A series of student texts in

CONTEMPORARY BIOLOGY

A Biologist's Physical Chemistry

Second Edition

J. Gareth Morris

B.Sc., D.Phil., F.I.Biol.

Professor of Microbiology, The University College of Wales, Aberystwyth

First published 1968
by Edward Arnold (Publishers) Limited
25 Hill Street, London W1X 8LL
Reprinted 1969, 1971, 1972

Second edition 1974
Reprinted 1974
Reprinted 1976

Boards edition ISBN: 0 7131 2413 x
Paper edition ISBN: 0 7131 2414 8

To Mary, Martha and Paul

Printed in Great Britain by
William Clowes & Sons, Limited
London, Beccles and Colchester

Foreword to the First Edition

Until fairly recently, the study of biology in British schools and Universities was characterized by two tacit assumptions. In the first place, it was believed that biological subjects are, in general, more descriptive than are their cousins of the physical (or "exact") sciences; scientifically-inclined pupils who, through lack of mathematical ability, would be debarred from the study of physics and chemistry might thus still settle for the anecdotal pursuit of biology. Secondly, it was held that the amount of detailed information that a student needed to acquire before he could describe himself as a biologist was so great that, possibly even before he came to University, he would do well to specialize in the study of either animals or plants. Even in the late 1960's, the great majority of biology graduates emerge from British Universities with specialist degrees in botany or zoology.

The revolutionary advances in biological knowledge, resulting principally from the researches of biologists working at the cellular and subcellular levels of organization, have shown both assumptions to be untenable. It is now conceded that one of the most important aims of biology is the description of life processes in physical and chemical terms: such a description can only be achieved by biologists thoroughly familiar with the principles of the physical sciences. It is also realized that the vertical divisions of biology into "Kingdoms", such as those of animals and plants, are both inadequate and misleading, and that such divisions obscure the underlying unity of the events which characterize the behaviour of living matter. It has become clear that the education of biologists must include both the molecular and organismic levels of enquiry. In consequence, Universities are instituting more-or-less loosely federated Schools of Biology, which offer first degrees in biology rather than in biological specialisms; due largely to the enlightened sponsorship of the Nuffield Foundation, the teaching of biology in schools is being similarly reorganized.

However, this desirable integration of biological specialisms in the schools and in the Universities may also have rather undesirable but inevitable consequences. Owing to difficulties of staffing, of time-tabling (and, of course, of inclination on the students' part), the pre-University education of students entering Schools of Biology may be gravely deficient in mathematics, and may also lack adequate preparation in physics and chemistry. Yet it is impossible, in a three-year degree course leading to a B.Sc. in biology, to allot sufficient time to pursue in depth these essential but non-biological topics. The provision of brief but intensive "crash courses" in physical chemistry, taught by physical chemists, often produces results opposite to those intended: such courses may frighten students, by their tendency towards "rigorous" (by which is usually meant, mathematical) treatment of the subject, and may bore them, by their lack of apparent relevance to biology. It may thus be incumbent upon the staff of a School of Biology to remedy the deficiencies of students' knowledge in these indispensable subjects in a manner which maintains (or even arouses) the students' interest, for, without understanding the basic concepts of physical chemistry as they apply to living systems, students cannot hope to progress beyond the purely anecdotal description of biological events.

The present book is intended to help students to understand those parts of physical chemistry which, though of immediate relevance to biology, are often also the most poorly understood. Dr. Morris has ensured, by his erudite and often witty treatment, that interest is sustained: his book may be read for pleasure as well as for profit. Numerous biological examples are cited and the solution of specimen problems is illustrated in detail. By working out the questions given at the end of each Chapter, students will acquire confidence that quantitative concepts can be used even without advanced mathematical knowledge; by using such concepts, students will also learn to understand them. I warmly commend this book and believe that, in writing it, Dr. Morris has rendered a real service both to teachers and students of biology.

H. L. KORNBERG

Preface to the First Edition

This book has been written in response to a need encountered when teaching elementary biochemistry to students reading for a degree in the School of Biology at Leicester. These students differ greatly in their background knowledge of the physical sciences, and many do not intend to specialize in biochemistry. It is our experience that the majority of them are unduly wary of the physical and quantitative aspects of biochemistry. In particular, they tend to assume that the principles of physical biochemistry can only be conveyed as an endless succession of mathematical equations whose comprehension and manipulation demand a greater knowledge of mathematics than they already possess. Often I have been told, 'Of course it's too mathematical for me, I learn things better when they are described in words—and anyway, it's all about frictionless pistons and the decomposition of nitrogen pentoxide'.

Accepting that this unreasonable fear of elementary mathematics remains a very real obstacle, I have attempted in three ways to help such students and to engage their interest in physico-biochemical matters:

1. by ensuring that no mathematical techniques are employed that are beyond their immediate comprehension; supplying an adequate number of quantitative examples that are painstakingly worked out, and offering a number of (answered) problems on which they may practise and gain in confidence;
2. by indicating the immediate relevance of all topics to the biological situation and, initially at least, examining only those topics and examples that they will encounter early on in their other studies in cell biology;
3. by clothing essential equations with a description of their meaning and implications which will help the student who likes to have principles 'described in words', and will ensure that the memorizer of equations is aware of their true worth.

These were the guide-lines that I followed in writing this small book and which are therefore largely responsible for its final form and content. It will be obvious from its limited coverage that it is not intended as a comprehensive textbook of physical chemistry, nor as the basis for a course intended to convey disciplinary instruction in physical chemistry. Nor does it cover all of the many areas in which physical chemistry impinges on biology, for a number of these (e.g. the properties of membranes, the behaviour of macromolecules in solution) are specifically studied in courses of molecular biophysics and are already well described in more advanced texts of biophysical chemistry. Rather, this book is offered to students and teachers of biology at the undergraduate level, as an aid to the understanding of the application of physico-chemical principles to biological situations, to be used in conjunction with other books of this Series, more particularly those which deal with cell biology and metabolism and with the physiology of plants and animals. It can also be profitably used together with less biologically-orientated and more complete textbooks of physical chemistry.

To this end, its content falls into 'natural' sections that are virtually self-contained. Where the subject matter bridges chapters, cross-references are included in the text so that sections may be selected at will. For example, it is hoped that the necessary background to a discussion of mechanisms of electron transport could be obtained from the account of oxidation-reduction processes in Chapter 12, allied with information about the energetics and coupling of reactions obtained from Chapters 7 to 9. Similarly, those studying the nature and function of enzymes might employ Chapters 10 and 11 to acquaint them with the aims and methods of reaction kinetics in general, and 'enzyme kinetics' in particular. Since pH is of prime interest to all biologists, Chapters 5 and 6 are devoted directly or indirectly to this topic, complemented (necessarily in my experience) by a section of Chapter 1 which describes the meaning and use of logarithms. The topics discussed in Chapter 6 to illustrate the usefulness to every biologist of a sound working knowledge of pH were chosen chiefly because they are those which a student of cell biology is likely to encounter in the first year of his course. For the same reason, manometry is described in the chapter on gases, and the chapter on solutions gives perhaps undue prominence to osmotic pressure.

Though primarily suited therefore to the needs of undergraduate (particularly 1st and 2nd year) students of biology and biochemistry, it is hoped that this approach will prove helpful to students in related schools, e.g. of medicine and pharmacy, and that it will be of interest to their more advanced colleagues.

The late Professor E. R. Redfearn first brought the need for this book to my attention, and during its preparation I have received much good counsel from many more of my colleagues. Much that you might find agreeable will have been derived from their helpful suggestions. I am especially grateful to Professor K. Burton of the University of Newcastle, Dr. D. G. Wild of the University of Oxford, and Dr. K. W. Morcom of the Department of Chemistry at this University. My colleague Dr. K. M. Jones has been unstinting in his help, and I have constantly drawn on his expert knowledge and good sense. I am also indebted to Mr. P. J. Price who, on behalf of Edward Arnold Ltd., has been a source of great patience and good advice. My thanks are due to Christopher Butler and John Payne who checked the answers to the problems, and to a host of past students who unwittingly drew my attention to sources of puzzlement which otherwise I should not have suspected. Errors and failings doubtless remain, and for these I am entirely responsible. I should be grateful to have them brought to my attention and to receive any suggestions as to how this book may be made more helpful to students.

Finally I wish to express my thanks to Professor Hans Kornberg for support and encouragement, and for his ensuring, by consenting to write the Foreword, that this book is not entirely without merit.

University of Leicester
1968

J. Gareth Morris

Preface to the Second Edition

The major innovation in this Edition is its 'conversion' to SI. Working biologists are understandably loth to exchange their familiar units, e.g. litre, °C, calorie, for the equivalents prescribed by this internationally agreed system of measurement which many physiologists consider less appropriate to their needs than to the requirements of the physical scientist. Yet teachers of biological chemistry in Universities, Polytechnics and other Colleges of Advanced Education will be aware that their students are now being instructed in school in the m.k.s. system, and are probably already more familiar with joules than with calories, and more at home with N m^{-2} than with mm Hg. Furthermore the biologist cannot (and should not) nurture his prejudices when in the related disciplines of Chemistry and Physics his students are being instructed through the medium of SI.

In this Edition therefore, Chapter 2 has been entirely rewritten to introduce the reader to SI, its units and the conventions adopted in their use. These rules and conventions are thereafter used consistently, but I trust sensibly, e.g. molarities are 'translated' into mol dm^{-3} rather than into kmol m^{-3} terms. At least during the lifetime of this Edition, many non-SI units will continue to appear both in the biological literature and in textbooks (expecially those originating in the U.S.A.). To aid the student in the task of converting such non-SI units into their SI equivalents a comprehensive conversion table is provided (pp. 20–2) and non-SI units have purposely been retained in the occasional *Example* to remind the reader of the need to acquire facility in this exercise.

I have resisted the temptation to enlarge the scope of this book since this would distort its primary objectives and inevitably lead to such an increase in its cost as to remove it from the possession of those it is most intended to help. I have however taken this opportunity to rewrite large sections on Real Gases, Background Thermodynamics, Coupling of Biochemical Reactions, Inhibition of Enzyme-Catalysed Reactions and Electrode & Redox Potentials, being much aided in this task by constructive comments from many colleagues. I continue to be indebted to all of these and to my publishers. It pleases me to learn that this small book has already proved helpful to some students whose main purpose in studying physical chemistry is to enhance their competence as biologists. I would be grateful to receive their views, and those of their teachers, of the changes made in this Edition.

The University College of Wales, J. GARETH MORRIS
Aberystwyth
1973

Table of Contents

Constants

The following values for constants have been used throughout this book:

1 atmosphere = 101 325 Pa = 101 325 N m^{-2}
Avogadro constant = 6.023×10^{23}
Faraday constant = 96 487 C mol^{-1}
Gas constant = 8.314 J K^{-1} mol^{-1}
Molar volume of ideal gas at s.t.p. = 22.414 dm^3
Molar f.p. depression constant of water = 1.86 K
0°C $\triangleq$ approx. 273 K

1

Mathematics Revision

Every student of biology is aware of the value of quantitative measurements, and of the need to make exact correlations of many of his experimental findings. Otherwise, biology would be limited to qualitative observations, and to the subjective appraisal of the structure and behaviour of living systems. Yet, although the need for a working knowledge of mathematics is recognized, every teacher of biology is soon made aware of the fact that many of his students are fearful of the simplest mathematical equation, and uneasy with the most straightforward graph.

In this book, three steps have been taken to counter this unwarranted fear of elementary mathematics:

1 As far as possible, mathematical 'short cuts' have been avoided in all worked examples, and all graphs and their implications are described in non-mathematical terms as well as by mathematical equations.
2 The number of mathematical concepts employed has purposely been kept to a minimum. Thus, virtually no use is made of calculus (except for brief mentions in Chapters 8 and 10). It was thought that, in this elementary book, the inconvenience of sometimes being prevented from pursuing the derivation of a given relationship was to be preferred to the danger of 'scaring off' students to whom dx/dy and $\int$ are anathema.
3 This chapter is offered as an aid to the revision of simple mathematical relationships and techniques. Even so, it is not necessary that you should read this chapter before proceeding with the remainder of the book; rather, you should turn to it only if you encounter purely mathematical difficulties in later chapters (particularly with exponents and logarithms in Chapter 5).

FRACTIONS, MULTIPLES AND POWERS

Common fractions

You will be familiar with the expression of fractional terms in the 'common' form of $\frac{\text{numerator}}{\text{denominator}}$. In handling such fractions remember the following points:

1 Common fractions with the same denominator can be added or subtracted by adding or subtracting their numerators.

To add or subtract common fractions possessing different denominators, the lowest common multiple of their denominators must be obtained.

2 To multiply common fractions, multiply the numerators to obtain the new numerator, and multiply the denominators to obtain the new denominator; then simplify if possible.

3 To divide common fractions, invert the divisor and multiply.

4 $\frac{0}{x}$ equals zero, provided that x does not itself equal 0; however, $\frac{x}{0}$ is meaningless, and division by 0 is 'banned'.

5 You will *retain* the value of a fraction if you multiply both its numerator and denominator (or divide both) by the same term,

$$\text{thus, } \frac{a}{b} = \frac{Ka}{Kb} \left(\text{but this does } \textit{not} \text{ equal } K\frac{a}{b}\right)$$

You *alter* the value of a fraction if you:

(i) add a constant to both numerator and denominator

$$\frac{a}{b} \neq \frac{a+K}{b+K}$$

(ii) raise both numerator and denominator to the same power or reduce each to the same root,

$$\frac{a}{b} \neq \frac{a^2}{b^2} \quad \text{and} \quad \frac{a}{b} \neq \frac{\sqrt{a}}{\sqrt{b}}$$

Decimal fractions

Addition and subtraction of decimal fractions present no difficulty if the figures are tabulated so that their decimal points fall in a vertical line. Multiplication and division are best carried out by the use of a slide rule, or logarithms (see later).

Expression of a number y as a multiple of another number x

If $y = ax$ where a is a constant, so long as the value of x is known, then y is defined by this equation as a function of x.

By expressing a series of very large numbers as multiples of a common denominator x which itself has a large value, the numbers are reduced to more manageable proportions. (Similarly for a series of very small numbers, when x would be made a small number.) In this way a series of numbers can be 'scaled up' or 'scaled down' by the factor x (see p. 16).

Expression of a number y as a power of another number x

If $y = x^i$, this is a shorthand way of stating that if x is multiplied by itself i times, the product will equal y. The term x^i (described as x to the power of i), consists of the ***base*** x and its ***index*** (or *power* or *exponent*) i. Since the index i need not necessarily be an integral number, the value of y can be expressed as an exponential term whatever is the value of y, and whatever number is chosen as the base x; that is, whatever is the defined value of x, it is possible by raising it to a certain power i (possibly negative or fractional) to equal the value of y.

Multiplication and division of numbers are much simplified by first converting them into exponential terms having the same base, for:

(i) to multiply exponential terms having a common base, one adds their indices,

$$x^a \times x^b = x^{(a+b)}$$

(ii) to divide exponential terms with a common base, one subtracts their indices,

$$x^a \div x^b = x^{(a-b)}$$

It makes no difference if some of the indices have a negative value, for they are added or subtracted algebraically,

$$x^a \times x^{-b} = x^{(a-b)}$$

Note the following relationships:

$$x^{-a} = \frac{1}{x^a} \qquad x^0 = 1$$

$$x^{1/a} = \sqrt[a]{x} \qquad x^{a/b} = \sqrt[b]{x^a}$$

Expression of any number as the product of two numbers, one of which is an integral power of 10

This method of expressing a number is usually employed when the number is very large or very small (having a large number of noughts immediately preceding or following its decimal point); e.g.

$$3\ 670\ 000\ 000 = 3.67 \times 10^9$$
$$0.000\ 044\ 3 = 4.43 \times 10^{-5}$$

The proper index to the base 10 is given by the number of places the decimal point must be shifted to the left (positive index), or to the right (negative index), to arrive at the number by which the power of 10 is multiplied.

Multiplication and division of large or small numbers are facilitated by handling them in this form; for example,

$$(3.67 \times 10^9) \times (4.43 \times 10^{-5}) = (3.67 \times 4.43) \times 10^{(9-5)} = 16.26 \times 10^4$$
$$= \underline{1.626 \times 10^5}$$

$$\frac{3.67 \times 10^9}{4.43 \times 10^{-5}} = \frac{3.76}{4.43} \times 10^{(9+5)} = 0.848 \times 10^{14} = \underline{8.48 \times 10^{13}}$$

The conventional way of tabulating values which are all multiples of the same product of 10 should be explained, for it will be used in later chapters both in Tables and for defining the scales employed in plotting graphs. According to this convention, the heading (or scale) reports the power of 10 by which the measured quantities *have been* multiplied to obtain the listed values. For example, consider the following:

Table 1.1

$10^4\ K_{eq}$	10^{-4} mass/g
1.67	2.2
3.32	3.6

This must be translated as meaning that the reported equilibrium constants are 1.67×10^{-4} and 3.32×10^{-4}, and that the masses are 22 000 and 36 000 g.

LOGARITHMS

If $y = x^i$, then so long as the magnitude of x is defined, the value of y can be represented by the index i. This indeed is what is done when a number is represented by its logarithm,† for a logarithm is another name for the index of an exponential term. Thus the relationship between y and x,

† A word of Greek origin; could mean 'ratio number' or 'reckoning number'.

defined by the equation $y=x^i$, may also be stated as follows, '*i* is the logarithm to the base *x* of the number *y*', and written in 'shorthand' as

$$i = \log_x y$$

The identity of the logarithm with the exponential index is sometimes obscured by the conventional way in which logarithms are written. For example, 0.02 equals $10^{-1.7}$ but $\log_{10}0.02$ equals $\bar{2}.3$.

The logarithm is written as a decimal number composed of two parts:

(i) the ***characteristic***, which precedes the decimal point and is therefore either zero or an integral number. It may have a positive or a negative value. When it has a negative value, this is indicated by a superscript bar, e.g. $\bar{2}$;

(ii) the ***mantissa***, which is that part of the logarithm that follows the decimal point. It is *always* positive.

Thus the logarithm to the base 10 of 2, is

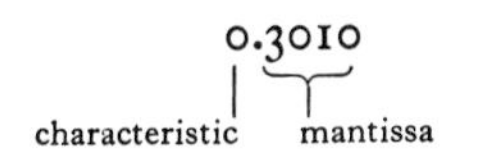

Because the mantissa of a logarithm is always positive, to represent $10^{-1.7}$ as a logarithm to the base 10, you cannot simply demote the index; i.e. log 0.02 cannot be written as -1.7, for this would mean that the mantissa would be negative ($-.7$) and this is forbidden. But -1.7 is 0.3 more positive than -2.0 (i.e. $-1.7=-2+0.3=\bar{2}.3$); thus, $\log_{10} 0.02=\bar{2}.3$.†

Bases for logarithms

Although *any* number could be made the base for a set of logarithms, three numbers are most often employed in practice:

(a) *10*—logarithms to the base 10 are the most frequently used of any and are called *common* logarithms (sometimes *Briggsian* logarithms after their inventor). In this book we shall symbolize these logarithms as log, e.g. log 2 = 0.3010;

(b) *2*—logarithms to the base 2 ($\log_2$) have a very evident application to processes wherein some property increases by doubling (e.g. binary fission effecting the multiplication of bacteria);

(c) '*e*' (*the* '*natural base*')—logarithms to the base e ($\log_e$ or ln) are also known as *Napierian* logarithms. Invented to describe processes that increase or decrease in a naturally exponential manner, these logarithms

† By keeping the mantissa positive (and using the superscript-bar system for the negative characteristic), a single table of logarithms serves for numbers which are greater than 1 as well as for numbers less than 1.

have as their base the number symbolized by e, which is the outcome of the infinite series,

$$1+\frac{1}{1}+\frac{1}{1\times 2}+\frac{1}{2\times 3}+\frac{1}{2\times 3\times 4}+\text{ etc.}$$

To five places of decimals e equals 2.71828 and,

$$\ln x = 2.303 \log x$$

Use of tables of common logarithms

A Table of Logarithms to the base 10 (Appendix) lists mantissas only; the characteristic of the common logarithm of a number must be obtained by inspection of that number.

To determine the common logarithm of a number, ignore any decimal point and from the Table of Logarithms read off that mantissa which corresponds to the first four figures of the number (reading from left to right). Insert a decimal point to the left of this mantissa and precede this with a characteristic calculated according to the following simple rules:

(a) for any number greater than 1, the characteristic is positive and is one *less* than the number of figures preceding the decimal point;

(b) for any number less than 1, the characteristic is negative and is one *plus* the number of noughts immediately following the decimal point. Alternatively, if the small number is written as a number between 1 and 10 multiplied by a negative power of 10, then the characteristic is identical with the negative index of 10.

For example, 2000 'looked up' in the Table of Logarithms gives the mantissa 3010. Therefore,

$$\log 200 = 2.3010 \qquad \log 0.2 = \bar{1}.3010$$

$$\log 2 = 0.3010 \qquad \log 2\times 10^{-5} = \bar{5}.3010$$

The number whose value is given as the logarithm to a stated base is itself called the *antilogarithm*, i.e. antilogarithm = $(\text{base})^{\text{logarithm}}$. Thus, 'the antilog of 0.3010' means 'that number whose log is 0.3010', i.e. 2.0.

To determine the antilogarithm of a given common logarithm, the number corresponding to its mantissa is read off from the Table of Antilogarithms (Appendix). The actual antilogarithm is then obtained by inserting a decimal point into this number in a position defined by the characteristic of the logarithm. Obviously, the rules that governed the assignment of the characteristic to the logarithm in the first place operate in reverse when the characteristic is used to locate the decimal point in the antilogarithm, so that:

(a) if the characteristic is *positive*, the number of figures that must precede the decimal point in the antilogarithm is one *greater* than the characteristic;

(b) if the characteristic is *negative*, then between the decimal point and the number 'read off' from the Table of Antilogarithms there must be inserted a number of noughts which is one *less* than the characteristic.

EXAMPLE

Determine the antilogarithms of (i) 2.5529, and (ii) $\bar{3}.5529$.

First 'read off' the number corresponding to the mantissa 5529 in the Table of Antilogarithms. This turns out to be 3572.

(i) If the logarithm is 2.5529, the characteristic is 2 (and positive), so the antilogarithm will have $(2+1)=3$ figures preceding its decimal point.

$$\therefore \text{antilog } 2.5529 = 357.2$$

(ii) If the logarithm is $\bar{3}.5529$, the characteristic is -3, and the antilog will have $(3-1)=2$ noughts immediately following its decimal point.

$$\therefore \text{antilog } \bar{3}.5529 = 0.003\,572 = 3.572 \times 10^{-3}$$

If you should ever be in doubt about the application of these 'rules', translate the logarithm into its exponential form. For example, if $\log x = \bar{3}.5529$

$$x = 10^{0.5529} \times 10^{-3}$$

$$\text{whence } x = (\text{antilog } 0.5529) \times 10^{-3} = 3.572 \times 10^{-3}$$

Multiplication and division using logarithms

Because logarithms are exponents, logarithms to the same base conform to the same rules as indices to the same base (p. 3). Therefore to multiply two numbers, one simply adds their common logarithms and determines the antilogarithm of the sum; i.e.

$$a \times b = \text{antilog} (\log a + \log b)$$

Similarly, to divide two numbers, one subtracts their common logarithms and determines the antilogarithm of their difference; i.e.

$$a \div b = \text{antilog} (\log a - \log b)$$

(Remember to subtract the log denominator from the log numerator.)

Powers and roots are similarly easily determined by multiplying and dividing their logarithms.

Remember therefore,

$$\log ab = \log a + \log b$$

$$\log \frac{a}{b} = \log a - \log b$$

$$\log \frac{1}{a} = -\log a$$

$$\log a^n = n \log a$$

$$\log \sqrt[n]{a} = \frac{\log a}{n}$$

EXAMPLE

(*i*) *What is the logarithm of* 6.35×10^{-4}?

$$\log ab = \log a + \log b$$

$$\therefore \log 6.35 \times 10^{-4} = (\log 6.35 + \log 10^{-4}) = 0.8028 + (-4)$$

$$= \underline{\bar{4}.8028}$$

(*ii*) *Calculate* $\dfrac{17.53 \times 13.76 \times 0.356}{5.41 \times 0.022}$ *using logarithms*

(a) *Add* the logs of the numerators

log 17.53	1.2437
log 13.76	1.1386
log 0.356	$\bar{1}$.5514
Sum	1.9337

(b) *Add* the logs of the denominators

log 5.41	0.7332
log 0.022	$\bar{2}$.3424
Sum	$\bar{1}$.0756

(c) *Subtract*: (summed logs of numerators) − (summed logs of denominators)

	1.9337
	$\bar{1}$.0756
Difference	2.8581

From the Table of Antilogarithms the mantissa 8581 = 7213

Since the characteristic is +2

Antilog 2.8581 = $\underline{721.3}$ = Answer

Arithmetic, geometric and logarithmic series

Consider the following series of numbers,

0, 10, 20, 30, 40, 50, 60, 70, etc.

The numbers form an ***arithmetic*** progression or series, in which the *difference* between two successive numbers is constant. In the present example, the increment is 10 and, irrespective of their location in the series, the difference between two successive numbers is always 10. (An arithmetic series will be obtained whatever is the size of the increment so long as this is constant.)

Now consider the following series,

1, 10, 100, 1000, 10 000, 100 000, etc.

These numbers form a ***geometric*** progression, or series, in which it is the *ratio* between two successive numbers that is constant. Thus in the example given, whatever is their place in the geometric progression one number is always 10 times as great as the preceding number. Note, however, that the arithmetic increment (i.e. the difference) between two successive numbers depends on their situation in the series, i.e. from 10 to 100 is an increase of 90, but from 10 000 to 100 000 is an increase of 90 000.

The geometric progression consists of numbers that increase 'exponentially' in magnitude. This is made plain by rewriting the numbers of a geometric series in their exponential form; i.e.

1, 10, 100, 1000, 10 000, 100 000, etc.

may be rewritten as,

10^0, 10^1, 10^2, 10^3, 10^4, 10^5, etc.

Employing the common logarithms of these numbers, we transmute the geometric series into the following arithmetic series:

logs are: 0, 1, 2, 3, 4, 5, etc.

Even if we selected a different base for the logarithms, the geometric series would in its logarithmic form be an arithmetic progression. For example, taking logarithms to the base 2,

the geometric series: 1, 10, 100, 1000, 10 000, 100 000, etc.
becomes: 0, 3.322, 6.644, 9.966, 13.288, 16.610, etc.

This $\log_2$ series is arithmetic, an increment of 3.322 in the value of $\log_2$ representing a tenfold increase in the value of the antilogarithm; i.e.

$$\log_2 10 = 3.322$$

This makes sense of the 'classical' but cryptic definition of logarithms as 'a series of numbers in arithmetic progression corresponding to another series of numbers (their antilogarithms) in geometric progression'.

Scales whose divisions are constructed so that they form an arithmetic progression are used when plotting parameters that increase arithmetically in value, e.g. time. Scales whose equal divisions mark steps in a geometric progression are useful when plotting parameters whose values increase exponentially.

In general, when any quantity y varies with another quantity x in such a way that the rate of change in y is always proportional to the value of y, it is said to vary in an *exponential* manner. The equation $y = Ze^{kx}$, where Z and k are constants and e is the 'natural base', defines the behaviour of systems that conform to what is known as the *law of continuous growth*, e.g. exponentially growing bacterial cultures. On the other hand, the similar exponential function which has a negative index $y = Ze^{-kx}$, describes systems that conform to the '*law of decay*', e.g. decrease of radioactivity of DNA labelled with ^{32}P.

GRAPHIC REPRESENTATION OF THE RELATIONSHIP BETWEEN TWO QUANTITIES x AND y

The graphs employed in this book possess two axes which are mutually perpendicular. It is conventional to refer to the horizontal axis as the *abscissa* (or x axis) and to the vertical axis as the *ordinate* (or y axis). The point of intersection of these axes is the *origin*.

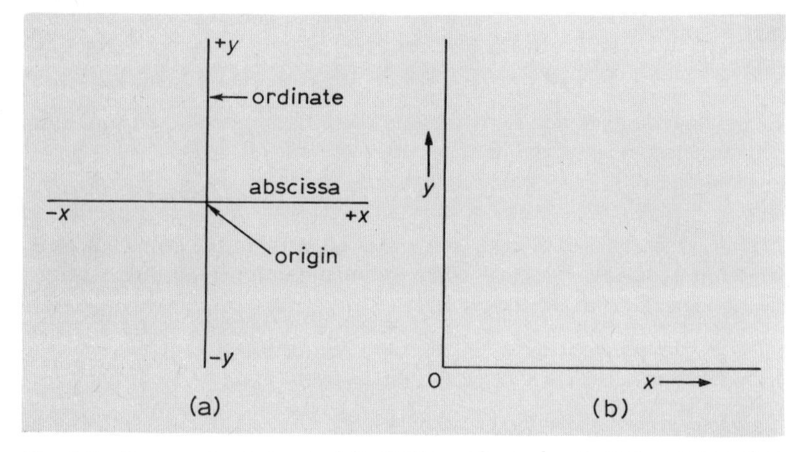

Fig. 1.1 Rectangular axes used in plotting values of y against corresponding values of x.

As shown in Fig. 1.1a, positive and negative values of x and y may be measured on the axes of such a graph, but when dealing solely with positive values of x and y we need use only its upper right-hand (positive) quadrant as in Fig. 1.1b.

The position of any point in the plane of the graph is defined by its perpendicular distances from the ordinate and abscissa. These distances, measured in the scale units into which the axes are divided, are the ***coordinates*** of the point. Any relationship between values of y and corresponding values of x can be represented by a line on this graph. The shape of the curve followed by this line defines the relationship between x and y, and can be expressed as an equation. With a little experience, it is possible from the form of the equation that relates x and y to predict the appearance of the curve that will be obtained when values of y are plotted against corresponding values of x.

Equation of a straight line

Whenever the relationship between x and y can be expressed in the following form

$$y = ax + b$$

where a and b are constants, one can be certain that if y is plotted against x, a straight line will be obtained whose slope is a, and whose intercept on the y axis equals b (for b is the value of y when x is 0).

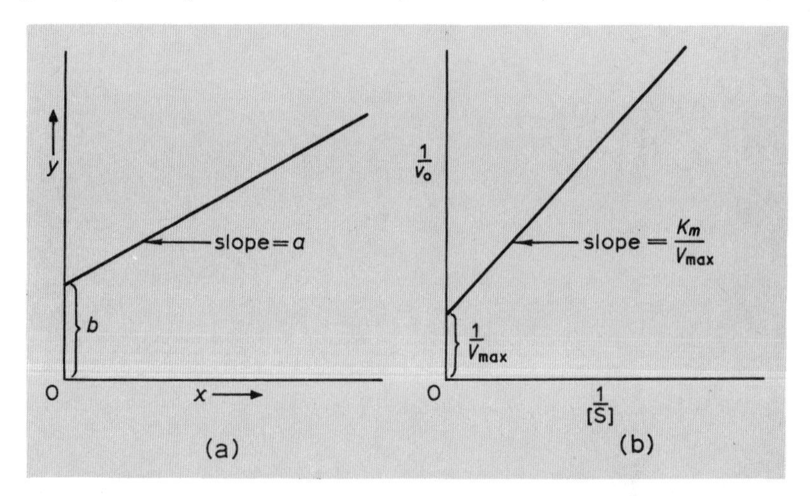

Fig. 1.2 Graphs of 'straight line' equations: (a) $y=ax+b$

and (b) $\dfrac{1}{v_0} = \dfrac{K_m}{V_{max}} \cdot \dfrac{1}{[S]} + \dfrac{1}{V_{max}}$

For example, in Chapter 11 (p. 285), we see that the Lineweaver–Burk equation relates the initial velocity of an enzyme-catalysed reaction (v_0) to the concentration of its substrate [S] as follows,

$$\frac{1}{v_0} = \frac{K_m}{V_{max}} \cdot \frac{1}{[S]} + \frac{1}{V_{max}} \qquad \text{(where } K_m \text{ and } V_{max} \text{ are constants)}$$

This equation has the form $y=ax+b$ (where $y=1/v_0$, $x=1/[S]$, $a=K_m/V_{max}$ and $b=1/V_{max}$). This means that when values of $1/v_0$ are plotted against values of $1/[S]$, a straight line is obtained of slope K_m/V_{max}, which intersects the $1/v_0$ ordinate to make an intercept which is equal to $1/V_{max}$ (see Fig. 1.2b).

Equation of a standard exponential curve

If we plot values of y against corresponding values of x when $y = e^x$, for positive values of x we obtain the standard exponential curve shown in Fig. 1.3a.

This natural exponential curve for the expression $y = e^x$ crosses the ordinate at $y = 1$ (when $x = 0$), and rises progressively more steeply as x increases.

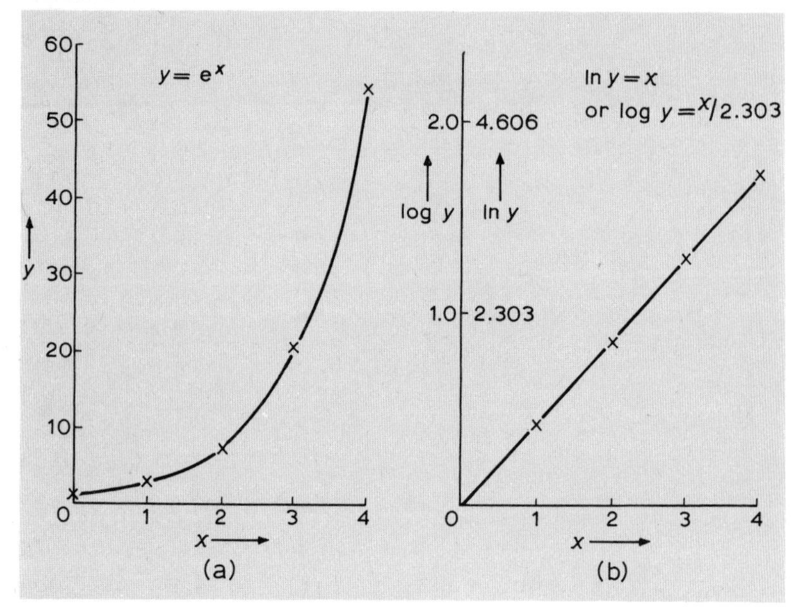

Fig. 1.3 Exponential curve (a) obtained for $y = e^x$ by plotting y versus x is transformed into a straight line (b) by plotting either ln y or log y versus x. (Note that the exponential curve (a) intersects the ordinate at $y = 1$.)

If we plot these values of y as their logarithms (to the base e or to the base 10 as shown in Fig. 1.3b, or for that matter, to any other base), we obtain a straight line. This comes about, since given that $y = e^x$, then $\ln y = x$, which is the equation of a straight line of slope $= 1$ stemming from the origin (when $x = 0$, $\ln y = 0$). If any other base (n) is used for the logarithmic expression of values of y,

$$\log_n y = \frac{x}{\text{constant}}$$

and a straight line is obtained which again passes through the origin but has a slope of 1/constant.

With a little thought, you will recognize that the use of a 'semi-log' plot to transform an exponential curve into a straight line is made possible by the fact that the logarithms of the numbers in a geometric (exponential) series are themselves members of an arithmetic series (p. 9).

Frequently, instead of calculating the logarithms of the 'y' term, and plotting these values of $\log y$ on 'ordinary' graph paper whose x and y axes are both subdivided arithmetically, values of y itself are directly plotted on 'semi-logarithmic' graph paper whose x axis is scaled arithmetically but whose y axis is scaled logarithmically. Inspection of this type of graph paper is sufficient to convince one of an important feature of any logarithmic scale, namely that it is in a sense distorting, in that it gives prominence to differences between lesser numbers but becomes progressively more 'squashed' as the scale is ascended.

The equations of many regular shaped curves could be listed here, e.g. hyperbolae, parabolae, etc., but this information is to be found in every elementary textbook of coordinate geometry.

The Calculus

Though, as stated at the outset, minimal use is made of calculus in this book, it would not be sensible to discuss rates (e.g. of reactions) without some reference to the differential calculus, whilst the evaluation of changes wrought at stipulated rates is aided by application of the integral calculus.

A very helpful account of the basic principles of the calculus is given in the useful textbook by Angela Crowe and Alan Crowe *Mathematics for Biologists*, Academic Press, London (1969).

In conclusion, I should like to stress once again that nowhere in the chapters that follow will you be asked to carry out difficult mathematical tasks. Never be alarmed by the first appearance of any quantitative problem; it is often helpful to appraise each problem quite calmly in the following three stages:

(i) what information is provided?
(ii) what is it that I am really being asked to determine?
(iii) what is the relationship between (ii) and (i)?

2

SI Units and their Usage

The magnitude of any quantity can be measured only in relation to a specific unit size that has been defined for that quantity. This means that the unit of measurement must be reported along with the numerical measurement, else the latter becomes meaningless. You will see in the following chapters that the only numbers to be written without defining units are logarithms, and ratio of quantities that are measured in identical units.

In calculations involving several measured quantities, it is essential that all are expressed in mutually consistent units; i.e. units based on the same fundamental system of measurement. The system that formerly was

Table 2.1

Physical quantity	Name of SI unit	Symbol
Base units :		
length	metre	m
mass	kilogramme	kg
time	second	s
electric current	ampere	A
thermodynamic temperature	kelvin	K
luminous intensity	candela	cd
amount of substance	mole	mol
Supplementary units :		
plane angle	radian	rad
solid angle	steradian	sr

most generally employed in chemical and biological work was the centimetre–gram–second (c.g.s.) system, but in the U.K. we are now adopting for all scientific work the Système International d'Unités, known to the initiated as SI. This is an internationally agreed form of the metre–kilogramme–second (m.k.s.) system of measurement which allots basic (primary) units to the seven physical quantities listed in Table 2.1. Two supplementary units for plane angle and solid angle are also provided.

Table 2.2 Special names and symbols for SI derived units

Physical quantity	Name of SI unit	Symbol for SI unit	Definition of SI unit	Equivalent in SI units
energy	joule	J	$m^2\ kg\ s^{-2}$	N m
force	newton	N	$m\ kg\ s^{-2}$	$J\ m^{-1}$
pressure	pascal	Pa	$m^{-1}\ kg\ s^{-2}$	$N\ m^{-2}$
power	watt	W	$m^2\ kg\ s^{-3}$	$J\ s^{-1}$
electric charge	coulomb	C	A s	$J\ V^{-1}$
electric potential difference	volt	V	$m^2\ kg\ s^{-3}\ A^{-1}$	$J\ C^{-1}$
electric resistance	ohm	Ω	$m^2\ kg\ s^{-3}\ A^{-2}$	$V\ A^{-1}$
electric conductance	siemens	S	$m^{-2}\ kg^{-1}\ s^3\ A^2$	Ω^{-1}
electric capacitance	farad	F	$m^{-2}\ kg^{-1}\ s^4\ A^2$	$C\ V^{-1}$
luminous flux	lumen	lm	cd sr	
illumination	lux	lx	m^{-2} cd sr	$lm\ m^{-2}$
frequency	hertz	Hz	s^{-1}	

The size of each of these primary (base) units is defined as an invariant, internationally agreed standard, and they are in turn employed to construct a variety of derived units. The derived SI units most relevant to biological studies are listed in Table 2.2.

Choice of units of appropriate size

For certain purposes, the primary and derived SI units could prove inconveniently large or inordinately small. In these circumstances, secondarily modified units might be used. The size of any secondary unit will be defined as a multiple of the primary or derived unit. To obtain a convenient range of secondary units from a primary or derived unit this is multiplied by different powers of 10. The size of the secondary unit is then indicated by attaching a modifying prefix to the name of the primary or derived unit. These prefixes (listed in Table 2.3) indicate the power of 10 by which the primary or derived unit has been multiplied.

Table 2.3 SI prefixes

Multiple	Prefix	Symbol	Multiple	Prefix	Symbol
10	deca	da	10^{-1}	deci	d
10^2	hecto	h	10^{-2}	centi	c
10^3	kilo	k	10^{-3}	milli	m
10^6	mega	M	10^{-6}	micro	μ
10^9	giga	G	10^{-9}	nano	n
10^{12}	tera	T	10^{-12}	pico	p
			10^{-15}	femto	f
			10^{-18}	atto	a

General observations on the usage of SI units and their symbols

Symbols will consist of capital letters when the units that they represent are named after persons, though when the names of these units are written in full they are not given initial capital letters.

Example: a force of 50 newton(s) = 50 N

Symbols must only be written in their singular form:

Example: 2.4 mol, *not* 2.4 mols

No point (i.e. full stop) should be inserted after the symbol (unless of course this happens to conclude a sentence).

Example: six milligrammes of NaCl = 6 mg NaCl, *not* 6 mg. NaCl

When two or more unit symbols are combined to create a derived unit symbol a space is left between them.

Example: 1 C = 1 A s

No space is left between a prefix (indicating a power of 10) and the symbol to which it applies. This rule is particularly important in guarding against confusion between m (meaning milli) and m (meaning metre).

Example: One millisecond = 1 ms
One metre per second = $1\ \mathrm{m\ s^{-1}}$

Compound prefixes should not be used; use only one multiplying prefix

Example: 10^{-9} m = 1 nm, *not* 1 mμm

A combination of prefix and symbol for a unit is regarded as a single symbol. Thus when a modified unit is raised to a power of 10 the power applies to the whole unit including the prefix.

Examples: 1 litre = 1 dm^3 = 1 $(dm)^3$ = 10^{-3} m^3, *not* 10^{-1} m^3
1 cm^3 = 1 $(cm)^3$ = 10^{-6} m^3, *not* 10^{-2} m^3
$\therefore$ 1 cm^3 = 10^{-3} dm^3

The decimal sign may either be a full point *on the line* or a comma on the line. In the U.K. we use the full point, but some other metric countries will continue to use the comma.

To facilitate easy reading, digits should be arranged in groups of three about the decimal sign. Because a comma can be employed as the decimal sign, commas must not be used to space groups of thousands and millions.

Example: write 1 352 670.47, *not* 1,352,670.47

Use of the solidus (i.e. stroke) is discouraged in favour of the negative index when writing symbols of reciprocal units.

Example: preferable to write N m^{-2} rather than N/m^2

The solidus must in any event never be used more than once in any unit.

Example: the molar gas constant $R = 8.314$ J K^{-1} mol^{-1}, *not* 8.314 J/K/mol.

It is in fact convenient to keep the solidus (a) for division of a physical quantity by another physical quantity, e.g. PV/RT, or (b) for the division of a physical quantity by its unit, e.g. $R/\text{J K}^{-1}\text{ mol}^{-1} = 8.314$.

It is recommended that the degree sign is omitted when recording temperature in the SI unit (kelvin).

Example: the f.p. of water is 273.15 K, *not* 273.15 °K

Some common non-SI units are likely to remain in use but should no longer be employed in a precise scientific context; e.g. min., hr., cal., Curie, atm., *etc.*

Some implications of the adoption of SI

Volume: The SI unit of volume is the cubic metre, m^3. The old definition of the litre (leading to the value 1.000 028 dm^3) was abolished in 1964, and the litre was redefined as being exactly equal to the cubic decimetre. But, although it is recognized that the litre will remain in common usage, it is still recommended that both the litre and millilitre are abandoned in exact scientific work, their respective volumes being represented as the corresponding fractions of a cubic metre;

i.e. 1 litre (symbol, l) = 1 dm^3 = 10^{-3} m^3
1 millilitre (symbol, ml) = 1 cm^3 = 10^{-6} m^3

Mass: The primary SI unit of mass is the kilogramme, kg. This means, somewhat anachronistically, that the SI rules regarding usage of prefixes confer a special status on the gramme.

Example: 10^{-6} kg cannot be written as 1 μkg but can quite properly be written as 1 mg.

[Note the agreed spelling: gramme, *not* gram.]

Amount of substance: The primary SI unit is the mole, given the symbol mol. The mole is defined as 'the amount of substance of a system which contains as many elementary entities as there are atoms in 0.012 kg of carbon 12'. The elementary unit must be specified and may be an atom, a molecule, an ion, an electron, a photon, etc., or a specified group of such entities.

Since 0.012 kg of carbon 12 contains the Avogadro number ($6.022\,52 \times 10^{23}$) of atoms of carbon, this is the number of elementary units (of a specified kind) contained in 1 mol. This means that terms such as gram-equivalents, gram-molecules, gram ions, etc., are all obsolete and are abandoned in favour of the mol.

Example: 1 mole of electrons has a mass of 5.4860×10^{-7} kg and carries a negative electric charge equal to 96 487 C.

Concentration:† Previously, molality (symbol, m) was used to indicate the number of moles of solute in 1000 g of solvent. Fortunately, this is easily transmuted into primary SI units; what was termed a 1 molal solution has a concentration of 1 mol kg^{-1}. Because the symbol m (for molality) could be confused with m (for metre), both the term molal, and use of the symbol m to represent molality, should be abandoned in favour of mol kg^{-1}.

We face greater difficulties in determining the appropriate SI unit in which to express concentrations in mole/unit volume terms. Previously, molarity (symbol, M) was used to indicate the amount of solute (in moles) dissolved in 1 litre of solution. Therefore what was termed a 1 M solution has a concentration of 10^3 mol m^{-3} = 1 kmol m^{-3} = 1 mol dm^{-3}.

† The use of square brackets around its chemical formula to denote the concentration of a substance is still common practice, and this device is employed in this text.

Because the symbol M is already employed in SI to represent the prefix mega, use of the term molarity and its representation by the symbol M should be abandoned in favour of kmol m^{-3} or mol dm^{-3}. Furthermore, the word 'molar' before the name of an extensive quantity is to be restricted to the meaning 'divided by amount of substance', where the amount of substance need not be expressed as a function of 1 mol. It would therefore be wrong to continue to speak of molar concentrations.

For ease of 'transliteration' of old-style molarities into their SI equivalents, it is certainly easiest to employ the mol dm^{-3} nomenclature, and I would recommend this procedure during the period of transition wherein much of the biological literature will continue to contain references to metabolite concentrations in μmoles/ml and to concentrations of reagents, etc., in M;

i.e. $1\ \mu\text{mole/ml} = 1\ \mu\text{mol cm}^{-3}$

$$1\ \text{M} = 1\ \text{mol dm}^{-3} = 1\ \text{mol l}^{-1}$$

Force: The derived SI unit of force is the newton, N (Table 2.2). The dyne ($=10^{-5}$ N) becomes redundant.

Pressure: The derived SI unit of pressure is the pascal, Pa (Table 2.2). Pressures should therefore be expressed either in Pa or in N m^{-2} (for 1 Pa $=$ 1 N m^{-2}). All other units of pressure are obsolete and should be abandoned;

e.g. 1 lbf/in^2 = 6894.76 Pa
1 mmHg = 133.322 Pa
1 millibar = 100 Pa
1 atm. = 101 325 Pa

Energy: The derived SI unit of energy is the joule, J (Table 2.2). All other units of energy should be abandoned,

e.g. 1 erg = 10^{-7} J
1 litre-atm. = 101.328 J
1 $\text{calorie}_{\text{(thermochemical)}}$ = 4.184 J

Radioactivity: The curie (symbol, Ci) is redundant,

$$1\ \text{Ci} = 3.7 \times 10^{10}\ \text{s}^{-1}$$

Frequency: The derived SI unit of frequency is the hertz, Hz (Table 2.2).

Example: an n.m.r. machine using 60 megacycles/sec. should now be rated as 60 MHz; i.e., 60 megahertz.

Temperature: The SI unit of temperature is the kelvin, K (1 K $=$ 1/273.16

of the thermodynamic temperature of the triple point of water). It is envisaged that although the Celsius (i.e. centigrade) scale will be retained for everyday, domestic purposes it will be discontinued for exact scientific work.

Example: $0°C = 273.15$ K, so that $T/K = (t/°C + 273.15)$

Luminous intensity: The primary SI unit is the candela, cd (Table 2.1).

Luminous flux: The derived SI unit is the lumen, lm (Table 2.2).

Illumination: The derived SI unit is the lux, lx (Table 2.2). Other units should be abandoned;

e.g. 1 foot candle = 10.7639 lx

During the period of transition to uniform adoption of SI, many non-SI units will continue to appear, both in the original scientific literature and in textbooks. You will therefore be called upon to convert these non-SI units into their SI equivalents. To aid you in this task, Table 2.4 lists the SI equivalents of most of the units encountered in biochemical texts, while useful physical constants (in SI units) are listed in Table 2.5. More complete lists are given in the pamphlets listed at the end of this chapter.

Table 2.4 Conversion table for translation of common units into their SI equivalents

Unit	SI equivalent
ampere, A	1 A
ångström, Å	100 pm $=10^{-10}$ m
atmosphere, standard ; atm.	101 325 Pa
bar, b	10^5 Pa
calorie (international table) ; cal	4.1868 J
calorie 15°C ; cal_{15}	4.1855 J
calorie, thermochemical	4.184 J
candela, cd	1 cd
centigrade (Celsius) degree, °C	$(t/°C+273.15)$ K
centimetre, cm	10^{-2} m
coulomb, C	1 C
cubic centimetre, cm^3	1 $cm^3 = 10^{-6}$ m^3
cubic decimetre, dm^3	1 $dm^3 = 10^{-3}$ m^3 = 1 litre
cubic foot, ft^3	0.028 316 8 m^3
cubic inch, in^3	16.3871 cm^3
cubic metre, m^3	1 m^3
curie, Ci	3.7×10^{10} s^{-1}
cycle/second, c/s	1 Hz

Table 2.4—*continued*

Unit	SI equivalent
degree (angle), °	$\pi/180$ rad
degree centigrade (degree Celsius), °C	$(t/°C+273.15)$ K
degree Fahrenheit, °F	$(t/°F+459.67)$ K
drachm (apothecaries)	3.887 93 g
drachm, fluid	3551.63 mm^3
dram (avoirdupois)	1.771 85 g
dyne, dyn	10^{-5} N
electron volt, eV	1.6021×10^{-19} J
erg	10^{-7} J
farad	1 F
fluid ounce, fl oz	28.4131 cm^3
foot, ft	0.3048 m
foot-candle, lm/ft^2	10.7639 lx
foot-lambert	3.426 26 cd m^{-2}
foot of water (pressure)	2989.07 Pa
foot pound-force, ft lbf	135 582 J
gallon, gal	4.546 09 dm^3
gramme, g	10^{-3} kg
henry, H	1 H
hertz, Hz	1 Hz
hour, h	3600 s
inch, in	25.4 mm
inch of water (pressure)	249.089 Pa
joule, J	1 J
kilowatt, kW	1 kW
kilowatt hour, kW h	3.6 MJ
litre, l	1 $dm^3=10^{-3}$ $m^3=1$ l
litre atmosphere	101.328 J
lumen, lm	1 lm
lumen/sq. ft, lm/ft^2	10.7639 lx
lumen/sq. metre, lm/m^2	1 lx
lux, lx	1 lx
micron, μ	1 μm
millibar	100 Pa
millilitre	1 $cm^3=10^{-6}$ $m^3=1$ ml
millimetre of mercury, mmHg	133.322 Pa
millimetre of water	9.806 65 Pa
minim	59.1939 mm^3

Table 2.4—*continued*

Unit	SI equivalent
molal, m	1 mol kg^{-1}
molar, M	1 mol dm^{-3} = 1 mol l^{-1}
mole	1 mol
newton, N	1 N
ohm	1 Ω
ounce, oz	28.3495 g
ounce, apothecaries	31.1035 g
ounce fluid	28.4131 cm^3
pascal, Pa	1 Pa = 1 N m^{-2}
pint, pt	0.568 261 dm^3
poise, P	0.1 kg m^{-1} s^{-1}
poiseuille, Pl	1 N s m^{-2} = 1 Pl
pound, lb	0.453 592 37 kg
pound-force, lbf	4.448 22 N
pound-force/sq. in, lbf/in^2	6894.76 Pa
pound/sq. in, lb/in^2	703.070 kg m^{-2}
rad (100 erg/g)	0.01 J kg^{-1}
radian	1 rad
siemens, S	1 S
square foot, ft^2	0.092 903 m^2
square inch, in^2	645.16 mm^2
stokes, St	10^{-4} m^2 s^{-1}
therm	105.506 MJ
ton of refrigeration	3516.85 W
torr	133.322 Pa
volt, V	1 V
watt, W	1 W

Table 2.5 Values in SI units of some useful physical constants

Physical constant	Symbol	Value
Avogadro constant	L (or, N_A)	$6.022\ 52 \times 10^{23}$ mol^{-1}
Boltzmann constant	k	$1.380\ 54 \times 10^{-23}$ J K^{-1}
Gas constant	$R = Lk$	8.3143 J K^{-1} mol^{-1}
charge of electron	e	$1.602\ 10 \times 10^{-19}$ C
Faraday constant	$F = Le$	9.6487×10^{4} C mol^{-1}
Planck constant	h	6.6256×10^{-34} J s
Molar volume of ideal gas at 273.15 K and 101 325 Pa		22.4136 dm^3 mol^{-1}

Conventions for labelling axes on graphs and heading columns in tables of quantities

To avoid repetitious use of defining units, all physical quantities, constants, etc., are tabulated as columns of pure numbers. This in turn means that the heading of each column must itself be a pure number, e.g. the quotient of the measured physical quantity and the symbol for the unit of measurement.

Examples: temperatures should be listed under the column heading T/K, pressures might be recorded under the heading P/N m^{-2} or P/Pa, and changes in Gibbs free energy would be tabulated under the heading ΔG/J mol^{-1}.

The same principle applies to the labelling of axes on graphs (for example, see Fig. 3.2).

Until one gets used to this convention it could prove confusing; just remember that the unit of measurement is always indicated by putting its symbol after a solidus (i.e. stroke) inserted immediately following the symbol for the physical quantity being tabulated or plotted.

To avoid terms containing an inconvenient number of digits, the actually tabulated or plotted numbers might be the measured values multiplied by a convenient (constant) power of 10. The power of 10 that is in this case written into the heading or legend (immediately preceding the symbol for the physical quantity) is that number by which the measured quantities have been multiplied to yield those more convenient numbers that are listed or plotted. This of course does not preclude the concurrent use of the most suitable size of unit obtained by modifying the SI unit with one or other of the prefixes listed in Table 2.3.

Examples: (i) An entry 3.6 under the heading 10^2k means that the value of k is 0.036; the same entry (3.6) under the heading $10^{-2}k$ means that the value of k is 360.
(ii) A substrate concentration of 0.000 12 mol dm^{-3} can be listed as 1.2 under a heading 10^4[S]/mol dm^{-3} or as 0.12 under a heading [S]/mmol dm^{-3}.
(iii) A value for 1/[S] given in the literature as 300 mM^{-1} could be plotted on a Lineweaver–Burk graph as 3 when the abscissa was labelled either as 10/[S] in dm^3 mol^{-1}, or as 10^{-2}/[S] in dm^3 mmol^{-1}.
(iv) As an illustration of the application of this convention, consider the table of quantities employed in the calculation on p. 270; viz:

t/°C	T/K	10^4k/s^{-1}	$10^3/T$ in K^{-1}	log k
15	288	2.51	3.472	$\bar{4}$.3997
20	293	4.57	3.412	$\bar{4}$.6599
25	298	8.22	3.356	$\bar{4}$.9149

The first (horizontal) line would be read as follows: At 15°C (i.e. ≏288 K) the rate constant k has the value 2.51×10^{-4} s^{-1}. The reciprocal of 288 K equals 3.472×10^{-3} K^{-1} and log k equals $\bar{4}$.3997.

HANDLING PHYSICAL QUANTITIES IN CALCULATIONS

The following rules must be obeyed:

(1) Every measured quantity must be represented as a *number* times a *unit*; although one could write in the multiplication sign, this is usually omitted.
(2) Only quantities that have the same dimensions can be added or subtracted from each other. For example, one can add two energy terms but one cannot sensibly add an energy term and a pressure term.
(3) Be consistent in the choice of units; wherever it appears, the one physical quantity must be measured in identical units. To a large extent this is ensured by strict adherence to the rules of SI, so that, for example, with all energy terms expressed in joules there can be no confusion as might occur when non-coherent energy terms were employed in the one calculation, e.g. ergs and calories.

In multiplying and dividing quantities, their units are treated in precisely the same way as the accompanying numbers, and the product or quotient of the unit must be appended to the final numerical answer.

The way in which units can be multiplied and divided can be illustrated by deriving the units of the molar gas constant R whose value is defined by the following ideal gas equation (Chapter 3, p. 31):

$$R = \frac{PV}{nT} \qquad \text{where} \begin{cases} R = \text{molar gas constant} \\ P = \text{pressure of gas in N m}^{-2} \\ V = \text{volume of gas in m}^3 \\ n = \text{quantity of gas in mol} \\ T = \text{temperature in K} \end{cases}$$

$$\text{i.e. } R = \left(\frac{PV}{T}\right) \text{mol}^{-1} \text{ and the units of } R = \frac{\text{N m}^{-2} \times \text{m}^3}{\text{K}} \text{mol}^{-1}$$

$$= \text{N m K}^{-1} \text{mol}^{-1}$$

But 1 N = 1 J m^{-1}, whence 1 N m = 1 J

$$\therefore R \text{ is measured in the units J K}^{-1} \text{mol}^{-1}$$

As a final caution, it should be emphasized that you must be painstaking in multiplying and dividing units during the course of a calculation. It is extremely hazardous to ignore the units until you have arrived at a numerical result and only then 'conjure up' an appropriate unit to define this value.

BIBLIOGRAPHY

Further information on SI, its units and their proper usage, can be obtained from the following sources. Since these are subject to regular revision, you should consult them in their latest editions.

1. *Quantities, Units and Symbols.* The Symbols Committee of the Royal Society, The Royal Society, 6 Carlton House Terrace, London SW1Y 5AG (1971).
2. *The International System (SI) Units.* (BS 3763, 1970). British Standards Institution, 2 Park Street, London W1A 2BS.
3. *Physicochemical quantities and units (the grammar and spelling of physical chemistry)* (2nd ed. 1971) by M. L. McGlashan. Royal Institute of Chemistry Monographs for Teachers, The Royal Institute of Chemistry, 30 Russell Square, London W.C.1.

3

The Behaviour of Gases

Some of the most important substrates and products of metabolism are gases: for example, oxygen, carbon dioxide, nitrogen and hydrogen. It is therefore important that we should understand some of their characteristic properties and, since most metabolic reactions take place in an aqueous medium, that we should also examine the behaviour of gases in solution.

The gaseous state is the simplest of the three fundamental states of matter (gas, liquid and solid). A gas differs from matter in the liquid or solid state in that it possesses no intrinsic volume, which means that, theoretically at least, it fully occupies any enclosed space into which it is introduced. This and other properties peculiar to the gaseous state can be interpreted in terms of what is known as the Kinetic Theory of Gases. At the outset it should be pointed out that when talking of a 'gas' we are usually discussing an 'ideal gas' whose behaviour is perfectly predicted by the various Gas Laws. All 'real gases' whether they be elemental, e.g. helium or chlorine, or compound, e.g. carbon dioxide or ammonia, differ to some extent from the imaginary ideal gas, but it is much more convenient to define the properties of an ideal gas and note particular deviations from this ideal, than to attempt an individual examination of the behaviour of every known gas as though these had no properties in common.

According to the Kinetic Theory, the ideal gas is composed of extremely small particles (its molecules) that are in continuous, random and independent motion. During their random motion the molecules of gas collide incessantly with the walls of the container, and it is this continuous bombardment of the confining walls that is recognized as the pressure of the gas. The component 'particles' of the ideal gas are completely elastic and rebound with an energy equal to that which they possessed when entering the collision. This appears to be very reasonable, for were it not so, the pressure of a gas kept at constant volume and temperature in any container would progressively decrease over the course of time. Furthermore, the molecules of an ideal gas must occupy nil volume (which confirms one's suspicion that the ideal gas is a useful fiction).

By virtue of the random, independent motion of its molecules, when a gas of a certain density is introduced into a larger space than that which it formerly occupied at the same temperature, the molecules redistribute themselves in such a way that each has maximum freedom of movement. The gas then fully occupies the new volume, with a corresponding decrease in its density. This tendency for gaseous molecules to move from a zone of high density to another of lower density, and so achieve a mean equilibrium density, is expressed in the force of diffusion. It follows that constraint must be placed upon a gas to increase its density—the force of compression.

The effect on a gas of changes in its temperature may also be interpreted in terms of the Kinetic Theory. Input of heat increases the kinetic energy of the molecules, enhances their tendency to move even further apart from one another, and thus provokes expansion of the gas at constant pressure. Decreasing the temperature decreases the mobility of the molecules and the tendency at constant pressure is for the gas to contract. In a sense, therefore, increasing the pressure and lowering the temperature tend towards the same end, namely decrease in the volume of the gas.

It follows that the condition of an ideal gas is affected by three interdependent variables: (i) volume, (ii) pressure and (iii) temperature. Examination of the effect of changes of pressure and/or temperature on the volume of a given mass of ideal gas has resulted in the establishment of certain fixed relations between these factors that are known as Ideal Gas Laws. For the most part these laws bear the names of their proponents.

1. Effect of changing pressure

If the temperature is not allowed to change, then by ***Boyle's Law***, 'the volume of a given mass of gas is inversely proportional to the pressure exerted upon it'. This means that an isothermal increase of pressure will proportionately decrease the volume of a quantity of gas, and vice versa.

$$V \propto \frac{1}{P} \quad \text{where} \begin{cases} V = \text{Volume} \\ P = \text{Pressure} \end{cases}$$

$$\text{or} \quad PV = \text{Constant}$$

EXAMPLE

A balloon, which was perfectly elastic up to its bursting volume of 1.68 dm^3*, was filled at sea level with 1 dm*^{3} *of a light ideal gas. To what atmospheric* *pressure may it rise before bursting? (Assume no change in temperature; atmospheric pressure at sea level was 101 kPa.)*

P_1 = Pressure at sea level = 1.01×10^5 Nm^{-2} $P_2 = ?$
$V_1 = 1.0\ \text{dm}^3$ $V_2 = 1.68\ \text{dm}^3$

By Boyle's Law, PV = Constant, for a given mass of gas at a constant temperature

$$\therefore \quad P_1V_1 = P_2V_2$$

$$\text{so that} \quad 1 \times 1.01 \times 10^5 = 1.68 \times P_2$$

$$\text{or} \quad \frac{1.01 \times 10^5}{1.68} = P_2$$

$$\text{whence} \quad P_2 = 6.01 \times 10^4\ \text{N m}^{-2}$$

The balloon will burst when the external pressure is 60.1 kPa.

2. Effect of changing temperature

If the pressure is maintained constant, then by ***Gay-Lussac's Law*** (sometimes called ***Charles' Law***), 'a given mass of gas will increase in volume by approximately 1/273rd of its volume at 0°C for every 1°C rise in its temperature' (and vice versa).

$$V_t = V_0 + t\left(\frac{V_0}{273}\right) \quad \text{where} \begin{cases} V_t = \text{Volume at } t°\text{C} \\ V_0 = \text{Volume at } 0°\text{C} \end{cases}$$

$$\text{or } V_t = V_0\left(1 + \frac{t}{273}\right)$$

Kelvin proposed that a new scale of temperature be adopted whose zero would be the temperature at which an *ideal* gas would occupy nil volume. If the plot of V versus t°C for an ideal gas is extrapolated to the point where $V=0$, it is found that this point is -273.15°C, so the zero point (0 K) on the Kelvin scale equals -273.15°C and the exact Gay-Lussac increment is 1/273.15 of the gas volume at 0°C per °C change in temperature. The degree Kelvin is equal in magnitude to the degree Celsius (i.e. 1 K = 1°C) so that,

$$t°\text{C} = (t + 273.15)\ \text{K}$$

and in all but the most exact calculations we can assume that t°C equals $(t+273)$ K.

Using the Kelvin scale of temperature we can express the Gay-Lussac relationship in another way. Let us consider the volumes V_1 and V_2 occupied by the same mass of gas at temperatures t_1°C and t_2°C.

Then, $$V_1 = V_0\left(1+\frac{t_1}{273}\right) \quad \text{and} \quad V_2 = V_0\left(1+\frac{t_2}{273}\right)$$

whence $$\frac{V_1}{V_2} = \frac{273+t_1}{273+t_2}$$

Expressing temperatures in K, when T_1 and T_2 are temperatures of the gas in K,

$$\frac{V_1}{V_2} = \frac{T_1}{T_2} \quad \text{and} \quad \frac{V_1}{T_1} = \frac{V_2}{T_2}$$

In general terms, for a given mass of gas maintained at constant pressure $V/T = \text{Constant}$.

EXAMPLE

Gas evolved during the fermentative growth of a bacterial culture had a volume of 580 cm³ when measured at a laboratory temperature of 17°C. What was the volume of this gas at the growth temperature of 37°C? (Assume that the gas volumes were measured at a constant pressure.)

$t_1 = 17°\text{C} \therefore T_1 = (17+273) = 290 \text{ K}$
$t_2 = 37°\text{C} \therefore T_2 = (37+273) = 310 \text{ K}$
$V_1 = 580 \text{ cm}^3$
$V_2 = ?$

By Gay-Lussac's Law, $\frac{V}{T} = \text{Constant}$

$$\therefore \frac{V_1}{T_1} = \frac{V_2}{T_2} \quad \text{and} \quad \frac{580}{290} = \frac{V_2}{310}$$

or, $$V_2 = \frac{580 \times 310}{290}$$

$$= \underline{620 \text{ cm}^3}$$

$\therefore$ the fermentation gas would have occupied a volume of 620 cm³ at 37°C.

3. Effect of changing temperature and pressure

The Laws of Boyle and Gay-Lussac can be combined to yield an expression which predicts the volume change that results from changes in the temperature and pressure of a given mass of ideal gas namely $PV/T = \text{Constant}$. This equation is called an *equation of state*; it describes a quantity of ideal gas solely in terms of its pressure, volume and temperature.

EXAMPLE

The gas pressure in a reaction vessel of fixed volume was to be reduced to 1 kPa. The available vacuum pump could lower the pressure only to 1.5 kPa at the laboratory temperature of 17°C. Could the required vacuum be obtained by then cooling the vessel in an ice-salt mixture at −25°C?

From the Laws of Boyle, and Gay-Lussac, $PV/T = \text{Constant}$

whence,
$$\frac{P_1 V_1}{T_1} = \frac{P_2 V_2}{T_2}$$

[*Initially*]	[*After cooling*]
$P_1 = 1.5$ kPa	$P_2 = ?$
$V_1 =$ volume of the reaction vessel	$V_2 = V_1 =$ volume of the reaction vessel
$T_1 = (17+273) = 290$ K	$T_2 = (-25+273) = 248$ K

Substituting these values in the above equation,

$$\frac{15 \times V_1}{290} = \frac{P_2 \times V_1}{248}$$

whence
$$\frac{15 \times 248}{290} = P_2 = \underline{1.28 \text{ kPa}}$$

Thus the cooling of the vessel to −25°C would not achieve the requisite partial vacuum.

Standard temperature and pressure (s.t.p.)

The volume occupied by a given quantity of a gas is dependent on the prevailing temperature and pressure. It therefore follows that if the volume of gas is to define its mass, its temperature and pressure must be defined. For this purpose, gas volumes may be recorded at a standard temperature and pressure (s.t.p.) defined as 273.15 K and standard atmospheric pressure (101 325 Pa). Of course, the volumes need not in practice be measured at 273.15 K and 101 325 Pa, for so long as the temperature and pressure of the gas are recorded when its volume is measured, the volume that it *would* have occupied at s.t.p. can always be calculated by applying the equation $PV/T = \text{Constant}$.

Avogadro's Law

The mole (symbol, mol) is a convenient unit in which to express the mass of any quantity of a compound (it equals the molecular weight of the compound in grammes). One mole of any compound contains 6.023×10^{23} molecules (the Avogadro number) and it has been discovered that one mole of any gas at s.t.p. will occupy a volume of 22.414 dm^3. This means that 22.414 dm^3 of any gas at s.t.p. contains 6.023×10^{23} molecules, and, since PV/T is constant for any given mass of an ideal gas, it follows that '*equal volumes of all ideal gases at the same temperature and pressure contain the same number of molecules*'. It is this statement that is known as ***Avogadro's Law.***

The Ideal Gas Law

For a given mass of any ideal gas PV/T is constant and when the 'given mass of gas' equals 1 mole the value of this constant is unique. Furthermore, as 1 mole of any one gas contains precisely the same number of molecules as are present in 1 mole of any other gas, this unique value of PV/T is a universal molar Gas Constant. If this molar Gas Constant is represented by the symbol R, then for any quantity of gas containing n mole of gas PV/T equals nR, or

$$PV = nRT$$

This expression is known as the ***ideal gas equation.***

Since real gases do not always behave even approximately in accordance with the predictions of the ideal gas equation, modifications of this equation (particularly that proposed by van der Waals) are sometimes employed (see p. 45). Yet most gases depart so little from ideality at low pressures that the ideal gas equation can then be applied to them in its unmodified form.

The value of the molar gas constant (R)

R equals PV/T for 1 mol of ideal gas which occupies a volume of 22.414 dm^3 at s.t.p.

$$\therefore \frac{PV}{T} = R \quad \text{when} \begin{cases} P = 101.325 \text{ kPa} = 101\,325 \text{ N m}^{-2} \\ V = 22.414 \text{ dm}^3 = 22.414 \times 10^{-3} \text{ m}^3 \\ T = 273.15 \text{ K} \end{cases}$$

$$\therefore R = \frac{101\,325 \times (22.414 \times 10^{-3})}{273.15} \text{ N m K}^{-1} \text{ mol}^{-1}$$

$$= \underline{8.3143 \text{ J K}^{-1} \text{ mol}^{-1}} \text{ (since 1 N m = 1 J)}$$

The units in which R is measured are units of energy per degree of temperature per mole, so that in older texts employing non-SI units you will find the value of R expressed in these units, e.g. $R = 1.987$ cal. degree^{-1} mole^{-1}, or $R = 0.082$ litre-atmospheres degree^{-1} mole^{-1}.

EXAMPLE

A 5-dm³ bomb calorimeter was to be filled with sufficient oxygen under pressure to support the complete combustion of 36 g of glucose. When filled with oxygen at room temperature from the only available cylinder, the final pressure was only 7.1×10^5 *Pa. Would this be sufficient to allow of complete combustion of the sugar? (Room temperature was 290 K, i.e. 17°C; glucose,* $C_6H_{12}O_6$*.)*

$P = 7.1 \times 10^5 \text{ N m}^{-2}$

$V = 5 \text{ dm}^3 = 5 \times 10^{-3} \text{ m}^3$

$R = 8.314 \text{ J K}^{-1} \text{ mol}^{-1}$

$T = 290 \text{ K}$

$n = ? \text{ mol}$

$$PV = nRT$$

$$\therefore n = \frac{PV}{RT} = \frac{(7.1 \times 10^5) \times (5 \times 10^{-3})}{8.314 \times 290} \text{ mol}$$

$$= \frac{3550}{2411} \text{ mol}$$

$$= \underline{1.47 \text{ mol}}$$

The complete combustion of glucose yields CO_2 and water;

$$C_6H_{12}O_6 + 6O_2 = 6CO_2 + 6H_2O$$

Six moles of O_2 would be required for the complete combustion of 1 mole of glucose. Thus for the 0.2 mol of glucose provided, 1.2 mol of oxygen would be required, and the 5 dm³ of O_2 at 7.1×10^5 Pa, being equal to 1.47 mol of O_2, would prove more than sufficient.

Partial pressures

It is characteristic of mixtures of gases that the component gases in many respects behave completely independently of one another. Thus it is possible to attribute the total pressure exerted by the gas mixture to the sum of the pressure contribution of each gas—this being the pressure that the gas would exert if it alone occupied the volume of the mixture at that temperature. This statement, that 'the pressure of the mixture is the sum of the partial pressures of the component gases', is known as ***Dalton's Law of Partial Pressures.*** From this and the Ideal Gas Law, it follows that the partial pressure of any gas in a mixture stands in the same ratio to the total pressure of the mixture as does its quantity in moles to the total number of moles of all gases present;

i.e. $$\frac{P_a}{P_t} = \frac{n_a}{n_t}$$

where
- P_a = partial pressure of gas (A)
- P_t = total pressure of gas mixture
- n_a = number of moles of gas (A)
- n_t = total number of moles of all gases in the mixture

EXAMPLE

In clinical studies involving gas mixtures, their composition is frequently expressed as the volume per cent (vol. %) contribution of each component gas, *which is that percentage of the total volume that is occupied by the stated gas (all volumes reduced to s.t.p. and referring to dry gas). Thus, alveolar air from the human lung contains nitrogen 80.5, oxygen 14.0 and carbon dioxide 5.5 vol.%. If the pressure in the lung is 1.01×10^5 Pa, and the vapour pressure of water is 6.25×10^3 Pa, calculate the partial pressures exerted by these major constituents.*

Actual pressure exerted by the dry gases = Total pressure − contribution of water vapour

$$= (1.01 \times 10^5) - (6.25 \times 10^3) \text{ Pa}$$
$$= \underline{94.75 \text{ kPa}}$$

By Avogadro's Hypothesis, the number of moles of gas is proportional to the volume it occupies at s.t.p. and since, from Dalton's Law plus the Ideal Gas Law, $\frac{P_a}{P_t} = \frac{n_a}{n_t}$,

then, $\frac{P_a}{P_t} = \frac{V_a}{V_t}$ where $\begin{cases} V_a = \text{volume at s.t.p of gas (A)} \\ V_t = \text{total volume at s.t.p. of mixture} \end{cases}$

Thus,

Partial pressure of N_2

$$P_{N_2} = \frac{V_{N_2} \times P_t}{V_t} = \frac{80.5 \times (94.75 \times 10^3)}{100} = \underline{76.27 \times 10^3 \text{ Pa}}$$

Partial pressure of O_2

$$P_{O_2} = \frac{V_{O_2} \times P_t}{V_t} = \frac{14.0 \times (94.75 \times 10^3)}{100} = \underline{13.27 \times 10^3 \text{ Pa}}$$

Partial pressure of CO_2

$$P_{CO_2} = \frac{V_{CO_2} \times P_t}{V_t} = \frac{5.5 \times (94.75 \times 10^3)}{100} = \underline{52.11 \times 10^3 \text{ Pa}}$$

Gas density

The density of a gas (as of any other state of matter) is its mass divided by its volume,† i.e.

$$\therefore d = \frac{10^{-3} w}{V} \quad \text{where} \begin{cases} d = \text{density in kg m}^{-3} \\ w = \text{mass in g} \\ V = \text{volume in m}^3 \end{cases}$$

† In SI, density is usually reported in kg m^{-3} or g cm^{-3}.

Since the volume occupied by a given mass of gas is affected by changes in pressure and temperature, its density will also be affected (though inversely). In the Ideal Gas Equation, $PV=nRT$, the term n represents the number of moles of gas considered,

whence, $n = \frac{w}{M}$ where $\begin{cases} w = \text{mass in g} \\ M = \text{molecular weight} \end{cases}$

or $w = nM$

Substituting for w in the density equation (above):

$$d = \frac{10^{-3}nM}{V}$$

For a given mass of gas at a fixed temperature and pressure, both n and V are constant. Therefore, under these conditions the density of a gas is directly proportional to its molecular weight.

The variable density of a gas must not be confused with its *vapour density* which is a constant and is defined as being equal to the fraction $\frac{\text{mass of a given volume of gas}}{\text{mass of an equal volume of hydrogen}}$, both masses being measured under identical conditions of pressure and temperature.

$$\text{Thus, vapour density of a gas} = \frac{\text{mass of } n \text{ mol of gas}}{\text{mass of } n \text{ mol of } H_2}$$

$$= \frac{\text{mass of 1 mol of gas}}{\text{mass of 1 mol of } H_2}$$

But molecular weight of $H_2 = 2$,

$$\therefore \text{vapour density} = \tfrac{1}{2} \text{ molecular weight of the gas}$$

EXAMPLE

A 250 cm³ container filled with argon at 273 K and 101.3 kPa weighed 48.30 g. Filled at 303 K and 105.9 kPa with the gas evolved by a photosynthesizing culture of green algae, it weighed 48.19 g. Assuming the unknown gas to be both dry, and a pure compound, attempt to identify its composition from these findings. (Density of argon at s.t.p. = 1.78 kg m⁻³; molar gas constant, R = 8.314 J K⁻¹ mol⁻¹.)

Mass of argon-filled container at 273 K and 101.3 kPa (i.e. s.t.p.) = 48.30 g. Density of argon at s.t.p. = 1.78 kg m^{-3} = 1.78×10^{-3} g cm^{-3}

$$\therefore \text{mass of 250 cm}^3 \text{ of argon at s.t.p.} = 250 \times (1.78 \times 10^{-3}) \text{ g} = 0.445 \text{ g}$$

whence, mass of empty container = (48.30 − 0.445) g = 47.855 g

∴ mass of 250 cm^3 of unknown gas sampled at 303 K and 105.9 kPa

$$= (48.19 - 47.855) \text{ g} = 0.335 \text{ g}$$

By the Ideal Gas Equation,

$$PV = nRT \qquad \text{whence,} \quad PV = \frac{wRT}{M} \quad \text{and} \quad M = \frac{wRT}{PV}$$

Here,

$$\begin{cases} P = 105.9 \text{ kPa} = 1.059 \times 10^5 \text{ N m}^{-2} \\ V = 250 \text{ cm}^3 = 2.5 \times 10^{-4} \text{ m}^3 \\ w = 0.335 \text{ g} \\ T = 303 \text{ K} \\ R = 8.314 \text{ J K}^{-1} \text{ mol}^{-1} = 8.314 \text{ N m K}^{-1} \text{ mol}^{-1} \\ M = \text{molecular weight in g mol}^{-1} \end{cases}$$

Substituting these values in the equation given above,

$$M = \frac{0.335 \times 8.314 \times 303}{(1.059 \times 10^5) \times (2.5 \times 10^{-4})} \text{ g mol}^{-1} = \underline{31.87 \text{ g mol}^{-1}}$$

The only gas with a molecular weight of approx. 32 is oxygen, thus it may be suggested that the gaseous product of algal photosynthesis was O_2.

Diffusion of gases

It has been mentioned that the expansion of a gas to fill a newly available space is the consequence of diffusion, which impels molecules of gas to quit a region of high density for one of lower density until homogeneity has been achieved. ***Graham's Law of Diffusion*** states that 'the rate of diffusion of a gas is inversely proportional to the square root of its density',

$$D \propto \frac{1}{\sqrt{d}} \qquad \text{where} \begin{cases} D = \text{rate of diffusion} \\ d = \text{density} \end{cases}$$

But the density of a gas at a fixed temperature and pressure is directly proportional to its molecular weight. Hence the rate of diffusion of a gas is inversely proportional to the square root of its molecular weight. For two gases whose rates of diffusion are measured at the same pressure and temperature,

$$D_1 \propto \frac{1}{\sqrt{M_1}} \quad \text{and} \quad D_2 \propto \frac{1}{\sqrt{M_2}}$$

$$\therefore \frac{D_1}{D_2} = \sqrt{\frac{M_2}{M_1}} \qquad \text{where} \begin{cases} D_1 = \text{rate of diffusion of gas 1} \\ M_1 = \text{M. Wt. of gas 1} \\ D_2 = \text{rate of diffusion of gas 2} \\ M_2 = \text{M. Wt. of gas 2} \end{cases}$$

The process of effusion (issuance of a gas through a small hole in its container) obeys the same laws as simple diffusion. By either means the molecular weight of a gas may be experimentally determined if a 'reference' gas of known molecular weight is available. In practice it is usual to measure the time (t) taken by a given volume of gas to effuse through a small hole in its container. This will be inversely proportional to its rate of diffusion; i.e.

$$t \propto \frac{1}{D}, \quad \text{and} \quad \frac{D_1}{D_2} = \sqrt{\frac{M_2}{M_1}}, \quad \therefore \frac{t_1}{t_2} = \sqrt{\frac{M_1}{M_2}}$$

EXAMPLE

A certain volume of the gas evolved by a photosynthesizing algal culture took 231 s to stream through a small hole. Under precisely the same conditions, an equal volume of argon took 258 s. Calculate (a) the molecular weight, and (b) the vapour density of the unknown gas (M. Wt. of argon = 40).

From Graham's Law $\dfrac{t_1}{t_2} = \sqrt{\dfrac{M_1}{M_2}}$

For the unknown gas:

$t_1 = 231$ s
$M_1 = ?$

For argon:

$t_2 = 258$ s
$M_2 = 40$

$$\frac{231}{258} = \sqrt{\frac{M_1}{40}}$$

$$\text{whence, } \left(\frac{231}{258}\right)^2 \times 40 = M_1$$

$$\therefore 0.802 \times 40 = M_1$$

$$\therefore \underline{32.08} = M_1$$

Thus, (a) molecular weight of the unknown gas = 32
(b) vapour density of the unknown gas = $\frac{1}{2}$ molecular weight
= $\underline{16}$

Vapour pressure

Liquids and solids in a closed space are in equilibrium with their vapour. The partial pressure of this vapour (known as the *vapour pressure* or, more correctly, the *saturation vapour pressure*) is dependent on the temperature, but is always significant in the case of liquids. This phenomenon is discussed more fully in Chapter 4 (p. 59). When considering the composition of gases maintained above a liquid phase, it must be remembered that the liquid will contribute its vapour pressure as a component of the total measurable gas pressure.

SOLUBILITY OF GASES IN LIQUIDS

Any gas will be to some extent soluble in every liquid. The rate at which it dissolves will depend on several factors, e.g. temperature, pressure and the surface area of the gas-liquid interface, but equilibrium will be established between a given volume of liquid and an excess of the gas only when the liquid is fully saturated with gas. The amount of gas that will then be in solution will again depend on the prevailing temperature and pressure, but will also depend upon the degree of solubility of the gas in that liquid. A useful term that is employed to express the degree of solubility of a gas in a liquid, at a fixed temperature and pressure, is the ***Bunsen absorption coefficient***. This is defined as 'that volume of gas in dm^3 at s.t.p. which saturates 1 dm^3 of the liquid when it is presented to the liquid at the reported temperature and a partial pressure of 1 atmosphere (i.e. 101 325 Pa). Since 1 mol of gas occupies 22.414 dm^3 at s.t.p., the solubility of the gas in mol dm^{-3} at standard atmospheric pressure and the prevailing temperature equals 1/22.414 times the Bunsen absorption coefficient of the gas at that temperature.

Partial pressure and solubility of a gas

By ***Henry's Law***, 'the *mass* of any gas which will dissolve in a given volume of liquid is directly proportional to the pressure of the gas' (provided the temperature remains unchanged). In this respect also, components of a gas mixture behave independently of each other, and the mass of each gas which at constant temperature dissolves in a given volume of liquid will be directly proportional to *its* absorption coefficient and to *its* partial pressure in the mixture.

Since the volume occupied by a gas at s.t.p. is a more convenient measure of its quantity than is its mass, we may express Henry's Law in the form of the following equation,

$$S = K\alpha P$$

where
- S = solubility of gas at the prevailing temperature expressed as volume at s.t.p. dissolved in unit volume of liquid
- K = a constant
- α = absorption coefficient for the gas in that liquid at the prevailing temperature
- P = partial pressure of the gas

Henry's Law is yet another example of an Ideal Gas relationship applicable to gases at low pressures and therefore to dilute solutions of many real gases. Gross deviations from the Law at low partial pressures of a gas are generally indicative of chemical interaction between gas and liquid, or of association or dissociation of the gas molecules in solution.

EXAMPLE

Calculate the quantity of nitrogen (in g) dissolved by 100 cm³ of blood plasma when this is aerated at 311 K (i.e. 38°C) and 102.7 kPa. (Absorption coefficient of N_2 in plasma at 311 K = 0.012; air consists of 78% N_2 by volume.)

'Absorption coefficient of N_2 in plasma at 311 K = 0.012' means that 0.012 cm³ (at s.t.p.) of N_2 dissolves in 1 cm³ of plasma at 311 K and standard atmospheric pressure (i.e. 101.325 kPa).

∴ 100 cm³ of plasma at 311 K and 101.325 kPa will dissolve 1.2 cm³ (at s.t.p.) of N_2

Partial pressure of N_2 in air at 102.7 kPa = 78/100 × 102.7 kPa

From Henry's Law, $S = K\alpha P$ or $\dfrac{S_1}{S_2} = \dfrac{P_1}{P_2}$

Solution of pure N_2 at 101.325 kPa	*Solution of N_2 from air at 102.7 kPa*
$S_1 = 1.2$ cm³ N_2/100 cm³ plasma	$S_2 = ?$ cm³/100 cm³ plasma
$P_1 = 101.325$ kPa	$P_2 = \dfrac{78 \times 102.7}{100}$ kPa

$$\therefore \frac{1.2}{S_2} = \frac{101.325 \times 100}{1.027 \times 78}$$

$$\text{whence } S_2 = \frac{1.2 \times 1.027 \times 78}{101.325 \times 100}$$

$$= 0.949 \text{ cm}^3 \text{ of } N_2/100 \text{ cm}^3 \text{ plasma}$$

But 24.414 dm³ (at s.t.p.) of N_2 has a mass of 28 g

∴ 0.949 cm³ (at s.t.p.) of N_2 will have a mass of $\dfrac{0.949 \times 28}{24.414 \times 10^3}$ g

$$= \underline{1.186 \times 10^{-3}} \text{ g } (= 1.186 \text{ mg})$$

Thus, 100 cm³ of blood plasma will dissolve 1.186 mg N_2 when aerated under the given conditions.

Temperature and solubility of a gas

Heat is generally liberated when gases dissolve in water. Since most gases have a positive heat of solution, it follows (p. 221) that the higher is the temperature the lower will be the solubility of a gas in an aqueous medium, provided that no chemical interaction occurs. The decrease in solubility that will result from a given rise in temperature can be predicted from the quantity of heat that is liberated when 1 mole of the gas is added to such a large volume of the solution that it causes no appreciable change in its concentration. This heat is called the *differential heat of solution* of the gas, and its value can be determined experimentally. However, in most biochemical calculations it is only necessary to select the appropriate values of the absorption coefficient (α) from reference tables that list values of α in water for most gases, over a wide range of temperatures.

REAL GASES

For a gas to conform precisely at all pressures and temperatures to the predictions made by Boyle's and Gay-Lussac's Laws, its molecules would have to be devoid of volume and exert no mutual forces of attraction or repulsion. It is not surprising therefore that the *ideal gas* is a notional concept, of value because it approximately describes the actual properties of *real gases* at low pressures and moderate temperatures, and facilitates interpretation of their aberrant behaviour as the pressure is increased or the temperature lowered.

We have seen that according to the ideal gas law,

$$PV = nRT \quad \text{or} \quad \frac{PV}{nRT} = 1$$

Fig. 3.1 shows the manner in which three real gases are experimentally observed to deviate from this prediction. Whereas for 1 mol of ideal gas at 273.15 K, PV/nRT should be unity irrespective of the prevailing pressure, we see that both 'positive' and 'negative' deviations from this prediction are evidenced by different real (i.e. actual) gases. The nature and extent of these deviations should in turn be interpretable in terms of those molecular properties of these real gases which are the root cause of their non-ideal behaviour. Consider neon, whose value of PV/nRT becomes increasingly greater than 1.0 as the pressure is increased (Fig. 3.1). The low boiling point of this gas (27 K) indicates that the force of attraction between its molecules is extremely weak, so that the 'positive' deviation in PV/nRT is likely to be almost entirely attributable to the non-ideality of its molecules in possessing a far from negligible volume. Intermolecular attraction is evidently more important in the case of oxygen, whose boiling point (90 K) is substantially higher, while carbon dioxide condenses to the solid state at 195 K and the greater attraction between its molecules is mirrored in the even more pronounced 'negative' deviation of PV/RT (see Fig. 3.1).

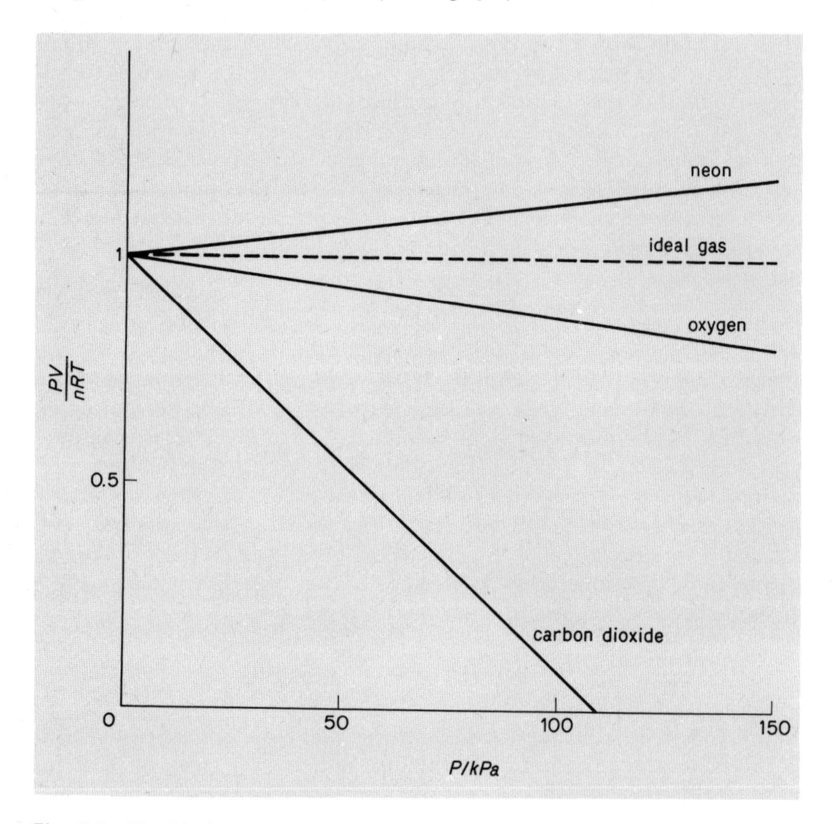

Fig. 3.1 'Positive' and 'negative' deviations from the ideal gas law (at 273 K).

Boyle's Law predicts that at a given temperature a given increase in pressure will proportionately decrease the volume of a quantity of ideal gas, thus if one plots P *versus* V for an ideal gas maintained isothermally (i.e. at constant temperature) one predictably obtains a hyperbolic graph (Fig. 3.2)—predictably, since whenever y changes in inverse proportion to x the plot of x versus y is a hyperbola of this type. Series of corresponding values of P and V obtained for an ideal gas at different temperatures will plot as a family of these hyperbolas. Each of these is called an

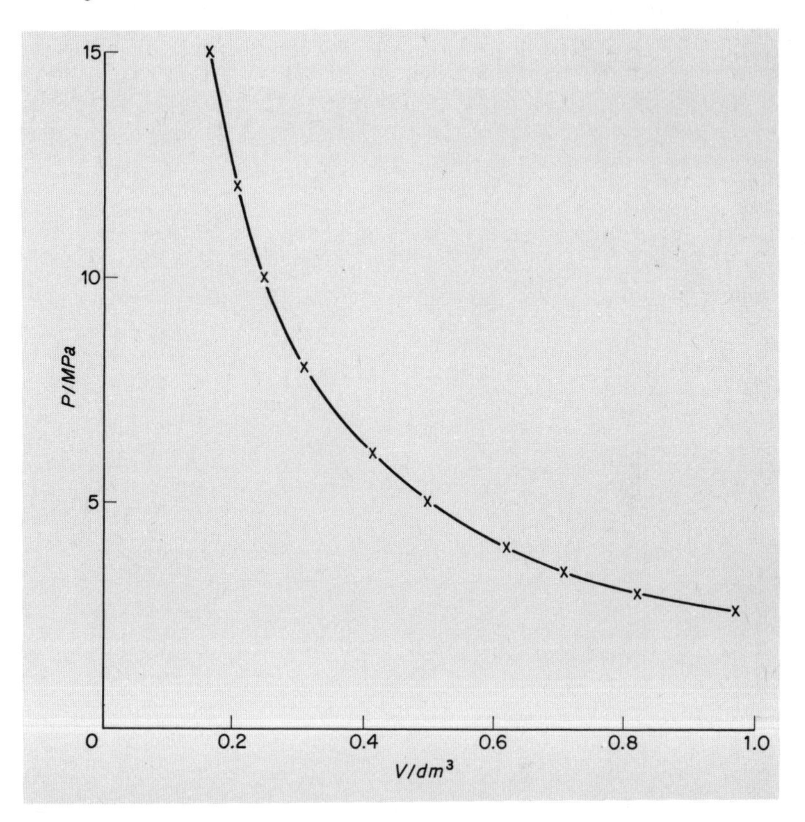

Fig. 3.2 Isotherm of the ideal gas at 298 K.

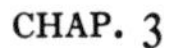

isotherm (or *isothermal*) since it represents the P vs. V behaviour of the gas at a given constant temperature.

In contrast, Fig. 3.3 shows the actual isotherms obtained when 1 mole of carbon dioxide was held at six different temperatures between 273 K and 323 K. The isotherm obtained at 323 K most nearly resembles the ideal hyperbolic plot. At lower temperatures the isotherms are manifestly aberrant, mirroring the existence of potent forces of intermolecular attraction, and, when the temperature is sufficiently low, the phenomenon of liquefaction. Consider the isotherm obtained at 273 K. As the pressure is increased (from A) the volume decreases until at B there is a marked discontinuity in the shape of the isotherm. From B to C there is considerable decrease in volume with imperceptible increase in pressure, but thereafter (C to D) considerable increase in pressure produces relatively little change in volume. The peculiar shape of this 273 K isotherm is explained by the fact that at point B the pressure is such that liquefaction is initiated, this then proceeds (B to C) until at point C the gas has been totally condensed into the liquid state. Thereafter, the steep

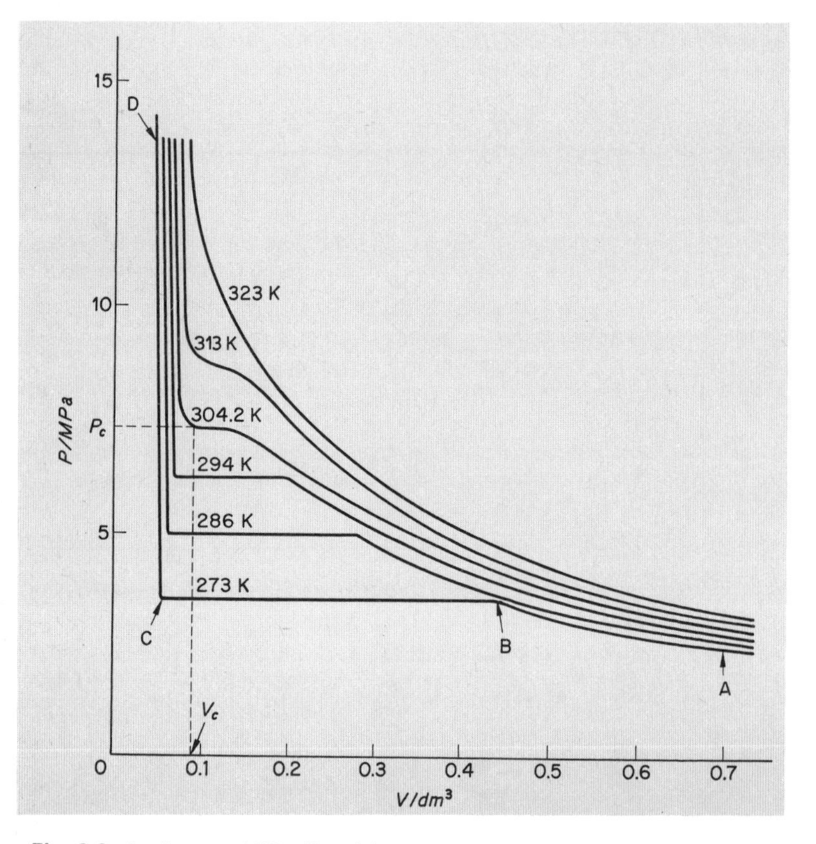

Fig. 3.3 Isotherms of CO_2 (1 mole).

rise in the isotherm (C to D) reflects the liquid's resistance to compression, accounting for the sharp discontinuity in the isotherm at point C. Similar sharp discontinuities in the isotherms obtained at other temperatures accurately identify the situation in which liquid and gas phases coexist (i.e. throughout the horizontal portions of these isotherms). The higher the temperature the shorter is this horizontal segment of the isotherm; that is to say, the smaller is the change in volume at constant pressure as gas isothermally condenses to yield liquid. If V_g is the volume of gas at the start of this liquefaction and V_l is the volume of liquid produced, then $(V_g - V_l)$ diminishes as the temperature is increased, until at the *critical temperature* it becomes zero. Thus we can determine the critical temperature (T_c) by plotting values of $(V_g - V_l)$ against the values of T at which these were obtained, and then extrapolating to $(V_g - V_l) = 0$. In the case of carbon dioxide the critical temperature emerges as 304.2 K (i.e. 31.1°C), and as shown in Fig. 3.3 the isotherm of carbon dioxide obtained at 304 K is the first (as the temperature is increased) not to display a recognizable horizontal segment. The isotherm obtained at the critical temperature therefore has zero slope only at a single point (the *critical point*); the pressure at the critical point is the *critical pressure* (P_c) and the molar volume is the *critical volume* (V_c). Below the critical temperature, application of pressure to a real gas decreases its volume and increases its density until intermolecular cohesive forces cause condensation of a portion of the gas to produce liquid. The two distinct phases are simultaneously present (and at a given temperature and pressure are in equilibrium with each other) over a visible boundary which we know as the fluid meniscus. If the temperature were increased to the critical temperature this meniscus would disappear, for the density of the gas would then precisely equal that of the liquid. Thus, above the critical temperature (as at the critical point) it is invidious to talk of liquid or gas at all, one can only talk of a homogeneous hyperfluid state.

It might be thought that acquaintance with isotherms and an understanding of the concept of critical point and critical temperature are scarcely likely to benefit the practising biologist. Yet this topic provides an apt example of how the informed biologist is also the better-equipped experimentalist, for knowledge of the critical properties of carbon dioxide enabled Anderson to suggest the so-called 'critical point drying procedure' as a method of retaining the three-dimensional structures of biological specimens during their preparation for electron microscopic examination.[15]

Since electron microscopy requires that the specimen be held in a vacuum, it must be pre-dried. Dehydration is generally accomplished by passage through a series of ethanol solutions of progressively increasing concentration, but whatever is the terminal solvent, it must ultimately be removed by evaporation. Normally as the solvent evaporates, the shrinking liquid/air meniscus (phase boundary) passes through the specimen, distorting and flattening many of its more flimsy components. To obviate such artefactual distortion of the specimen, Anderson in 1951 suggested that the ethanol-dehydrated specimen be transferred via amyl acetate into liquid CO_2 which might then be converted into its hyperfluid state and removed in that condition at a temperature just higher

A

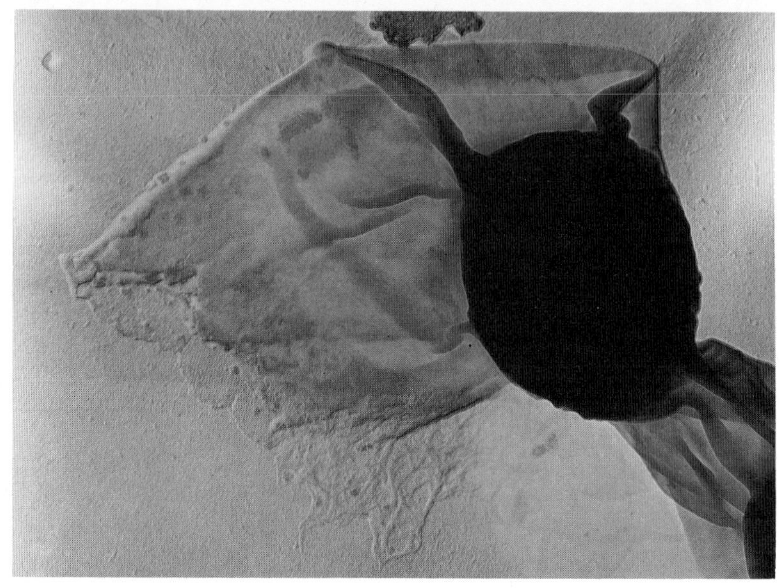

B

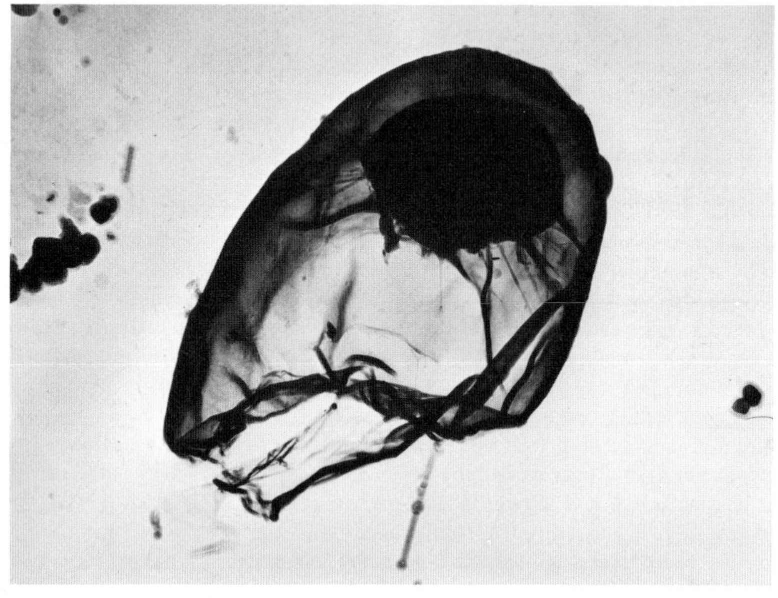

Fig. 3.4. A platinum shadowed spore of *Clostridium pasteurianum* demonstrating how the flimsy outer integument (exosporium) is flattened by the procedures usually employed for specimen preparation. B Critical point-dried spore; the true nature of the exosporium is now revealed (see Mackey & Morris, *J. gen. Microbiol.* **73**, 325 [1972]).

than its conveniently moderate critical temperature of 304 K. In practice, the specimen is removed from amyl acetate into a steel pressure vessel at room temperature and liquid CO_2 from a gas cylinder is introduced. Some liquid CO_2 with gaseous CO_2 and residual amyl acetate is allowed to escape before the vessel is closed and its temperature is raised to 45°C (i.e. 318 K). The enclosed liquid CO_2 immediately passes into the hyperfluid state (accompanied by a substantial rise in pressure) so that when the exit valve is again opened the CO_2 escapes in this state, leaving a dry specimen which has not suffered damage by contact with a phase boundary. The value of this technique is illustrated in Fig. 3.4.

The van der Waals equation

If the real gas behaves non-ideally because its molecules (a) occupy a significant volume, and (b) exert upon each other a significant force of attraction, it might be possible to modify the ideal gas equation by introducing terms which compensate for these factors, so as to produce a somewhat more complex equation which better predicts the behaviour of real gases. In 1873, van der Waals proposed such a modified gas equation which he derived as follows. He argued that if the volume occupied by the gas molecules themselves ($V_{\text{molecules}}$) was subtracted from the measured volume occupied by a given quantity of gas (V_{observed}) the resultant volume (V_{free}) was that space which the molecules could freely enter and in which they could execute their movements. Strictly speaking, in the ideal gas equation $PV = nRT$, the volume term V is actually V_{free}, since the molecules of an ideal gas occupy nil volume. For a real gas however,

$$V_{\text{free}} = (V_{\text{observed}} - b), \text{ where } b \text{ is the allowance made for } V_{\text{molecules}}$$

Unfortunately it is not possible to measure or calculate $V_{\text{molecules}}$ since the molecule is not a rigid entity with a well-defined and inflexible boundary. Thus the value of b must be deduced from measurements of the magnitude of the deviation from ideal behaviour that is displayed by the real gas at high pressures and relatively high temperatures where the possession of a significantly large $V_{\text{molecules}}$ will be the predominant cause of aberrant behaviour.

The pressure term P in the ideal gas equation is the ideal kinetic pressure exerted by the gas molecules. Its value equals the observed pressure only because the molecules of an ideal gas exert no mutual force of attraction. Thus the observed pressure exerted by a real gas will be less than the ideal kinetic pressure by an amount equal to that pressure dissipated in overcoming the actual intermolecular attraction,

$$\text{i.e.} \quad P_{\text{kinetic}} = (P_{\text{observed}} + P_{\text{against intermolecular attraction}})$$

Since the force of intermolecular attraction should be inversely proportional to the square of the volume, $P_{\text{against intermolecular attraction}}$ should similarly decrease in magnitude as the volume increases,

$$\text{i.e.}\quad P_{\text{against intermolecular attraction}} = \frac{a}{V^2}$$

where *a* is an empirical constant characteristic for each real gas. The value of *a* is obtained by measuring the extent by which the gas deviates from ideal behaviour at relatively low temperatures where the effect of intermolecular attraction predominates.

The effective kinetic pressure of a real gas is therefore given by the equation,

$$P_{\text{kinetic}} = \left(P_{\text{observed}} + \frac{a}{V^2}\right)$$

Van der Waals concluded that it was the product of the kinetic pressure and the free volume which was constant for a given quantity of gas at a set temperature, i.e. for 1 mol of real gas,

$$P_{\text{kinetic}} \times V_{\text{free}} = RT$$

Thus when the observed pressure is P and the observed volume is V,

(i) *for an ideal gas* (1 mol)

$$\left.\begin{aligned} P_{\text{kinetic}} &= P \\ V_{\text{free}} &= V \end{aligned}\right\} \qquad \therefore\ PV = RT$$

and for a quantity of n mol, $PV = nRT$

(ii) *for a real gas* (1 mol)

$$\left.\begin{aligned} P_{\text{kinetic}} &= \left(P + \frac{a}{V^2}\right) \\ V_{\text{free}} &= (V-b) \end{aligned}\right\} \qquad \therefore\ \left(P + \frac{a}{V^2}\right)(V-b) = RT$$

while for n mol of a real gas this van der Waals equation becomes

$$\left(P + \frac{an^2}{V^2}\right)(V-nb) = nRT$$

Values of the van der Waals constants *a* and *b* have been determined for most gases and are listed in reference books. Generally those gases with the larger molecules have the larger values of *b*. Similarly, since the larger molecule will have the larger surface, it is to be expected that it will exert a more powerful attractive force on its neighbour, so that a larger value of *a* might also be anticipated. But complications can arise from the fact that the shape of the molecule will also be a determinant of its surface area and other forces such as hydrogen bonding will unduly increase the magnitude of *a*.

Van der Waals' equation only approximately describes the behaviour of a real gas, but, even so, it represents a considerable advance over the attempt to impose the ideal relationship on real gases at high pressures or low temperatures. Fortunately, since the biologist is generally interested in the behaviour of gases at low pressures and at near room temperature, he can often assume that any difference between real and ideal behaviour may be disregarded.

MEASUREMENT OF GAS UPTAKE OR OUTPUT BY BIOLOGICAL SYSTEMS

The measurement of gas volumes at various temperatures and pressures, and the conversion of these volumes into the molar quantities of gas that they represent, is a common task in biological investigations. Sometimes comparatively large amounts of gas are involved; for example, when the yield of gaseous products is measured during the fermentative growth of microbial cultures. More frequently, the evolution or consumption of gases by biological materials is studied on a micro-scale using manometric methods. These remain among the most sensitive methods of following the course and extent of a biochemical reaction, and some reactions that do not involve gases are frequently coupled to others that do, merely so that they can be studied by manometric techniques, e.g. a reaction producing acid could be made to liberate CO_2 from a CO_2/bicarbonate buffer (see also reference[35]).

Two manometric methods may be used:

1 ***Constant pressure manometry***, which measures a change in the *volume* of a gas at constant pressure and constant temperature. The Haldane–Barcroft manometer operates on this principle, as does the Van Slyke 'volumetric' apparatus for gas analysis.

2 ***Constant volume manometry***, which measures a change in the *pressure* of a gas whose volume is kept constant at a constant temperature. The Warburg manometer which operates on this principle is the most frequently used form of 'micro-respirometer' in biochemical laboratories. The Van Slyke & Neill 'manometric' apparatus for gas analysis also assays amounts of gas in terms of their partial pressures in a fixed volume at constant temperature.

The Warburg constant volume manometer

The apparatus (Fig. 3.5) is essentially a capillary U-tube bearing at its base a reservoir of manometer fluid. The left-hand limb of the U-tube is left open to the atmosphere, while the right-hand limb may be closed by a

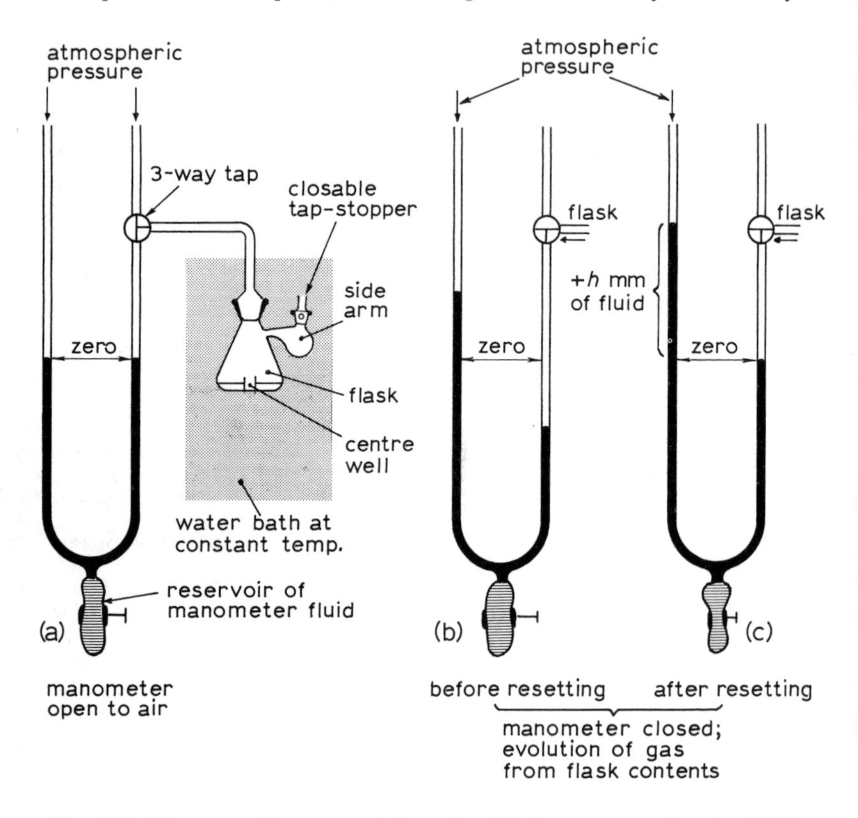

Fig. 3.5 Diagrammatic representation of a Warburg manometer. In (b) the increase in pressure resulting from gas evolution in the flask has forced the manometer fluid 'down' in the right-hand (closed) limb, and 'up' in the left-hand (open) limb. The resultant increase in pressure is measured at constant volume in (c) as $+h$ mm manometer fluid, by returning the meniscus in the right-hand limb to its initial, zero level. This is done by squeezing additional fluid into the manometer from the basal reservoir.

3-way stop tap. High up on the right-hand limb is a side arm terminating in a 'male' ground glass joint onto which a small manometer flask (about 20 cm^3 capacity) can be attached. A scale marked in millimetres is fixed behind the U-tube so that a rise or fall of the fluid meniscus in the open (left-hand) limb can be accurately measured. The zero point of this scale is approximately half way up the U-tube, and the fluid in the right-hand limb is always returned to this zero point before any reading is taken of the level of the meniscus in the left-hand limb. This maintains a constant volume of gas (V_g) in the manometer and flask. The value of V_g for any manometer and its flask is given by ($V_0 - V_f$), where V_0 is the overall volume enclosed by the flask *plus* manometer side arm *plus* that portion of the right-hand manometer limb between the stop tap and the zero calibration point, while V_f equals the volume of the fluid contents of the flask.

The gas-producing or gas-consuming reaction proceeds in the liquid contents of the manometer flask. This liquid is in equilibrium with the enclosed gas phase so that utilization of dissolved gas is reflected as a decrease in gas pressure, and vice versa. If gas is released during the reaction in the flask, the level of the manometer fluid falls in the right-hand limb of the manometer U-tube, causing a rise in the level of fluid in the left-hand limb. To read the actual increase in pressure 'at constant volume', the meniscus in the right-hand limb must be returned to the zero point. This is done by introducing more manometer fluid from the reservoir. The rise in the level of fluid in the left-hand limb of the manometer can be quantitatively related to the amount of gas released in the flask. Thus if such readings are taken at frequent intervals during the course of the reaction, an almost continuous record of the course of the gas output is obtained. Exactly the reverse occurs when gas is consumed in the flask, for now the level of fluid in the right-hand limb rises and must be returned to the zero point by withdrawal of manometer fluid into the reservoir. The level of the meniscus in the left-hand limb falls proportionately.

The shape of the manometer vessel is dictated by the need to provide (1) one or more side limbs to hold reactants which may be 'tipped in' following equilibration of the contents of the vessel with the gas phase, (2) a closable, side-arm tap-stopper through which, (and via the manometer 3-way stop tap), the gas space may be flushed with different gas mixtures, and (3) a compartment within the flask (a centre well) which can accommodate various gas-absorbing materials—usually KOH to remove CO_2. The manometer flask contains a comparatively small volume of reaction mixture, and it is oscillated or reciprocally shaken in a constant-temperature water bath. By this means a large liquid-gas surface is established in order to promote rapid diffusion and equilibration of 'free' and dissolved gases at the strictly maintained temperature. Concurrent readings, taken with a 'blank' manometer (a 'thermobarometer') whose flask contains only water or buffer solution, are used to compensate for minor fluctuations in atmospheric pressure and bath temperature during the course of an experiment.

The pressure changes in the manometer are thus translated into distances moved by the manometer fluid in the left-hand limb in either an upward (positive) or downward (negative) direction. Movements of the similar meniscus in the thermobarometer are subtracted from those recorded in the experimental manometers. The resulting distances (in mm) can be translated into mm^3 of dry gas at s.t.p. by multiplication by a constant (the manometer constant). The value of this constant must be calculated afresh for each manometer whenever one of its determinants is changed. These include:

(a) the nature of the gas whose partial pressure changes in the manometer and flask;
(b) the prevailing temperature;
(c) the volume of liquid in the flask (V_f).

Calculation of the manometer constant

The value of the manometer constant (K) relating to a specific gas in a particular manometer at a single temperature, is obtained by substitution in the following equation:

$$K = \frac{V_g \frac{273}{T} + V_f . \alpha}{P_0}$$

where V_g = volume of available gas space in mm^3
T = experimental temperature in K
V_f = volume of fluid in manometer flask in mm^3
α = absorption coefficient of the exchanged gas in the liquid contents of the manometer flask at temperature T
P_0 = normal atmospheric pressure, expressed in mm of manometer fluid. (This fluid is usually so constituted that $P_0 = 10\,000$ mm manometer fluid.)

Note that:

1. the effect of temperature on the solubility of the exchanged gas is taken into account by using the value of its absorption coefficient (α) that is relevant to the experimental temperature;
2. in both solubility and partial pressure the exchanged gas is wholly independent of the concurrent presence of other gases, and this independence extends to the manometer constant. Thus the manometer constant (K_{O_2}), to be used when following the utilization of oxygen by respiring tissues in a given manometer, has the same value irrespective of whether the atmosphere in the manometer flask consists of pure oxygen, air, or an oxygen plus carbon dioxide mixture.

EXAMPLE

A Warburg manometer flask (23.0 cm^3 volume when attached to its manometer) contained 3.0 cm^3 of a washed suspension of bacteria supplied with suitable substrate. It was to be used to follow (a) uptake of oxygen from air when the suspension was incubated at 37°C, and (b) evolution of nitrogen from nitrate at 30°C.

Calculate the values of the manometer constant appropriate to these experiments.

$$(\alpha_{O_2} \text{ at } 310 \text{ K} = 0.024; \quad \alpha_{N_2} \text{ at } 303 \text{ K} = 0.013)$$

(a) *Uptake of oxygen at 310 K (i.e. 37°C)*

The value of the flask constant K is given by the equation,

$$K = \frac{V_g \times \frac{273}{T} + V_f \alpha}{P_0}$$

where $V_g = (23-3) \text{ cm}^3 = 20 \times 10^3 \text{ mm}^3$
$T = 310 \text{ K}$
$V_f = 3 \text{ cm}^3 = 3 \times 10^3 \text{ mm}^3$
$\alpha = 0.024$
$P_0 = 10 \times 10^3$ mm of manometer fluid

$$\therefore K = \frac{\left(20 \times 10^3 \times \frac{273}{310}\right) + (3 \times 10^3 \times 0.024)}{10 \times 10^3}$$

$$= \frac{\left(\frac{20 \times 273}{310}\right) + (3 \times 0{\cdot}024)}{10}$$

$$= \frac{17.60 + 0.072}{10} = \underline{1.767}$$

$\therefore$ flask constant for uptake of oxygen at 310 K = 1.767

(b) *Evolution of nitrogen at 303 K (i.e. 30°C)*

Values of the relevant terms are as in (a) above, except that now $\alpha = 0.013$ and $T = 303$ K

$$\therefore K = \frac{\left(\frac{20 \times 273}{303}\right) + (3 \times 0.013)}{10}$$

$$= \frac{18.02 + 0.039}{10} = \underline{1.806}$$

$\therefore$ flask constant for evolution of nitrogen at 303 K = 1.806

Use of the Warburg manometer

The amount of gas produced or consumed in the flask of a Warburg manometer during a certain time interval, is calculated by applying the following simple equation:

$V = h \times K$ where
- V = volume of gas exchanged in mm^3 at s.t.p.
- h = distance in mm by which the meniscus moves in the left-hand limb of the manometer
- K = relevant flask (manometer) constant

If h has a negative value (i.e. the level of the manometer fluid falls in the left-hand limb), then V also has a negative value equal to the volume of gas in mm^3 at s.t.p. 'taken up' within the flask. If h has a positive value, then V has a positive value equal to the volume of gas in mm^3 at s.t.p. that is produced within the manometer flask.

Since 1 μmol of any gas at s.t.p. occupies 22.4 mm^3, $V/22.4$ equals the number of μmol of gas produced or consumed within the manometer flask.

EXAMPLE

The following readings were obtained in three manometers incubated at 30°C and containing in their flasks (under air):

1. *3 cm^3 buffer (i.e. thermobarometer);*
2. *bacterial suspension in 2.8 cm^3 buffer (20 mg dry weight of bacteria) plus 0.2 cm^3 of 4 mol dm^{-3} KOH in the centre well;*
3. *as (2), but with 0.1 cm^3 of an aqueous solution of sodium acetate introduced from the side arm at zero time, and with the bacterial suspension containing the same quantity of bacteria made up in 2.7 cm^3 of buffer.*

Time/min	*Manometer*		
	1	*2*	*3*
0	29	130	145
5	27	123	115
10	29	120	86
15	30	116	55
20	28	109	24
25	28	104	19
30	27	99	14

Calculate, (a) Q_{O_2} for acetate oxidation by the bacterial suspension, (b) the quantity of acetate added (assuming total oxidation).

(Q_{O_2} = mm^3 of oxygen utilized mg^{-1} dry wt of organisms h^{-1}. The relevant manometer constants (K_{O_2}) at 303 K were (1) = 2.07, (2) = 1.98, (3) = 2.23.)

The manometer readings can be interpreted in the following way (Table 3.1):

(a) calculate the amount by which each thermobarometer reading differs from that immediately previous to it;

(b) calculate the 'interval changes' in the readings of all other manometers as in (a): the results are entered in column i for each manometer (see below);

Table 3.1

Time/min	Thermobarometer Reading	Thermobarometer Change	Manometer 2 (K = 1·98) Reading	i	ii	iii	iv
0	29	—	130	—	—	—	—
5	27	−2	123	−7	−5	−10	−10
10	29	2	120	−3	−5	−10	−20
15	30	1	116	−4	−5	−10	−30
20	28	−2	109	−7	−5	−10	−40
25	28	0	104	−5	−5	−10	−50
30	27	−1	99	−5	−4	− 8	−58

Time/min	Thermobarometer Reading	Thermobarometer Change	Manometer 3 (K = 2·23) Reading	i	ii	iii	iv
0	29	—	145	—	—	—	—
5	27	−2	115	−30	−28	−63	−63
10	29	2	86	−29	−31	−69	−132
15	30	1	55	−31	−32	−71	−203
20	28	−2	24	−31	−29	−65	−268
25	28	0	19	− 5	− 5	−10	−278
30	27	−1	14	− 5	− 4	− 8	−286

(c) subtract the thermobarometric interval changes from the corresponding interval changes for the remaining manometers (column ii);

(d) multiply the true interval changes in the manometric readings (column ii) by the relevant manometer constants to give the gas volume interval changes in mm^3 at s.t.p. (column iii);

(e) sum the gas volume interval changes (column iii) to give the total change in the gas volume at s.t.p. (column iv);

(f) represent graphically the course of the gas exchange in each manometer, by plotting the *total gas volume change* (mm^3 at s.t.p.) against *time* in min as in Fig. 3.6.

From Fig. 3.6 we see that the consumption of oxygen in manometer 2 was slight but progressive, due presumably to the endogenous respiration of the bacterial suspension. The very rapid uptake of oxygen in manometer 3 ceased after 20 min when presumably all of the acetate had been oxidized; the oxygen consumption then fell to the same slow rate as that in manometer 2. Assuming that the endogenous respiration proceeded

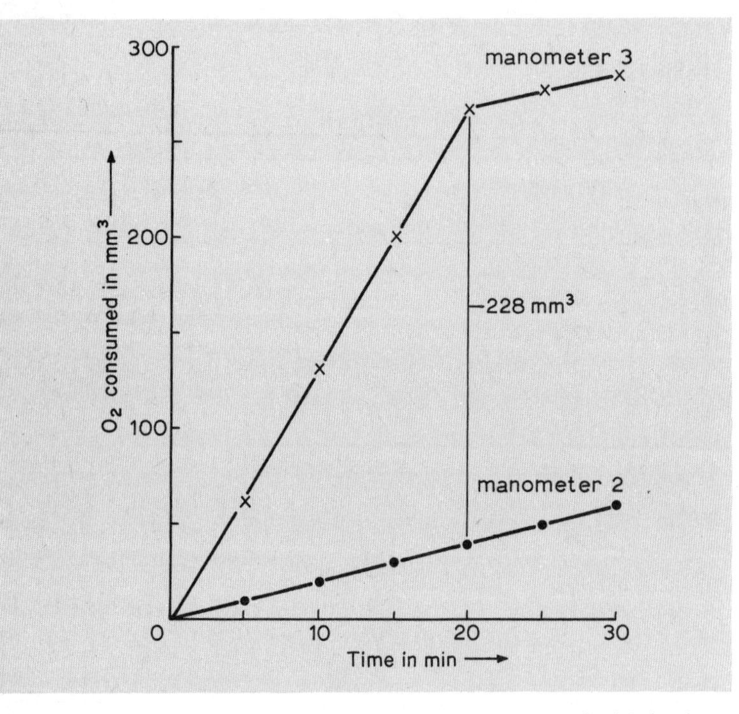

Fig. 3.6 Plot of oxygen uptake versus time in the absence of added substrate (●—●), and in the presence of acetate (×—×).

unchanged during acetate utilization, the actual uptake of oxygen that is attributable to the oxidation of the acetate is equal to the difference between the volumes of gas consumed at the end of the first 20 min in manometers 2 and 3. This equals 228 mm^3 at s.t.p. (from Fig. 3.6). (Note: CO_2 produced in the manometer flasks is removed by the KOH in their centre wells.)

As the complete oxidation of acetate proceeds as follows,

$$CH_3COOH + 2O_2 \longrightarrow 2CO_2 + 2H_2O$$

the total oxidation of 1 μmol of acetate consumes 2 μmol of oxygen.

(a) Q_{O_2} *for acetate oxidation by the bacterial suspension*

20 mg dry wt of bacteria consumed 228 mm^3 of O_2 over a period of 20 min for the oxidation of acetate;

$$\therefore Q_{O_2} = mm^3\ O_2 \text{ at s.t.p., consumed for acetate oxidation } mg^{-1} \text{ dry wt } h^{-1}$$

$$= 228 \times \frac{1}{20} \times \frac{60}{20} = \underline{34.2}$$

(b) *Amount of acetate added*

1 μmol of O_2 occupies a volume of 22.4 mm^3 at s.t.p.; thus 2 μmol of O_2 occupy a volume of 44.8 mm^3 at s.t.p.

Now, 2 μmol of O_2 (44.8 mm^3 at s.t.p.) are consumed for every 1 μmol of acetate that is completely oxidized. Since, in all, 228 mm^3 at s.t.p. of O_2 were consumed for acetate oxidation, assuming complete oxidation of the acetate,

$$\text{quantity of acetate oxidized} = \frac{228}{44.8}\ \mu\text{mol} = 5.09\ \mu\text{mol}$$

Therefore approximately 5 μmol of sodium acetate was added to the bacterial suspension in the flask of manometer 3.

PROBLEMS

(*Assume where necessary that s.t.p. is 273 K and 101.3 kPa, and that the gas constant* $R = 8.314\ J\ K^{-1}\ mol^{-1}$.)

1. A perfectly elastic, hydrogen balloon of 2 m diameter was released at sea level. What would be its diameter when it had risen to 3050 m above sea level? (Assume no change in temperature. Atmospheric pressures are: 101.3 kPa at sea level, and 68.1 kPa at 3050 m above sea level. Volume of a sphere is $4/3\ \pi r^3$.)

2. Convert the following volumes of gas into volumes at s.t.p.:

(i) 450 cm^3 at 303 K and 102.7 kPa;
(ii) 25 cm^3 at 310 K and 12.156×10^5 Pa;
(iii) 25 cm^3 at 256 K and 72 kPa.

3. What volume would be occupied by 1 dm^3 (at s.t.p.) of an ideal gas at:

(i) 303 K and 102 kPa?
(ii) 288 K and 2.026×10^6 Pa?
(iii) 258 K and 2.026×10^5 Pa?

4. A small cylinder contained 500 cm^3 of gas at 1.52 MN m^{-2} and 291 K. If this gas were dissolved in 10 dm^3 of water, what would be the molarity of the solution?

5. An insoluble gas, produced during fermentation by a bacterial culture, was collected above water at 30°C and 750 mm Hg pressure. If it occupied

430 cm^3 under these conditions, calculate the volume of the dry gas at s.t.p. (Vapour pressure of water at 303 K = 4.266 kPa; 1 mmHg = 133.32 Pa.)

6. The bacterium *Azotobacter chroococcum*, growing aerobically in a nitrogen-free medium, obtains all of its nitrogen by the 'fixation' of atmospheric N_2. What volume of air at standard atmospheric pressure and 303 K would supply the nitrogen requirement of 1 dm^3 of a culture of this bacterium, which grows to a density of 0.84 mg dry weight cm^3 of organisms, having a nitrogen content of 7% of the dry weight? (Air contains 78 vol.% N_2.)

7. What pressure of air at 303 K would be required to support the complete combustion of 1.5 g of lactic acid in a vessel of 1 dm^3 capacity? (Lactic acid, $CH_3CHOHCOOH$, has a molecular weight of 90; air contains 21% by volume of oxygen.)

8. It was proposed to manufacture a small quantity of a gas mixture containing 95% N_2 and 5% CO_2 (% by volume). A watch glass carrying a weighed amount of anhydrous sodium carbonate was floated on the surface of concentrated sulphuric acid in the base of a vacuum desiccator. By successively evacuating the desiccator and flushing with pure nitrogen gas, it was filled with 2 dm^3 of N_2 at 98.6 kPa. The desiccator was now tipped so as to react the carbonate with the acid. What weight of sodium carbonate should have been taken to yield the correct gas mixture? (Temperature = 290 K; molecular weight of anhydrous sodium carbonate, Na_2CO_3 = 106.)

9. Workers in underwater caissons necessarily breathe air at greater than normal pressures. If returned too rapidly to the surface, N_2, dissolved in their blood at the previously higher pressure, comes out of solution and may cause emboli (gas bubbles in the blood stream), severe pains (bends), and a general malaise (decompression sickness). Slow return to the surface, or the equivalent use of a decompression chamber, allows time for the gradual dissipation of this dissolved nitrogen gas.

Calculate the volume of N_2 likely to be liberated from the blood plasma of a caisson worker who is returned to an atmospheric pressure of 101.3 kPa after prolonged exposure to an air pressure in excess of this by 270 metres of water. (Absorption coefficient of N_2 at body temperature = 0.012; total blood plasma volume of the average adult male = 3.2 dm^3; 1 m of water = 9.807 kPa; air contains 78% nitrogen.)

10. Under anaerobic conditions, a suspension of the bacterium *Micrococcus denitrificans* will quantitatively reduce nitrate to nitrogen gas if it is also supplied with excess of an oxidizable substrate, e.g. succinate.

$$2NO_3^- \longrightarrow N_2 + 2OH^- + 4H_2O$$

Two hundred cm^3 of a washed suspension of *M. denitrificans*, supplied with excess succinate and 0.25 mol of nitrate, were incubated at 303 K in a 2 dm^3 closed bottle which initially contained an atmosphere of oxygen-free N_2 at 100 kPa. What would be the final pressure of gas in this bottle after complete reduction of the nitrate? (Assume negligible solubility of N_2 in the culture medium.)

11. To prepare tetrahydrofolic acid (PtH_4G), pteroylglutamic acid (PtG) is catalytically hydrogenated at room temperature, i.e.

$$PtG + 2H_2 \longrightarrow PtH_4G$$

A 5 g sample of pteroylglutamic acid of unknown purity was treated in this fashion and absorbed 507 cm^3 of H_2 at 290 K and 102.3 kPa. Assuming that the impurities do not react with H_2, calculate the approximate percentage purity of the sample. (Molecular weight of pteroylglutamic acid = 441.)

12. An anaerobic culture of a bacterium, isolated from sewage, liberated an inflammable gas during growth. A pure sample of this gas took 491 s to stream through a minute hole. Under identical conditions of temperature and pressure, an equal volume of nitrogen took 650 s to stream through the same hole. Calculate the molecular weight of the inflammable gas and suggest what it might be.

13. The main compartment of a Warburg manometer flask contained 0.4 cm^3 of 0.1 mol dm^{-3} potassium ferricyanide, 0.4 cm^3 of 4 mol dm^{-3} sodium hydroxide and 1.2 cm^3 of water. In the side arm of this flask were placed 0.5 cm^3 of a saturated solution of hydrazine sulphate and 0.5 cm^3 of 4 mol dm^{-3} sodium hydroxide. After preliminary equilibration at 303 K and 'setting' of the manometer in preparation for a gas output, the contents of the side arm were tipped into the main flask compartment. Reaction was complete in a very few minutes when the manometer reading had increased by 103 mm. In the same period the thermobarometer reading had decreased by 2 mm.

Calculate, (a) the manometer constant for nitrogen at 303 K, (b) the manometer constant for carbon dioxide at 298 K.
(Ferricyanide oxidizes hydrazine in alkaline solution to yield N_2 gas:

$$4Fe(CN)_6{}^{3-} + N_2H_4 \longrightarrow N_2 + 4H^+ + 4Fe(CN)_6{}^{4-}$$

Absorption coefficient α for N_2 at 303 K = 0.0134; α for CO_2 at 298 K = 0.759; standard atmospheric pressure = 10 000 mm of manometer fluid.)

[*Note:* This reaction is sometimes used to obtain approximate values for the constants of Warburg manometers.]

14. Pyruvic acid may be assayed manometrically by enzymic decarboxylation at pH 5,

$$CH_3COCOOH \xrightarrow{\text{yeast carboxylase}} CO_2 + CH_3CHO$$

0.5 cm^3 of a pyruvate solution, of unknown concentration, was introduced into a manometer flask containing 2.5 cm^3 of yeast carboxylase in acetate buffer pH 5. During incubation under nitrogen at 303 K, the output of CO_2 caused an increase of 110 mm in the manometer reading. The thermobarometer reading had decreased by 7 mm in the same period.

Calculate the concentration of pyruvate in the solution used. (Volume of the manometer flask plus closed limb of the manometer = 22.0 cm^3; absorption coefficient of CO_2 at 303 K = 0.665.)

4

Some Properties of Aqueous Solutions

In general terms, a solution consists of a wholly homogeneous, single phase mixture of two or more substances. The component which determines the phase of the solution, and which is usually present in the highest proportion, is called the *solvent*; the remaining components are known as *solutes* and are considered to be dissolved in the solvent. The physico-chemical behaviour of any solution is determined not only by the nature of its solvent and solutes, but also by their proportions in the solution.

Though this definition of a solution includes mixtures of gases, and solutions of substances in a solid, we are more usually concerned with the properties of solutions which have a liquid solvent. In this chapter we shall examine some of the properties of aqueous solutions since these are of most interest to biologists; we shall defer, until Chapter 5, any discussion of the special amphiprotic role of water when it acts as a solvent for acids, bases and salts.

Water as a liquid

The reference condition (or standard state) of a substance is considered to be its state at 298 K and standard atmospheric pressure (i.e. 101 325 Pa). At this temperature and pressure, water is a liquid with a density much greater than any gas, but with no rigid shape. The fluid properties of a liquid suggest that its molecules possess greater liberty than do the molecules of a solid, but undertake considerably less translational movement than do the molecules of a gas which quickly distribute themselves throughout any space into which they are introduced. That the liquid state of water is in many respects intermediate between its solid and gaseous states is simply demonstrated by changing its temperature at standard atmospheric pressure, when it freezes to form ice at approx. 273 K, and boils to form steam at 373 K. There is good reason to believe that in its liquid state the molecules of water are structurally associated, and there is further evidence (Chapter 5) that a proportion of them are ionized. These, and other unique characteristics, are reflected in the physical properties of water including its density, viscosity, high boiling point, high dielectric constant and electrical conductance; they also affect its behaviour as a solvent. However, for the elementary purposes of this chapter, we need not concern ourselves with many of these special properties of liquid water. Instead we shall be particularly concerned with one property that water shares with all liquids and solids, namely that at a given temperature in a closed vessel it exists in equilibrium with its vapour.

VAPOUR PRESSURE

Whenever a liquid (or solid) possesses an interface with a gaseous phase, a proportion of its molecules are lost by vaporization into this gaseous phase. When a quantity of liquid is introduced into a closed vessel, the rate at which its molecules quit the liquid for the gaseous phase will at first exceed the rate at which they recondense from the gaseous into the liquid phase. Consequently, the partial pressure exerted by the vapour molecules progressively increases and promotes their recondensation. Eventually the partial pressure is such that the rate of evaporation exactly equals the rate of condensation. The partial pressure of its vapour at this equilibrium is characteristic of a liquid at a given temperature and external pressure, and is called its ***vapour pressure***.†

The influence of temperature on the vapour pressure of a liquid

Increasing the temperature of a liquid causes its vapour pressure to rise until it eventually equals the external pressure; then, the liquid boils. It follows that the boiling point of a liquid depends on the prevailing external pressure. Thus, the *normal boiling point* of a liquid, which is the temperature at which it boils under an external pressure of 101 325 Pa, can also be defined as 'that temperature at which its vapour pressure is 101 325 Pa'. It follows that a liquid boils at a lower temperature than its normal boiling point when the external pressure is less than standard atmospheric pressure, a fact that is made use of to distil heat-labile materials under reduced pressure. Conversely, at pressures greater than

† Though the vapour pressure of a liquid is very sensitive to changes in temperature, it is relatively insensitive to changes in applied pressure, and unless this is very different from 'atmospheric pressure' the external pressure is generally not reported.

the standard atmospheric pressure, the boiling point of a liquid is higher than normal. Thus water, whose normal boiling point is 373 K, boils at 283 K at a pressure of 1226 Pa, and at 394 K under a pressure of 203 kPa. This shows that at 283 K the vapour pressure of water is 1226 Pa; at 373 K it is 101 325 Pa, while at 394 K it is 203 kPa.

If we consider the freezing of water at a fixed pressure, we find that it takes place at that temperature (approx. 273 K at standard atmospheric pressure) at which the vapour pressure of water is exactly equal to the vapour pressure of ice. The normal freezing point of a liquid may in fact be defined as 'that temperature at which, under standard atmospheric pressure, its solid and liquid states are in equilibrium'. By increasing the applied pressure, the freezing point of most liquids is raised, but it is a peculiarity of water (and of a few other liquids) that its freezing point is *lowered* by enhanced pressures.

When the temperature of water is increased at standard atmospheric pressure from its normal freezing point (273 K) to its normal boiling point (373 K), its vapour pressure concurrently rises in an exponential manner. In fact, if the logarithm of the vapour pressure of any liquid is plotted against the reciprocal of the temperature in K, a straight line of negative slope is obtained. Clapeyron and Clausius showed that the magnitude of this slope was directly proportional to the quantity of heat required to vaporize 1 mole of the liquid (slope $= -L/2.303R$ where $L=$ latent heat of vaporization of the liquid). This means that, by following the way in which the vapour pressure of a liquid alters with temperature, we can obtain some information concerning the magnitude of the forces that tend to maintain its molecules in the liquid state.

Effect of solutes on the vapour pressure of water

When water acts as a solvent, its vapour pressure is lowered by an amount which, in extremely dilute solutions, is directly proportional to the number of solute 'particles' introduced per unit volume of the aqueous solution, but is independent of the specific nature of these particles (i.e. their size, shape, chemical constitution, electrical charge). Coincident with this lowering of the vapour pressure the solution exhibits other properties to a proportional degree. All of these properties, in a sufficiently dilute solution, are determined by the concentration of solute particles regardless of their nature; they are called ***colligative properties*** of the solution and include,

(a) lowering of the vapour pressure of the solvent;

(b) elevation of the boiling point of the solvent;

(c) depression of the freezing point of the solvent;

(d) exhibition of an osmotic pressure.

SOLUTIONS OF NON-ELECTROLYTES

Raoult's Law

If the vapour pressure of a pure solvent at a given temperature and pressure is p^0, and if this is decreased to p by the addition of a quantity of solute,† then the solute is responsible for a *relative lowering of the vapour pressure of the solvent* equal to $(p^0-p)/p^0$. Raoult's Law states that 'this relative lowering of the vapour pressure of the solvent is equal to the mole fraction of solute in the solution'. This means that if n_2 mol of any solute is dissolved in N_1 mol of solvent at a given temperature and pressure, Raoult's Law predicts that,

$$\frac{(p^0-p)}{p^0} = \frac{n_2}{n_2+N_1} = \text{mole fraction of solute}$$

In this same solution, the mole fraction of solvent equals $N_1/(n_2+N_1)$, so that if we represent the mole fraction of solute by x_2, then $x_2=(1-x_1)$, for in the one solution the sum of the mole fractions of solute(s) and solvent must equal unity. If we substitute $(1-x_1)$ for x_2 in the Raoult's Law equation given above,

$$\frac{(p^0-p)}{p^0} = 1-x_1 \qquad \text{whence,} \quad \frac{p^0}{p^0}-\frac{p}{p^0} = 1-x_1$$

$$\therefore\ 1-\frac{p}{p^0} = 1-x_1 \quad \text{and} \quad p = p^0x_1$$

It follows that Raoult's Law predicts that the vapour pressure of the solvent above a solution is directly proportional to the mole fraction of *solvent* present in that solution. The proportionality constant is equal to the vapour pressure of pure solvent at the same temperature.

Just as the Ideal Gas Laws are only precisely obeyed by ideal gases, so too is Raoult's Law only perfectly obeyed by ideal solutions. Most real solutions depart somewhat from its dictates, though, as we shall see later, the more dilute is the solution the more likely is the solvent to behave according to Raoult's Law. It is interesting to compare the views of ideality that are taken (a) by Raoult's Law which defines the ideal behaviour of a solvent, and (b) by Henry's Law which defines the ideal behaviour of a gaseous, or other, solute.

Correlation of Raoult's Law with Henry's Law

Suppose that n_2 mol of a gas are dissolved in N_1 mol of a liquid solvent. Raoult's Law predicts (1) that the vapour pressure of solvent above the solution, at a given temperature, is directly proportional to the mole

† If the solute is appreciably volatile, p is the *partial* vapour pressure of the solvent.

fraction of the solvent, and (2) that the proportionality constant is equal to the vapour pressure of the pure solvent at the same temperature; i.e.

$$p = p^0 x_1 \qquad \text{where} \begin{cases} p = \text{vapour pressure of solvent above the solution} \\ p^0 = \text{vapour pressure of pure solvent at the same temperature and pressure} \\ x_1 = \text{mole fraction of solvent in the solution} \\ \quad = \dfrac{N_1}{n_2+N_1} \end{cases}$$

Henry's Law (p. 37) states that 'the mass of gas, dissolved in a certain volume of solvent, is directly proportional to the partial pressure of the gas that is in equilibrium with the solution'. Therefore, applying Henry's Law to this same solution, we obtain the following equation,

$$\frac{n_2}{N_1} \propto p_2$$

where p_2 is the partial pressure of the gas in equilibrium with the solution (i.e. the vapour pressure of the gas as solute).

For a sparingly soluble gas that yields a very dilute solution, n_2/N_1 must be approximately equal to $\dfrac{n_2}{n_2+N_1}$ (since n_2 is now insignificant in comparison with N_1). The Henry's Law prediction for such a solution is that,

$$\frac{n_2}{n_2+N_1} \propto p_2$$

where $\dfrac{n_2}{n_2+N_1} = x_2 =$ mole fraction of solute, and therefore,

$$p_2 = kx_2$$

In principle, Henry's Law applies to any solute, so that while Raoult's Law conceives of the ideal solvent as one whose vapour pressure is directly proportional to its mole fraction, Henry's Law similarly sees the ideal solute as one whose vapour pressure is directly proportional to *its* mole fraction. In a sense, therefore, Raoult's Law can be thought of as a 'special' application of Henry's Law in which the proportionality constant has a unique value (p^0) equal to the vapour pressure of the pure component at the same temperature and pressure. In practice, in the solution in which the solute behaves ideally according to Henry's Law, the solvent also behaves ideally according to Raoult's Law (though the reverse is not necessarily true and Raoult's Law ideality on the part of the solvent does not mean that the solute must obey Henry's Law).

Both Raoult's Law and Henry's Law are *limiting* laws in that they define the ideal behaviour of the components of a solution as their behaviour in an infinitely dilute solution; as a real solution approaches the limit of infinite dilution so will its components behave more ideally. The ideal state of the solvent envisaged by Raoult's Law is 'solvent at a mole fraction of unity', which is of course pure solvent. As a solute behaves ideally only in an infinitely dilute solution, its ideal state is more difficult to define. It is therefore more usual to define a hypothetical, reference solution in which, although the solute is present at unit concentration, it still behaves as if it were present in an infinitely dilute solution. This reference solution, which is taken as the standard state of the solute and in which the solute is said to be present at *unit activity*, is defined on the mol dm^{-3} scale of concentrations as 'the hypothetical 1 mol dm^{-3} solution of the solute in which it behaves as if it were actually present in infinitely dilute solution'.

Deviations from Raoult's Law exhibited by aqueous solutions of non-electrolytes

For the present, we shall consider only those solutes which, when dissolved in water, contribute only their intact molecules as solute 'particles' in the solution (i.e. non-electrolytes). We shall later (p. 79) examine the different behaviour of another class of solutes (electrolytes), a proportion of whose molecules 'disintegrate' in aqueous solution.

You will recall that Raoult's Law predicts that at a given temperature and pressure the vapour pressure (p) of solvent above a solution is determined (a) by the vapour pressure of the pure solvent at the same temperature (p^0), and (b) by the mole fraction of solute that is present in the solution (x_2), so that,

$$\frac{(p^0-p)}{p^0} = x_2 \quad \text{or,} \quad p = p^0 - p^0 x_2$$

It follows that if values of p at a given temperature and pressure are plotted against the corresponding values of x_2, a straight line is obtained if the solution behaves ideally at all concentrations of solute (dotted line in Fig. 4.1). The intercept that this line makes on the vapour pressure axis (when x_2 is zero) equals p^0.

In practice, although very dilute solutions may behave almost ideally, as the concentration of solute is increased so the measured value of p increasingly differs from the ideal value predicted by Raoult's Law. As shown in Fig. 4.1, this deviation can be 'positive', when increase in the concentration of the solute causes a smaller decrease in the vapour pressure of the solvent than is predicted by Raoult's Law. Alternatively, the deviation will be 'negative', when addition of solute causes a greater decrease in the solvent vapour pressure than is explicable by Raoult's Law. Such non-ideal behaviour suggests interaction between the molecules of solvent and solute. The existence of intermolecular forces of attraction (or repulsion) could explain why the vapour pressure of the solvent above the solution is

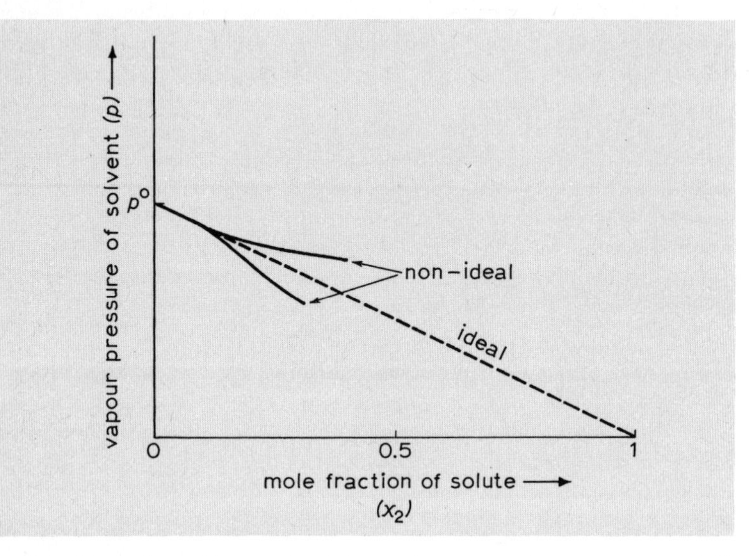

Fig. 4.1 Illustration of how real solutions of a non-volatile solute in a volatile solvent may deviate from Raoult's Law. (Note that only as the mole fraction of the solute approaches zero do the solutions behave ideally.)

smaller (or greater) than 'normal'. It follows that the extent by which solvent and solute interact, and so behave non-ideally in the solution, may be deduced from the vapour pressure of the solvent above the solution (see also p. 219).

Activity of solvent in a solution

At a given temperature and pressure, the mole fraction of solvent in an ideal solution equals the partial vapour pressure of the solvent above this solution divided by the vapour pressure of the pure solvent, i.e.

$$\text{mole fraction of solvent} = \frac{p}{p^0}$$

For a non-ideal solution this ratio of vapour pressures does not equal the actual mole fraction of solvent known to be present in the solution. Instead, it measures the *effective mole fraction* of solvent in the solution. So, whatever is its actual mole fraction, the solvent in a non-ideal solution behaves as the equivalent of ideal solvent present in a mole fraction of p/p^0.

As we have seen, Raoult's Law points to 'solvent at a mole fraction of unity' (pure solvent) as being solvent in its ideal state. By assigning to this ideal solvent an *activity* of unity, we define a standard against which we can measure the 'behavioural concentration' of the solvent in any other solution. On this 'activity scale', the activity of the solvent in any solution is measured as a proportion of its value (unity) in the defined ideal state of the solvent. In practice, this means that the activity of the solvent in a solution equals its effective mole fraction in that solution, measured as the ratio of the vapour pressures p/p^0.

The more a solvent differs in its behaviour in a solution from the ideal behaviour anticipated by Raoult's Law, the greater will be the difference between its activity and its actual mole fraction in that solution. The ratio of these terms is called the *rational activity coefficient* of the solvent in the solution, and is a measure of the extent to which the solvent behaves non-ideally in the solution:

$$\frac{\text{activity of solvent in the solution}}{\text{actual mole fraction of solvent in the solution}} = \text{rational activity coefficient of the solvent}$$

EXAMPLE

At 310 K the vapour pressure of glycerol is negligible compared with that of water. If the vapour pressure at 310 K above 56% glycerol solution in water was 4772 Pa, calculate the activity of water in this solution. (Vapour pressure of water at 310 K is 6275 Pa.)

The vapour pressure above the glycerol solution at 310 K is virtually entirely due to water vapour.

Vapour pressure of water above the solution $= p = 4772$ Pa

Vapour pressure of pure water at the same temperature $= p^0 = 6275$ Pa

$$\therefore \text{ Activity of water in the solution} = \frac{p}{p^0} = \frac{4772}{6275} = \underline{0.76}$$

Activity of the solute

By ascribing an activity of unity to the solute in the hypothetical 1 mol dm^{-3} solution in which it behaves ideally (as though it were present in infinitely dilute solution), the activity of the solute in a solution is equated with its *effective (mol dm*$^{-3}$*) concentration.* A practical activity coefficient can now be calculated whose value for a solute in a particular solution represents the extent by which it departs from ideality (according to Henry's Law) in its behaviour in this solution; i.e.

$$\frac{\text{activity of solute in solution}}{\text{actual concentration of solute in the solution in mol dm}^{-3}} = \text{concentration-based activity coefficient of solute} = y$$

Thus, the activity coefficient approaches 1 as the concentration of the solute in the solution approaches zero.† (Note: if unit activity is ascribed to a 1 mola*l*, ideal solution, of the solute, then a mola*l* activity coefficient γ for the solute is obtained by dividing its activity by its actual molal concentration in the solution. So long as only very dilute aqueous solutions are considered, there is negligible difference between the values of the mol dm^{-3} concentration and mola*l* activity coefficients.)

We shall consider activities and activity coefficients at greater length when we examine the thermodynamics of reactions in solution (Chapters 7 and 8, especially p. 217). In fact, 'activity' is a thermodynamic concept, and the most satisfactory explanation of its meaning, and of the significance of ideal behaviour, is based on thermodynamic principles. Meanwhile, just as we avoided the need to consider the non-ideal behaviour of 'real' gases by confining our attention to gases at low pressures, so we can largely avoid the necessity of considering activities and activity coefficients by restricting our discussion to the behaviour of extremely dilute solutions. In these very dilute solutions, both solvent and solute can be considered to act ideally and to possess activity coefficients of unity. Only when considering fairly concentrated solutions of non-electrolytes (>0.2 mol dm^{-3}), or solutions of electrolytes of significantly great ionic strength (p. 82, generally, electrolyte concentration $>10^{-4}$ mol dm^{-3}), will we have to be careful to assess their behaviour in terms of activities rather than in terms of concentrations.

Calculation of the molecular weight of a solute from the magnitude of the colligative properties of its dilute solutions

One of the reasons for making a detailed study of colligative properties is that they can be used for the experimental determination of the molecular weights of solutes.

If n_2 mol of solute is dissolved in N_1 mol of solvent to give a very dilute solution, the mole fraction of solute in this solution is insignificantly less than n_2/N_1. Assuming that this solution, being dilute, will behave ideally according to Raoult's Law, then

$$\frac{(p^0-p)}{p^0} = \frac{n_2}{N_1} \quad \text{or} \quad \frac{\Delta p}{p^0} = \frac{n_2}{N_1}$$

where p^0 is the vapour pressure of the solvent at the same temperature at which its vapour pressure in the solution (p) is less that p^0 by Δp. If the

† It is possible to determine the activity coefficients of both solvent and solute in a solution by following the way in which the vapour pressure of the solvent changes at a given temperature in response to the addition of increasing amounts of the solute.

solution was prepared by dissolving w_2 g of solute of molecular weight M_2 in w_1 g of solvent of molecular weight M_1, then,

$$\frac{n_2}{N_1} = \frac{w_2/M_2}{w_1/M_1} = \frac{\Delta p}{p^0}$$

It is therefore possible to obtain the molecular weight of a solute by observing what decrease in the vapour pressure of the solvent results from dissolving a known weight of the solute in a measured weight of the solvent (whose molecular weight and initial vapour pressure are both known). The requirement that the solution should behave ideally according to Raoult's Law limits the application of this method to dilute solutions. In turn this means that very small changes in vapour pressure must be measured accurately. Though very sensitive methods for measuring vapour pressure differences are available, it is more usual to make use of the fact that any decrease in the vapour pressure of the solvent is mirrored in changes in the other colligative properties of the solution.

One might measure the solute-induced elevation of the boiling point of the solvent, though when the solute is heat labile, or significantly volatile, this would not be the method of choice. More usually, with instruments capable of measuring very small changes in temperature, one would measure the extent by which a certain quantity of solute depresses the freezing point of a known mass of solvent. In a dilute solution, the solute lowers the freezing point of the solvent by an amount that is directly proportional to the relative lowering of its vapour pressure, i.e.

$$\Delta T_f = \frac{k.\Delta p}{p^0} = k.\frac{w_2/M_2}{w_1/M_1}$$

whence the depression in freezing point of the solvent,

$$\Delta T_f = k.\frac{w_2 M_1}{w_1 M_2}$$

The *molal depression constant* (K_f) for a solvent is the amount by which the freezing point of that solvent is lowered in a hypothetical 1 mol kg^{-1}, ideal solution (in which the solute behaves as if it were present in an infinitely dilute solution). For this 1 mol kg^{-1} ideal solution, $w_2/M_2 = 1$, and $w_1 = 1000$,

$$\text{so that} \quad K_f = \frac{kM_1}{1000}$$

The value of its molal depression constant (K_f) is characteristic of a solvent, and can either be determined experimentally using a solute of known molecular weight, or can be calculated from known thermodynamic properties of the solvent. (It can be shown that K_f equals $RT_f^2/1000l_f$ where $R = 8.314$ J K^{-1} mol^{-1}, T_f = freezing point in K and l_f = heat of fusion per gramme of solvent, which is the heat in joules required to melt 1 g of solid solvent at its freezing point and standard atmospheric pressure.)

Once the molal depression constant of a solvent is known (and values of K_f for most common solvents are now available in reference tables), the molecular weight of a solute can be calculated from the depression in the freezing point of the solvent that results from dissolving w_2 g of solute in w_1 g of solvent, for,

$$\Delta T_f = K_f \frac{1000 w_2}{w_1 M_2}$$

or molecular weight of solute $= M_2 = K_f \dfrac{1000 w_2}{\Delta T_f w_1}$

EXAMPLE

A neutral, nitrogenous compound, obtained from human urine, was recrystallized from ethanol. A solution, prepared by dissolving 90 mg of the purified compound in 12 g of distilled water, had a freezing point 0.233 K lower than the freezing point of pure distilled water. Calculate the molecular weight of the compound. (Molal depression constant of water = 1.86 K.)

$$\text{Molecular weight of the solute} = K_f \frac{1000 w_2}{\Delta T_f w_1}$$

where K_f = molal depression constant of water = 1.86 K

w_2 = weight of solute = 0.09 g

ΔT_f = depression of freezing point of water = 0.233 K

w_1 = weight of water in which w_2 of solute is dissolved = 12 g

$$\therefore \text{molecular weight of solute} = \frac{1.86 \times 1000 \times 0.09}{0.233 \times 12} = 59.8 \text{g mol}^{-1}$$

Therefore the probable molecular weight of the solute is 60.

Two points must be emphasized concerning this cryoscopic, or freezing point depression, method of determining the molecular weight of a solute.

1. It can only be applied to dilute solutions and, in order to make the depression of the freezing point as large as possible, there is some advantage to be gained by using a solvent which has a larger molal depression constant than water, e.g. K_f for water = 1.86 K, K_f for camphor = 40.0 K.
2. It can only be applied to a solution from which pure solid solvent separates on freezing. If a 'mixed' solid phase (i.e. a solid solution) containing a proportion of the solute appears on freezing, Raoult's Law is not applicable to the solvent vapour pressure at the freezing point.

Though the cryoscopic method is of great use in determining the molecular weights of some metabolites of relatively small molecular weight (and, as we shall see later, in predicting the osmotic pressure of a solution), it is impracticable as a means of determining the molecular weight of a macromolecular material such as a protein. In the first place, the depression of the freezing point of water caused by even a relatively large weight of protein dissolved in 1 dm^3 of water would be so small as to defy accurate measurement. For example, if 5 g of a protein of molecular weight 500 000 were dissolved in 1000 g of water, the freezing point of the solution would be only 1.86×10^{-5} K less than that of pure water. More importantly, relatively small but frequently unavoidable contamination of the protein with a salt of small molecular weight (e.g. traces of ammonium sulphate) would mean that the minute depression of freezing point attributable to the protein would be masked by the large depression caused by the contaminating salt. Yet, as we shall see later, use *can* be made of another colligative property (osmotic pressure) to determine the very large molecular weights of these substances.

OSMOSIS

Osmosis is the process whereby solvent spontaneously moves from one region in a solution where its activity is high, to another region where its activity is lower. In practice, osmosis takes place only when a semi-permeable membrane separates a solution, either from pure solvent or from a solution with the same solvent but having a different concentration of solute. The membrane that is used is in theory perfectly semi-permeable, which means that it is freely permeable to solvent but impermeable to all solutes. But such a completely semi-permeable membrane is very rare, and the membranes employed in practice to demonstrate osmosis between aqueous solutions are only *selectively* impermeable; they generally allow free passage of water and of solutes of small molecular weight, but are completely impermeable to substances of comparatively high molecular weight. The occurrence of osmosis is easily demonstrated with the simple apparatus shown in Fig. 4.2, where the semi-permeable membrane may be of cellophane, or visking, or of copper ferrocyanide deposited in porous porcelain.

Pure water, or a very dilute aqueous solution, is placed in the beaker, B, and a more concentrated aqueous solution whose solute cannot pass through the dividing membrane, is placed in A. The volumes of the solutions in A and B are adjusted so that at the outset their levels are identical. Over a period of time, at room temperature, the volume of liquid in A increases, causing the liquid meniscus to rise in the capillary tube A′. Eventually this meniscus remains stationary at a level which is significantly higher than the level of the ‘outer’, less concentrated solution in B. At this terminal equilibrium, the solution in A exerts a pressure on the semi-permeable membrane which is greater than the contrary pressure exerted by the less concentrated solution in B. The magnitude of this pressure difference is measured as the height at equilibrium of the column

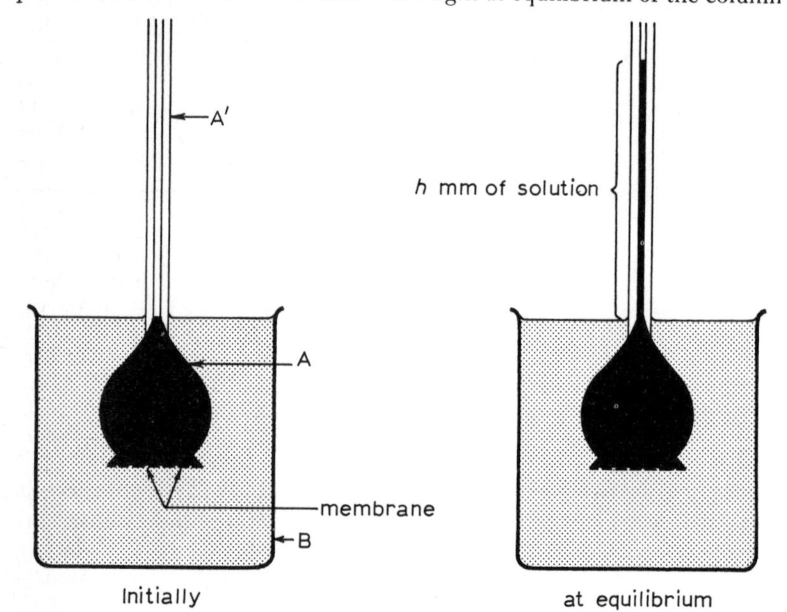

Fig. 4.2 Simple apparatus for demonstrating osmosis.

of liquid in A (*h* mm in Fig. 4.2); it is just sufficient to prevent any further net movement of solvent from B into A at equilibrium.

The ***osmotic pressure*** of a solution, at a given temperature, is defined as 'that pressure which must be exerted on the solution to prevent any net movement of solvent between the solution and its pure solvent, when these are separated by a perfect semi-permeable membrane'.† Thus, to measure

† Although it is quite usual to talk of 'the osmotic pressure of a solution', this is in fact an irrational statement. In the first place, no isolated solution can 'possess' an osmotic pressure since the phenomenon of osmosis is only demonstrable in a system in which pure solvent and solvent in solution (or two solutions in which the solvent is at different activities, see p. 64) are separated by a semi-permeable membrane. In the second place, since the osmotic pressure is the pressure that must be imposed upon the solution to maintain its solvent in equilibrium with pure solvent at the same temperature (p. 74), it is disconcerting to have to refer to this pressure as if it were exhibited *by* the solution (in the same sense that a gas exhibits a pressure). There appears to be a need for a new and agreed term to denote the 'osmotic potential' of a solution—meanwhile in common with current usage we shall continue to refer to the osmotic pressure of a solution, whilst trusting that you will understand the true meaning of this rather unsatisfactory phrase.

the osmotic pressure of a solution using the simple apparatus of Fig. 4.2, a semi-permeable membrane must be employed that is completely impermeable to the solute but is completely permeable to the pure solvent. The solution is then placed in vessel A and pure solvent is placed on the other side of the membrane in vessel B. If the solution is diluted by the inflow of solvent, the excess pressure measured at equilibrium will not be its true osmotic pressure; therefore the bore of the capillary tube A is made as fine as possible. Better still, an apparatus is used in which a measured external pressure can be exerted on the solution. Using a sensitive means for detecting the net flow of solvent across the semi-permeable membrane, the magnitude of this pressure is adjusted until it precisely prevents net flow of solvent in either direction. This pressure equals the osmotic pressure of the solution at the prevailing temperature.

Osmotic pressure as a colligative property of a solution

Early experiments performed with simple apparatus, of the type described above, showed that:

1. osmosis always consists of the net movement of solvent from the dilute to the more concentrated solution;
2. at a given temperature, the osmotic pressure of a dilute solution is directly proportional to its solute concentration;
3. the osmotic pressure of a solution of given solute concentration is directly proportional to its temperature.

These findings can be summarized in the equation,

$$\pi \propto \frac{n_2}{V} . T \quad \text{where} \begin{cases} \pi = \text{osmotic pressure of the solution in N m}^{-2} \text{ (i.e. in Pa)} \\ n_2 = \text{number of moles of solute present in } V \text{ m}^3 \text{ of solution} \\ T = \text{temperature in K} \end{cases}$$

whence

$$\pi V \propto n_2 T$$

This equation bears a striking resemblance to the ideal gas equation $PV \propto nT$, where the proportionality constant is R and equals 8.314 J K^{-1} mol^{-1}. Van't Hoff pointed out this analogy in 1887. He imagined that the solute in a dilute solution might behave similarly to an ideal gas since its molecules would be very widely dispersed throughout the large volume of the solvent. What came to be known as the van't Hoff theory of solutions can be summarized as follows: 'The osmotic pressure of a dilute solution is identical with the pressure that its solute would exert if it were an ideal gas occupying the same volume as the solution at the same temperature.' Accordingly, the osmotic pressure of a dilute solution (π) should be given by the equation,

$$\pi V = n_2 RT$$

In practice, it is found that this equation actually does represent the osmotic behaviour of ideal, dilute solutions. Yet, rather than imagining it as the 'solute bombardment pressure' it is more profitable to consider osmotic pressure as that pressure which must be exerted upon a solution to maintain it in equilibrium with its pure solvent at the given temperature. Solvent transfer by osmosis is then clearly seen to be yet another instance of spontaneous change motivated by the tendency to establish thermodynamic equilibrium.

Accepting the validity of the van't Hoff equation for ideal dilute solutions, we can employ it to calculate the molecular weight (M_2) of a compound from the osmotic pressure of the dilute solution in which the compound is present at a concentration of C g m^{-3}. In this solution, n_2/V equals C/M_2, and substituting this term in the osmotic equation, we obtain,

$$\pi = \frac{CRT}{M_2} \quad \text{or} \quad M_2 = \frac{CRT}{\pi}$$

EXAMPLE

An aqueous solution containing 5 g dm^{-3} of a soluble polysaccharide has an osmotic pressure of 3.24 kPa at 278 K. Assuming ideal behaviour, calculate the molecular weight of the polysaccharide (R=8.314 J K^{-1} mol^{-1}).

Assuming ideal behaviour we can apply the equation,

$$M_2 = \frac{CRT}{\pi} \quad \text{where} \left\{ \begin{array}{lll} \pi & = & \text{osmotic pressure} = 3.24 \text{ kPa} \\ & = & 3.24 \times 10^3 \text{ N m}^{-2} \\ C & = & \text{solute concentration in g m}^{-3} \\ & = & 5 \times 10^3 \text{ g m}^{-3} \\ R & = & 8.314 \text{ J K}^{-1} \text{ mol}^{-1} \\ & = & 8.314 \text{ N m K}^{-1} \text{ mol}^{-1} \\ T & = & 278 \text{ K} \\ M_2 & = & \text{molecular weight of solute in} \\ & & \text{g mol}^{-1} \end{array} \right.$$

$$\therefore M_2 = \frac{(5 \times 10^3) \times 8.314 \times 278}{3.24 \times 10^3} = 3567 \text{ g mol}^{-1}$$

This means that if the solute is homogeneous, its molecular weight is 3567. (If, as frequently happens with macromolecular solutes, the solute is not homogeneous (e.g. a polysaccharide may contain some 'molecules' of shorter chain length than others), then the value obtained by this method for its molecular weight is in fact an average molecular weight for the polydispersed solute.)†

Since the van't Hoff osmotic equation is only truly applicable to ideal (infinitely dilute) solutions, it is usually necessary to measure the osmotic

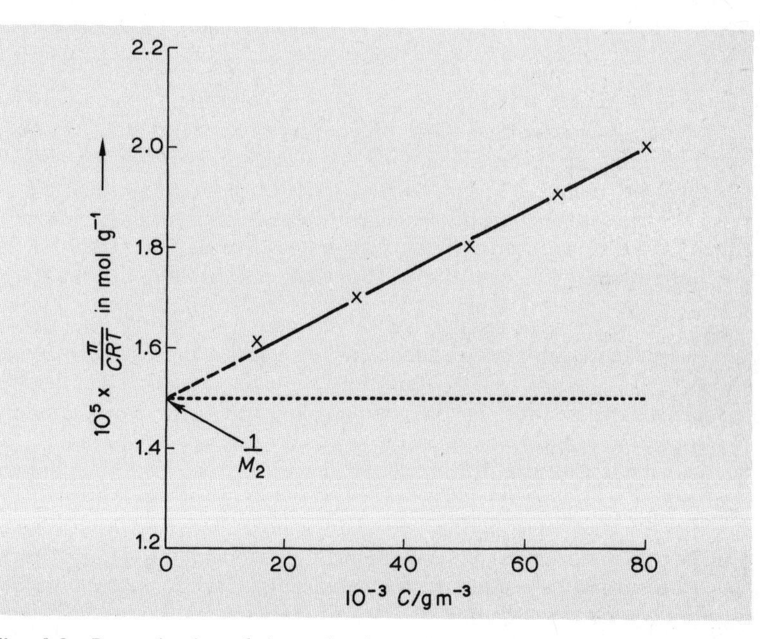

Fig. 4.3 Determination of the molecular weight of a protein from its osmotic behaviour in aqueous solutions. (The data are taken from the example on p. 75.)

pressure of a number of dilute solutions containing different small concentrations of the solute, and use these findings to obtain the osmotic pressure at infinite dilution. In practice, the equation $\pi = CRT/M_2$ is rearranged to give

$$\frac{\pi}{CRT} = \frac{1}{M_2}$$

and values of π/CRT are obtained at a constant temperature for several solutions containing different concentrations of the one solute. When the curve obtained by plotting these values against the corresponding values of C is extrapolated to infinite dilution ($C = 0$), it intercepts the π/CRT ordinate at a value equal to $1/M_2$ (see Fig. 4.3). Only if the solution behaves

† Actually the *number-average* molecular weight (in contrast to the *weight-average*).[10]

ideally at all values of C will a straight line be obtained which runs parallel to the concentration (C) axis (dotted line in Fig. 4.3).

Theoretical explanation of osmosis

To interpret the phenomenon of osmosis, we need not concern ourselves with the mechanism whereby the semi-permeable membrane discriminates between solute and solvent so long as we can assume that this is fully effective. One simple, empirical explanation of osmosis makes use of the following facts:

(i) a solute decreases the vapour pressure of a solvent;
(ii) increasing the pressure applied to a solvent increases its vapour pressure (though it takes a comparatively large increase in pressure to raise the vapour pressure by a significant amount);
(iii) at 'osmotic equilibrium' the vapour pressure of the solvent is the same on each side of the semi-permeable membrane.

Initially (as in Fig. 4.2), the pure solvent of vapour pressure p^0 is separated by the semi-permeable membrane from the solution in which its vapour pressure (p) is less than p^0. Solvent therefore tends to distil from the pure solvent into the solution. By the external application or internal development of an increased pressure on the solution's side of the membrane, the solvent vapour pressure of the solution is raised from p to p^0. Thus, at equilibrium, the solvent vapour pressure equals p^0 on each side of the membrane and net 'distillation' of solvent is no longer possible. It follows that the osmotic pressure of a solution at a given temperature is the pressure which must be applied to the solution to make its solvent vapour pressure equal to the vapour pressure of pure solvent at the same temperature.†

The most satisfactory answer to the question 'Why does osmosis occur?', involves thermodynamic concepts which we shall not examine until later (Chapter 7, p. 191). Briefly, it concludes that a solute decreases the chemical potential of a solvent, and that the osmotic pressure of a solution is that pressure which must be exerted on the solution to raise the chemical potential of its solvent so that it equals that of the pure solvent at the same temperature. By a quantitative thermodynamic analysis of this situation it is possible to derive, and so confirm, the osmotic equation.

According to this thermodynamic viewpoint, the lowering of the chemical potential of the solvent is the 'fundamental' colligative effect whose magnitude is faithfully mirrored in the magnitudes of all other colligative properties of the solution. This means, in practice, that from the extent of the

† According to this explanation, it is because the solvent vapour pressure is relatively insensitive to changes in pressure that the osmotic pressure of a solution is a magnified expression of the extent by which p is less than p^0.

depression of freezing point of the solvent in a solution, we can predict the magnitudes of:

1. the depression of vapour pressure of the solvent in the solution at a given temperature;
2. the elevation of boiling point of the solvent in the solution;
3. the osmotic pressure of the solution at a given temperature.

EXAMPLE

Calculate the osmotic pressure at 303 K (i.e. 30°C) of an aqueous solution of sucrose which at standard atmospheric pressure has a freezing point 0.093 K lower than water. (Molal depression constant of water = 1.86 K.)

To obtain the molecular weight of a solute from the depression in freezing point of a solvent which it causes, we use the equation,

$$M_2 = K_f \frac{1000 w_2}{\Delta T_f w_1}$$

where $1000 w_2 / w_1 = C' =$ g solute/1000 g solvent.

In dilute aqueous solutions C' approximately equals $10^{-3} C$ where C is the concentration in g m^{-3},

$$\therefore\ M_2 = \frac{K_f C'}{\Delta T_f} = \frac{K_f (10^{-3} C)}{\Delta T_f}$$

According to the van't Hoff osmotic equation, $M_2 = CRT/\pi$

$$\therefore\ \frac{K_f (10^{-3} C)}{\Delta T_f} = \frac{CRT}{\pi}$$

$$\text{whence} \qquad \pi = \frac{10^3\, \Delta T_f R T}{K_f}$$

In the example given,

π = osmotic pressure in N m^{-2} = ?
ΔT_f = 0.093 K
R = 8.314 J K^{-1} mol^{-1} = 8.314 N m K^{-1} mol^{-1}
T = 303 K
K_f = 1.86 K

Substituting these values in the above equation,

$$\pi = \frac{10^3 \times 0.093 \times 8.314 \times 303}{1.86}\ \text{N m}^{-2} = 1.26 \times 10^5\ \text{N m}^{-2} = \underline{126\ \text{kPa}}$$

Determination of molecular weights of proteins by osmometry

It has been mentioned (p. 69) that it is impracticable to deduce the molecular weight of a protein from the extent of the depression of the vapour pressure (or freezing point) of the water in its dilute aqueous solution. The chief objection stems from the fact that even a highly purified protein is likely to be contaminated with inorganic salts which make a significant contribution to the colligative properties of the solution. The same objection attaches to the determination of the extent by which the boiling point of water is elevated in the protein solution, though here the greater hazard is the likelihood that the molecular integrity of the protein will be destroyed by heating (resulting in denaturation). Although it is possible to obtain the molecular weight of proteins by other methods, e.g. centrifugal, gel filtration and light-scattering techniques, it is also possible to measure the osmotic pressure of an aqueous solution of protein at a temperature at which the protein retains its normal molecular structure, and in such a way that the osmotic effect of contaminating salts is nullified.

The protein is dissolved at its isoelectric pH in a fairly concentrated salt solution (about 0.2 mol dm^{-3}) which is separated by a semi-permeable membrane from an identical salt solution (minus the protein). At its isoelectric pH, the protein behaves as nearly as possible as though it were a non-electrolyte (p. 165), and the high ionic strength of the salt-containing solution minimizes the errors that would otherwise be introduced by the Gibbs–Donnan effect (p. 170). The system takes a comparatively long time to attain equilibrium, during which time the protein might deteriorate due to denaturation or microbial infection. Therefore the osmotic pressure contribution of the protein is usually determined by measuring the rate at which water flows into, or out of, the protein-containing solution when this is subjected to various external pressures. When the rate of flow of the water is plotted against the externally applied pressure, the osmotic pressure attributable to the protein is given by the intercept on the pressure axis at zero rate of flow. From the osmotic pressures of solutions containing different concentrations of the protein, the osmotic pressure of the ideal (infinitely dilute) solution of the protein can be calculated as in the following example (see also p. 73).

EXAMPLE

The following osmotic pressure data were obtained for a protein dissolved at its isoelectric point and at 278 K in 0.1 mol dm^{-3} buffer

Protein concentration/g dm^{-3}	*15*	*32.5*	*50*	*65*	*80*
Osmotic pressure/kPa	*0.557*	*1.277*	*2.076*	*2.856*	*3.697*

Calculate the molecular weight of the protein ($R=8.314$ J K^{-1} mol^{-1})

Values of π/CRT in mol g^{-1} are calculated for these solutions, where,

π = osmotic pressure in Pa (i.e. in N m^{-2})
C = protein concentration in g m^{-3}
R = 8.314 J K^{-1} mol^{-1} = 8.314 N m K^{-1} mol^{-1}
T = 278 K

Thus:

$10^{-3}C$/g m^{-3}	$10^{-5}CRT$/N m^{-2} g mol^{-1}	π/N m^{-2}	$10^5\pi/CRT$ in mol g^{-1}
15	347	557	1.604
32.5	752	1277	1.698
50	1155	2076	1.798
65	1502	2856	1.901
80	1849	3697	1.998

When the calculated values of π/CRT in mol g^{-1} are plotted against the corresponding values of C in g m^{-3}, a straight line is obtained which intersects the π/CRT ordinate at 1.5×10^{-5} mol g^{-1} (Fig. 4.3, p. 73). This is therefore equal to the reciprocal of the molecular weight of the protein (see p. 73).

∴ Molecular weight of the protein = $1/1.5 \times 10^{-5}$ = 66 670 g mol^{-1}

The osmotic behaviour of living cells

The osmotic pressure of an ideal solution of non-electrolytes is directly proportional to the sum of the mol dm^{-3} concentrations of all its solutes. It follows that ideal solutions of equal mol dm^{-3} concentration are *isosmotic* (i.e. possess the same osmotic pressure). This 'osmotic potential' is only fully displayed when, at a given temperature, the solution is separated from its pure solvent by a completely semi-permeable membrane. If the membrane is selectively impermeable, allowing free passage of solvent and of certain solutes while restraining other solutes, the solution can exhibit only that fraction of its total osmotic pressure that is due to the solutes to which the membrane is impervious. It is this fraction of the total osmotic pressure of a solution that is known as its *tonicity*. Evidently, therefore, the tonicity of a solution cannot be predicted solely from its known composition (as can the osmotic pressure), for the distinctive properties of the bounding membrane are also involved. Suppose, for example, that an aqueous solution containing 0.5 mol dm^{-3} sucrose and 0.5 mol dm^{-3} urea is separated from a 1.0 mol dm^{-3} aqueous solution of sucrose by a membrane that is permeable to water and to urea, but is completely impermeable to sucrose. The mixed urea plus sucrose solution is isosmotic with the 1 mol dm^{-3} sucrose solution, yet its tonicity is less than that of this solution, since tonicity is here determined solely by the sucrose concentration (as the urea is freely able to pass through the membrane). Because the mixed solution is hypotonic to the 1 mol dm^{-3} sucrose solution, it loses water to it by osmosis. As a further example, consider the situation (Fig. 4.4b) in which the same mixed solution of urea and sucrose (0.5 mol dm^{-3} each) is separated by the same selectively impermeable membrane from a 0.5 mol dm^{-3} solution of sucrose. Assuming ideal behaviour of the solvent in both these solutions, no net transference of water by osmosis would occur, for although

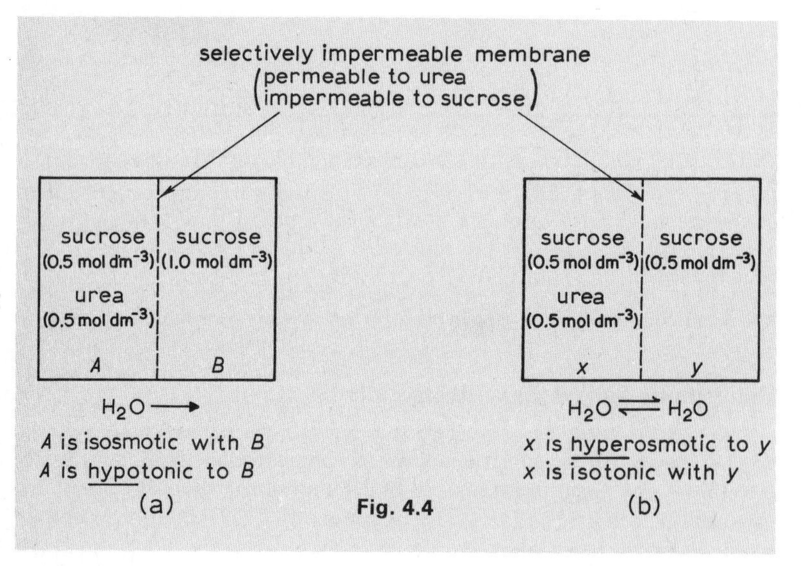

Fig. 4.4

the mixed urea plus sucrose solution is initially hyperosmotic to the 0.5 mol dm^{-3} sucrose solution, it is isotonic with it.

It is important that this distinction between 'osmotic potential' and tonicity is made in the case of the contents of a living cell, since the cell membrane is not truly semi-permeable but is only selectively impermeable to the enclosed solutes. Even so, the tonicity of the cell contents can be quite large; for example, a Gram-positive bacterium may have an intracellular tonicity of 0.8 to 1.2 MPa (i.e. 8 to 12 atmospheres) at room temperature. If such a cell were bounded only by its cell membrane, it would rapidly burst in any hypotonic medium due to its osmotic uptake of water; this explains the lysis of 'protoplasts' by osmotic shock. However, most plant cells and many micro-organisms are encased in a rigid cell wall of great mechanical strength, laid down external to the cell membrane. This cell wall acts as a corset, conferring a characteristic shape on the cell and enabling it to withstand considerable internal pressures. Should the construction of the cell wall be impeded (e.g. by treatment of susceptible bacteria with penicillin), or should the completed cell wall be weakened (e.g. by partial dissolution following treatment with certain lytic enzymes such as lysozyme or cellulase), the cell would again explode when it was immersed in a hypotonic medium. It follows that a cell which does not possess a rigid cell wall (such as a red blood cell) will best retain its normal shape and size in isotonic media. It is therefore not surprising to find throughout the animal kingdom numerous examples of osmoregulatory mechanisms which operate either to control the tonicity of extracellular fluids or to cause excess water (and certain solutes) to be expelled from cells which are bathed in hypotonic media.

As any student botanist is aware, the relatively high tonicity of protoplasm can be of direct advantage to a living cell; for example, it is responsible for the turgor of plant cells and is involved in the absorption of water and solutes by a plant's roots. It would be foolish, though, to suppose that the water and solute relationships of a living cell can be wholly explained by analogy with the osmotic behaviour of ideal solutions bounded by entirely 'passive', selectively impermeable membranes. Cell membranes are highly complex structures capable of utilizing metabolically derived energy (p. 242) to transport specific materials by a non-osmotic mechanism between the protoplasm and the extracellular medium. This can result in the concentration of certain solutes within a cell and the expulsion of others from its cytoplasm, both against normal concentration gradients.

SOLUTIONS OF ELECTROLYTES

Anomalous colligative properties of aqueous solutions of electrolytes

So far we have considered only those solutes (non-electrolytes), whose molecules remain intact when they are dissolved in water. Yet not all substances behave in this manner; a large number, including acids, bases and salts, yield aqueous solutions whose colligative properties are so great as to suggest that when they are dissolved in water their molecules disintegrate into submolecular particles. Solutions of these compounds in various solvents are distinguished by their marked ability to conduct an electric current, and by the consequent electrolysis of the solute. These substances are called ***electrolytes*** and their behaviour in aqueous solution is explained by the ionic theory. This theory is based on the original proposals of Arrhenius, whose theory of electrolytic dissociation (1887) first correlated the anomalous colligative properties of solutions of electrolytes with their ability to conduct an electric current and to be disrupted by its passage.

Ionization of electrolytes in aqueous solution

The ionic theory can be summarized as follows:

1 A molecule of an electrolyte when dissolved in water disintegrates (dissociates) into a fixed number of submolecular, electrically charged particles called *ions*. The number of unit charges (positive or negative) carried by an ion is called its *valency* or *valence*. Since the molecule of an electrolyte bears no net charge, its dissociation in water (or in a similarly ionizing solvent) must produce equal numbers of positive and of negative charges, though these are not necessarily borne by equal numbers of positive and negative ions.

2 Passage of a direct electric current through a solution of an electrolyte consists of the movement of its positive ions (cations) to the negative electrode (cathode), and the concurrent movement of its negative ions (anions) towards the positive electrode (anode). The products of electrolysis arise from the discharge of these ions at the electrodes.

3 Each ion acts as a 'particle of solute' making a contribution to the colligative properties of the solution equal to the contribution made by any uncharged molecule. If each molecule of an electrolyte were to dissociate completely in aqueous solution, to yield one positive and one negative ion, the colligative properties of its solution should ideally have twice the magnitude predicted from the molar concentration of the solute. Yet, the mutual attraction of ions of opposite charge ensures that all save extremely dilute aqueous solutions of electrolytes are likely to behave non-ideally.

The two main types of electrolytes are *electrovalent* and *covalent* compounds respectively. The ions of an electrovalent compound already exist in its solid (crystalline) state. The 'molecular unit' of such a compound consists of a characteristic number of anions and cations held together by electrostatic forces. Salts such as sodium chloride ($Na^{+}Cl^{-}$), or barium sulphate ($Ba^{2+}SO_4^{2-}$), are electrovalent compounds whose crystals are composed of electrically nullifying numbers of anions and cations held in a characteristic lattice structure. When these substances dissolve in water, their ions become hydrated, the electrostatic attraction between ions of opposite charge is weakened, and they are dispersed. In contrast to the behaviour of electrovalent compounds, covalent electrolytes yield ions only when they undergo chemical reaction with water molecules (or with the molecules of some other suitable solvent). The ionization of acids and bases in aqueous solution exemplifies the behaviour of covalent electrolytes and will be discussed in Chapter 5.

Electrovalent and covalent electrolytes are thus distinguishable by the origins of their ions. The extent to which the molecules of an electrolyte are ionized in an aqueous solution is of considerable practical interest, and is reported as the ***degree of dissociation*** or ***degree of ionization*** of the compound in that solution. Those substances that are completely or nearly completely ionized in dilute aqueous solution are called *strong* electrolytes, while those that ionize only to a small extent in dilute aqueous solution are called *weak* electrolytes.

Colligative properties of aqueous solutions of a strong electrolyte

Electrovalent salts, strong acids and strong bases are all strong electrolytes, and because ideally they are completely ionized in water, the magnitude of any one of the colligative properties of their aqueous solutions should indicate the number of ions formed from each of their molecules.

The measured value of the colligative property of a solution of any electrolyte will be greater than the value predicted on the basis of the mol dm^{-3} concentration of the solute. The ratio of the measured value to the value calculated on the mol dm^{-3} concentration basis is known as the *van't Hoff factor* of the solution, and is usually represented by the symbol i. Thus if the depression of freezing point of water is measured in an aqueous solution of an electrolyte,

$$\frac{\text{measured value of } \Delta T_{\text{f}}}{\text{value of } \Delta T_{\text{f}} \text{ calculated on a mol dm}^{-3} \text{ concentration basis}} = i$$

$$= \text{van't Hoff factor of the solution}$$

For an ideal solution of a non-electrolyte the van't Hoff factor equals 1, and for an ideal solution of a strong electrolyte (whose molecules completely dissociate in water to yield an integral number of ions) the van't Hoff factor is a simple multiple of 1. It is found that such a value for i is obtained only with extremely dilute solutions of strong electrolytes. For example, the value of the van't Hoff factor for solutions of sodium chloride approaches its limit of 2 as the solutions are made progressively more dilute ($Na^+Cl^- \rightarrow Na^+ + Cl^-$ is virtually complete at $< 10^{-3}$ mol dm^{-3}); similarly, solutions of barium chloride demonstrate values of i that approach 3 only when they are very dilute ($Ba^{2+}Cl_2^- \rightarrow Ba^{2+} + 2Cl^-$ is almost complete below 10^{-5} mol dm^{-3}). In reasonably concentrated solutions the strong electrolyte behaves non-ideally due to interionic effects, and the values of i for these solutions are less than the limiting value which is approached as the solution is made more and more dilute.

This non-ideal behaviour of reasonably concentrated solutions of electrolytes means that the solute ions behave as if they are present at effective concentrations (activities) which are conspicuously different from their actual concentrations in the solution. The extent by which an ionic species behaves non-ideally in a solution in which it is present in known concentration can therefore be expressed as an activity coefficient,

$$y = \text{activity coefficient of ion} = \frac{\text{activity}}{\text{actual concentration (mol dm}^{-3})},$$

where the activity of the ion is the ratio of the measured activity in the solution to the unit activity that it displays in the ideal solution (defined as the 1 mol dm^{-3} solution in which the ion behaves as if it were present in infinitely dilute solution).

It is not possible to determine experimentally the activity coefficient of a single ionic species; one can only obtain a value for the ***mean ion activity coefficient*** of an electrolyte. If a binary electrolyte dissociates to give v_+ positive ions and v_- negative ions per molecular unit, then its mean ion activity coefficient $y_\pm$ is defined by the equation,

$$y_\pm = (y_+^{v+} \times y_-^{v-})^{\frac{1}{v_+ + v_-}}$$

where y_+ is the activity coefficient of the cation, and y_- is the activity coefficient of the anion. For example, the mean ion activity coefficient of sodium sulphate in an aqueous solution of given concentration and temperature is $y_\pm$ Na_2SO_4, which equals $(y^2_{Na^+} \times y_{SO_4{}^{2-}})^{\frac{1}{3}}$

The value of $y_\pm$ for an electrolyte in aqueous solution, at a given concentration and at a fixed temperature, can be obtained by measurements of colligative properties, but might also be derived from solubility or e.m.f. measurements. You should consult a standard textbook of physical chemistry for a description of these methods.[8]

How the ionic strength of a solution of an electrolyte affects the value of the mean ion activity coefficient

The behaviour of any one ion in a solution is influenced by the number and charge of all other ions in its vicinity, regardless of their source. From a theoretical appraisal of the forces of interionic attraction in solutions of electrolytes in various solvents, Debye and Hückel in 1923 derived an expression which relates the mean ion activity coefficient of an electrolyte in very dilute solution to the nature of the solvent, the temperature, and the valences and concentrations of all other ions present in that solution.

For dilute aqueous solutions at 298 K, they predicted that

$$\log y_{\pm} = -0.51 z_{+} z_{-} \sqrt{\mu} \quad \text{where} \begin{cases} y_{\pm} = \text{mean ion activity coefficient of the electrolyte} \\ z_{+} = \text{valence of the electrolyte's cation} \\ z_{-} = \text{valence of the electrolyte's anion} \\ \mu = \text{ionic strength of the solution} \end{cases}$$

Or, theoretically, for a single ionic species in aqueous solution at 298 K,

$$\log y = -0.51 z^2 \sqrt{\mu}$$

The ionic strength (μ) is defined as being equal to half the sum of the terms obtained by multiplying the concentration of an ion (usually measured on the molality scale) by the square of its valence, i.e.

$$\mu = \tfrac{1}{2} \sum m z^2 \quad \text{where} \begin{cases} \sum \text{ indicates 'sum of all terms of the following description'} \\ m = \text{molality} \equiv \text{mol of ion/1000 g solvent} \\ z = \text{valence} \end{cases}$$

The Debye–Hückel equation given above is applicable only to aqueous solutions whose ionic strength is less than about 0.01. These solutions are so dilute that concentrations of ions in mol dm^{-3} can be substituted for m when calculating the ionic strength of an aqueous solution when this term is to be used in the Debye–Hückel equation.†

Since μ equals the total ionic strength of a solution (to which all its ions contribute), we conclude that at a given temperature the mean activity coefficient of an electrolyte will have the same value in all aqueous solutions of identical ionic strength, regardless of the concentration in which the electrolyte is itself present in these solutions.

EXAMPLE

Calculate the activities of sodium and sulphate ions in an aqueous solution containing 0.005 mol dm^{-3} sodium chloride and 0.001 mol dm^{-3} potassium sulphate at 298 K.

The salts are completely ionized in the solution, i.e.

$$Na^{+}Cl^{-} \rightarrow Na^{+} + Cl^{-}$$
$$K_2{}^{+}SO_4{}^{2-} \rightarrow 2K^{+} + SO_4{}^{2-}$$

† The Debye–Hückel relationship is a 'limiting relationship' which becomes more and more accurate as the value of μ approaches 0.

The ionic strength of the solution is given by the equation,

$$\mu = \tfrac{1}{2}\sum mz^2$$

Calculating the value of mz^2 for each type of ion in the solution,

$$\begin{aligned} Na^+ &= 0.005 \times 1^2 &&= 0.005 \\ Cl^- &= 0.005 \times 1^2 &&= 0.005 \\ K^+ &= 2 \times 0.001 \times 1^2 &&= 0.002 \\ SO_4^{2-} &= 0.001 \times 2^2 &&= 0.004 \\ & && \overline{0.016} = \sum mz^2 \end{aligned}$$

$$\therefore \mu = \tfrac{1}{2}(0.016) = \underline{0.008}$$

(a) *Activity of Na^+ ions*

The activity coefficient of these ions in the solution is y_{Na^+} where

$$\log y_{Na^+} = -Kz^2\sqrt{\mu} \quad \text{and} \quad \begin{aligned} K &= 0.51 \text{ at } 298\text{ K} \\ z &= 1 \text{ (for } Na^+) \\ \mu &= \text{ionic strength} = 0.008 \end{aligned}$$

$$\begin{aligned} \therefore \log y_{Na^+} &= -0.51 \times 1^2 \times \sqrt{0.008} \\ &= -0.51 \times 1 \times 0.0894 \\ &= -0.0456 = \bar{1}.9544 \end{aligned}$$

Thus $\log y_{Na^+} = \bar{1}.9544$ and $y_{Na^+} = \text{antilog } \bar{1}.9544 = 0.9$

But $a_{Na^+} = y_{Na^+} \times [Na^+] = 0.9 \times 0.005 = 0.0045$

$\therefore$ activity of Na^+ ions in the solution $= \underline{4.5 \times 10^{-3}}$

(b) *Activity of sulphate ions*

At 298 K, $\log y_{SO_4^{2-}} = -0.51 \times z^2\sqrt{\mu}$ where $z = 2$ and $\mu = 0.008$

$$\begin{aligned} &= -0.51 \times 4 \times 0.0894 \\ &= -0.1824 \\ &= \bar{1}.8176 \end{aligned}$$

$$\therefore y_{SO_4^{2-}} = \text{antilog } \bar{1}.8176 = 0.657$$

But $a_{SO_4^{2-}} = y_{SO_4^{2-}} \times [SO_4^{2-}] = 0.657 \times 0.001 = 0.000\,657$

$\therefore$ activity of SO_4^{2-} ions in the solution $= \underline{6.57 \times 10^{-4}}$

(Note the relatively large discrepancy between the values of the activity and concentration of this bivalent ion in a quite dilute solution.)

Summary of the colligative properties of a solution of a strong electrolyte

A strong electrolyte is fully ionized in all aqueous solutions, yet only in infinitely dilute solutions (of zero ionic strength) do its ions behave ideally.

In such a solution the van't Hoff factor (i) has its maximum integral value and the mean activity coefficient of the ions is unity. More concentrated aqueous solutions ($> 10^{-4}$ mol dm^{-3}) do not behave ideally; the van't Hoff factor for these solutions has less than its maximum value, and the mean activity coefficient of the electrolyte differs from unity. This non-ideal behaviour is not due to incomplete ionization of the strong electrolyte in more concentrated solutions, but is attributable to the existence of interionic forces whose magnitude at any given temperature is affected by the ionic strength of the solution.

Colligative properties of aqueous solutions of a weak electrolyte

A weak electrolyte is distinguished by being incompletely dissociated in even the most dilute aqueous solution. This means that, at any given temperature, when a known concentration of weak electrolyte is dissolved in water the resulting solution will contain undissociated solute molecules plus its ions.

Suppose that a weak electrolyte, of molecular formula XY, partially dissociates in water to give rise to the ions X^+ and Y^-. This dissociation is a chemically reversible reaction since, under appropriate conditions, ions X^+ and Y^- may reassociate to produce XY. The stoichiometric equation for the reaction should therefore be written as $XY \rightleftharpoons X^+ + Y^-$. We shall discuss such chemically reversible reactions in Chapter 8, p. 212; here we need only state that it is characteristic of these reactions that they proceed until a state of equilibrium is established in which the forward and back reactions proceed at identical rates. The ratio of activities of reactants and products that will coexist at this equilibrium is predictable from the Law of Mass Action (p. 213). The Law states that 'at a given temperature the rate of a chemical reaction is directly proportional to the 'active masses' of the reactants', which means that for a reaction proceeding between solutes the rate will be directly proportional to their activities. Thus for the reaction $XY \rightleftharpoons X^+ + Y^-$ in solution at a given temperature, if a_{XY} is the activity of the undissociated weak electrolyte at equilibrium, and a_{X^+} and a_{Y^-} are the activities of its ions also present in the equilibrium mixture, then at equilibrium,

$$\text{rate of dissociation of XY} \propto a_{XY}$$

$$\text{rate of the reverse association of its ions} \propto (a_{X^+} \times a_{Y^-})$$

Since at equilibrium these rates are equal,

$$\frac{a_{X^+} \times a_{Y^-}}{a_{XY}} = \text{Constant} = K$$

where K is the equilibrium constant (in this case also called the ionization constant), whose value at a given temperature is characteristic of the ionization of the weak electrolyte XY.

But activity = (activity coefficient × concentration) = $y.[XY]$ etc., and therefore the ionization constant $K = \frac{y_{X^+} \times y_{Y^-}}{y_{XY}} \times \frac{[X^+][Y^-]}{[XY]}$

In a very dilute aqueous solution of extremely low ionic strength, the activity coefficients of the undissociated weak electrolyte and of its ions will all equal 1 (approx.), and

$$\text{ionization constant} = \frac{[X^+][Y^-]}{[XY]}$$

If the dissociation of one molecule of weak electrolyte yields more than one ion of a single type, then the equation which gives the value of K contains the concentration of this ion raised to the power of n, where n is the number of these ions derived from a single molecule.

For example, if $$X_2Y \rightleftharpoons 2X^+ + Y^{2-}$$

then, $$K = \frac{[X^+]^2 \times [Y^{2-}]}{[X_2Y]}$$

It follows that the extent to which a weak electrolyte is ionized in aqueous solution, at a given temperature, can be reported as its ionization constant at that temperature.

Alternatively, the same information is given by the degree of ionization of the weak electrolyte in a dilute aqueous solution of reported concentration at a given temperature. The degree of ionization (α) is defined as 'the fraction of the weak electrolyte that is apparently totally ionized in the solution' (see also p. 107).

Calculation of the degree of ionization of a weak electrolyte

In practice, the degree of ionization of a weak electrolyte in dilute aqueous solution is most often determined from the electrical conductance of the solution.[2, 8] However, an approximate value for the degree of ionization (α) can be calculated from the van't Hoff factor (i) of a solution, when its ionic strength is sufficiently small for it to be possible to assume that the activity coefficients of the undissociated solute, and of its ions, are all equal to unity.

Let us assume that 1 mol of the weak electrolyte is dissolved in V dm^3 of water at a given temperature to yield a dilute solution whose freezing point is lower by ΔT K than the comparable freezing point of pure water. If α is the degree of ionization of the solute, the solution will contain $(1-\alpha)$ mol of unionized solute in V dm^3. Furthermore, if each of the molecules of solute that ionizes gives rise to n ions, this same volume of solution will contain $n\alpha$ ions. We have seen that the depression of freezing point of the solvent in an ideal solution is directly proportional to the number of 'particles' of solute present in unit volume of the solution; assuming no interionic association, ΔT_f is directly proportional to $\frac{(1-\alpha)+n\alpha}{V}$, which equals $\frac{1+\alpha(n-1)}{V}$.

Had the solute remained unionized, the solution would have exhibited a lesser depression of freezing point equal to Δt_f, where Δt_f is proportional to $1/V$, the proportionality constant being identical with that relating ΔT_f to the concentration of solute molecules plus its ions.

The van't Hoff factor (i) for the electrolyte solution (partially ionized) equals $\Delta T_f/\Delta t_f$, therefore,

$$i = \frac{\Delta T_f}{\Delta t_f} = \frac{1+\alpha(n-1)}{V} \div \frac{1}{V} = 1+\alpha(n-1)$$

Thus,

$$\alpha = \frac{i-1}{n-1}$$

Having determined the value of α for a weak electrolyte in a dilute aqueous solution of known concentration, the corresponding value of its ionization constant can now be calculated by applying ***Ostwald's Dilution Law*** (p. 107). This Law predicts that for a weak electrolyte that yields 2 ions/molecule, if α is its degree of ionization in the dilute aqueous solution in which it is present in a concentration of C mol dm^{-3} at a given temperature, then the value of its ionization constant K_i is given by the equation,

$$K_i = \frac{\alpha^2 C}{(1-\alpha)}$$

EXAMPLE

If the depression of freezing point of a 0.01 mol dm^{-3} aqueous solution of propionic acid at standard atmospheric pressure is -0.0193 K, calculate an approximate value for the ionization constant of propionic acid in this solution at approximately 273 K.

The van't Hoff factor for this solution of propionic acid is given by the equation,

$$i = \frac{\Delta T_f}{\Delta t_f} \quad \text{where} \begin{cases} \Delta T_f = \text{observed depression of f.p.} \\ \Delta t_f = \text{calculated depression of f.p. assuming} \\ \qquad\quad \text{no ionization of the solute} \end{cases}$$

Since the molal depression constant for water is 1.86 K, a 0.01 mol dm^{-3}

ideal, aqueous solution of a non-ionized solute would freeze at 0.0186 K lower than the f.p. of pure water.

$$\therefore \frac{\Delta T_f}{\Delta t_f} = \frac{0.0193}{0.0186} = 1.037 = i$$

$$\text{The degree of ionization } \alpha = \frac{i-1}{n-1}$$

$$\text{where} \begin{cases} i = 1.037 \\ n = \text{number of ions from 1 molecule} = 2 \end{cases}$$

$$\therefore \alpha = \frac{1.037-1}{2-1} = 0.037$$

Propionic acid is a weak electrolyte whose ionization forms 2 ions from 1 molecule; thus, according to Ostwald's Dilution Law, its ionization constant is related to its degree of ionization in a dilute solution (at a given temperature) by the following equation,

$$K_i = \frac{\alpha^2 C}{(1-\alpha)} \qquad \text{where} \begin{cases} \alpha = \text{degree of ionization} \\ C = \text{concentration in mol dm}^{-3} \\ K_i = \text{ionization constant at this temperature} \end{cases}$$

Thus since α is 0.037 for propionic acid in its 0.01 mol dm^{-3} solution at the given temperature,

$$K_i = \frac{(0.037)^2 \times 0.01}{1-0.037} = \underline{1.42 \times 10^{-5}}$$

Therefore, the ionization constant of propionic acid at the given temperature equals 1.42×10^{-5}

Summary of the colligative properties of a solution of a weak electrolyte

A weak electrolyte is incompletely ionized even in extremely dilute aqueous solution and the van't Hoff factor for these solutions does not possess an integral value. From the actual value of the van't Hoff factor of a dilute aqueous solution of a weak electrolyte, its degree of ionization can be determined; from this, one can in turn calculate its ionization constant. Although independent of the actual concentration of weak electrolyte in the dilute solution, the value of this ionization constant varies with temperature. In more concentrated solutions the activity of the unionized solute decreases, and the activities of the ions are different from their concentrations by an amount determined by the total ionic strength of the solution.

SOLUBILITY OF SALTS

Thermodynamic and apparent solubility products

We have seen that the mean ion activity coefficient of an electrolyte in solution is in large measure determined by the total ionic strength of the solution. Examination of the equilibrium between a solid salt and its ions in its saturated aqueous solution illustrates how important is this effect of the ionic strength.

Consider, for example, the saturated aqueous solutions of sodium chloride and barium sulphate at a given temperature. The saturated solution of sodium chloride contains the crystalline salt in equilibrium with its solute ions,

$$Na^+Cl^- \text{ (solid)} \rightleftharpoons Na^+ + Cl^-$$

Thus sodium chloride dissolves in water at 298 K to an extent that is determined by the value at 298 K of the equilibrium constant K, where

$$K = \frac{a_{Na^+} \times a_{Cl^-}}{a_{Na^+Cl^-}}$$

and the activity terms are the activities of the components of the saturated solution at 298 K. By definition, solid sodium chloride has unit activity (this being its standard state, p. 192), so that the equilibrium expression becomes,

$$K_{sp} = a_{Na^+} \times a_{Cl^-}$$

where the equilibrium constant is now represented by K_{sp} and is called the ***thermodynamic solubility product*** of sodium chloride in water at 298 K.

Similarly, because the following equilibrium exists in the saturated aqueous solution of barium sulphate,

$$Ba^{2+}SO_4^{2-} \text{ (solid)} \rightleftharpoons Ba^{2+} + SO_4^{2-}$$

the thermodynamic solubility product of barium sulphate in water at a given temperature is,

$$K_{sp} = a_{Ba^{2+}} \times a_{SO_4^{2-}}$$

where the activity terms refer to the activities of the Ba^{2+} and SO_4^{2-} ions in the saturated aqueous solution of barium sulphate at the given temperature.

If one 'molecular unit' of the salt gives rise to more than one ion of the same kind, then, when calculating its thermodynamic solubility product, the equilibrium activity of that ion must be raised to a power equal to the number of these ions produced from one molecular unit of the salt. Thus, since tricalcium phosphate ionizes as follows,

$$Ca_3^{2+}(PO_4)_2^{3-} \rightarrow 3Ca^{2+} + 2PO_4^{3-}$$

its aqueous thermodynamic solubility product is $(a_{Ca^{2+}})^3 \times (a_{PO_4^{3-}})^2$.

If a salt is only very slightly soluble in water, its saturated solution will have a very low ionic strength and its mean ion activity coefficient in that solution will approximately equal 1. This means that its thermodynamic solubility product will be very little different in value from its ***apparent solubility product*** which is derived from the *concentrations* of the ions of the salt in this saturated solution. Take, for example, barium sulphate, which is very slightly soluble in water at 298 K:

$$\text{apparent } K_{sp} = [Ba^{2+}][SO_4^{2-}]$$

$$\text{thermodynamic } K_{sp} = a_{Ba^{2+}} \times a_{SO_4^{2-}}$$

$$= y_{Ba^{2+}}[Ba^{2+}] \times y_{SO_4^{2-}}[SO_4^{2-}],$$

$$\text{and since } y_{\pm\,BaSO_4} = (y_{Ba^{2+}} \times y_{SO_4^{2-}})^{1/2},$$

$$\text{thermodynamic } K_{sp} = y^2_{\pm\,BaSO_4} \times [Ba^{2+}][SO_4^{2-}]$$

or thermodynamic K_{sp} = (mean ion activity coefficient)^2 × apparent K_{sp}

Because of the slight solubility of barium sulphate, its saturated solution is of low ionic strength and $y_{\pm BaSO4} \triangleq 1$

$$\therefore \text{ thermodynamic } K_{sp} \triangleq \text{apparent } K_{sp} \text{ for } BaSO_4$$

In the case of more soluble salts such as sodium chloride, the saturated solutions are of high ionic strength, the mean ion activity coefficient is usually very different from 1, and the values of the apparent and thermodynamic solubility products are markedly different. In Fig. 4.5 values of the mean ion activity coefficients of a few salts are plotted against their corresponding molal concentrations in aqueous solution at 298 K (i.e. 25°C).

This figure illustrates not only the possible extent of the deviation from unity of these coefficients, it also demonstrates how, as the molal concentration of the salt solution is increased, the values of the coefficients first fall but may pass through a minimum (at about 1 mol dm^{-3}), rising at higher concentrations and perhaps eventually exceeding unity. Figure 4.5

also emphasizes the importance of the valencies of a salt's ions in determining the ionic strength of its solution. This explains

(a) why there is a measurable discrepancy between the values of a 'thermodynamic' (activity based) constant which defines an ionic equilibrium in aqueous solution, and its 'apparent' (concentration based) counterpart;

(b) why this discrepancy is greater in the case of a fairly dilute solution of a salt whose ions are multivalent, than in the case of a solution of identical concentration of a salt whose ions are monovalent.

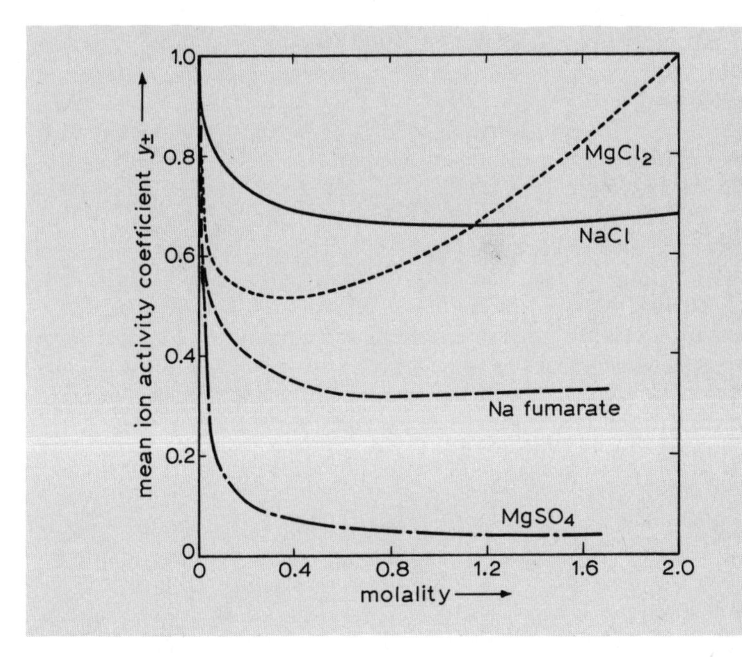

Fig. 4.5 How the mean ion activity coefficients of some salts vary with the molalities of their aqueous solutions (at 298 K).

We will again encounter this difference between 'apparent' and 'thermodynamic' constants for an ionic equilibrium in aqueous solution, when we discuss acidic and basic dissociation constants in Chapter 5. Remember, therefore, that the value of a thermodynamic (activity based) constant is truly constant, but the magnitude of an apparent constant changes with the ionic strength of the medium. This is demonstrated, as we shall now see, by the way in which the solubility of one salt in water is affected by the addition of another electrolyte.

The salt (or electrolyte) effect

Consider a saturated aqueous solution of silver chloride at a given temperature,

$$Ag^+Cl^- \text{ (solid)} \rightleftharpoons Ag^+ + Cl^-$$

$$\text{apparent } K_{sp} = [Ag^+][Cl^-] \text{ (at equilibrium)}$$

$$\text{thermodynamic } K_{sp} = y_{\pm Ag^+Cl^-} \times [Ag^+][Cl^-]$$

Because silver chloride is only very slightly soluble in water, its saturated solution is of low ionic strength and $y_{\pm Ag^+Cl^-}$ is approximately 1. This means that in the 'pure' saturated solution of silver chloride in water, the apparent value of K_{sp} equals the thermodynamic solubility product of silver chloride.

Suppose now that a quantity of sodium nitrate is added to the saturated aqueous solution of silver chloride. Sodium nitrate is very soluble in water and the ionic strength of the solution will be greatly increased by its addition; as a result, it is likely that the value of $y_{\pm Ag^+Cl^-}$ decreases. Yet the value of the thermodynamic solubility product remains constant, and since it equals ($y_{\pm Ag^+Cl^-} \times$ apparent K_{sp}), it follows that when $y_{\pm Ag^+Cl^-}$ decreases in value, the apparent solubility product must increase and more of the solid silver chloride must enter into solution to raise the concentrations of Ag^+ and Cl^- ions. These consequences of the increase in the ionic strength of a saturated salt solution, caused by the addition of a 'foreign' electrolyte, are collectively known as the ***salt or electrolyte effect.***

The common ion effect

The thermodynamic solubility product of silver chloride in water at 298 K is 1.7×10^{-10}. Because the ionic strength of the saturated solution is negligible, this is also the value of the apparent solubility product, and we deduce that so long as some silver chloride remains undissolved in water at 298 K the supernatant solution must contain 1.3×10^{-5} mol dm^{-3} each of Ag^+ and of Cl^- ions for,

$$\text{apparent } K_{sp} = 1.7 \times 10^{-10} = [Ag^+][Cl^-],$$

$$[Ag^+] = [Cl^-],$$

and at equilibrium each equals $= \sqrt{1.7 \times 10^{-10}}$

$$= 1.3 \times 10^{-5} \text{ mol dm}^{-3}$$

Suppose however that potassium chloride (10^{-3} mol dm^{-3}) is added to this saturated solution of silver chloride. The concentration of Cl^- ions is increased almost 100-fold and the value of the thermodynamic solubility product of silver chloride can only be kept constant by a corresponding decrease in the equilibrium concentration of Ag^+ ions. The net outcome is the precipitation of a considerable fraction of the previously dissolved silver chloride.†

The precipitation of one salt, by the addition to its saturated solution of a quantity of a soluble salt that possesses an ion in common with it, is known as the ***common ion effect***. It is usually put to practical use in the analytical and preparative precipitation of sparingly soluble salts.

EXAMPLE

Calcium oxalate is of interest to clinical biochemists since it is only slightly soluble in water (7 mg dm^{-3} at room temperature) and can be deposited in renal calculi.

Calculate values for the following terms at room temperature:

(*a*) *the concentration of oxalate ions in a saturated aqueous solution of calcium oxalate;*
(*b*) *the 'thermodynamic' and the 'apparent' solubility product constants of calcium oxalate in water;*
(*c*) *the concentration of oxalate ions in a saturated solution of calcium oxalate in 10^{-3} mol dm^{-3} aqueous calcium chloride.*

(*Molecular weight of calcium oxalate* = *128*)

The following equilibrium is established in a saturated solution of calcium oxalate:

$$\begin{matrix} COO^- \\ | \\ COO^- \end{matrix}\ Ca^{2+}\ (\text{solid}) \rightleftharpoons \begin{matrix} COO^- \\ | \\ COO^- \end{matrix} + Ca^{2+}$$

(a) In the saturated aqueous solution that contains 7 mg of calcium oxalate per dm^3, the concentration of solute $= 7\times 10^{-3}/128 = 5.47\times 10^{-5}$ mol dm^{-3}.

Since the dissolved calcium oxalate is completely ionized, the concentration of oxalate ions in this solution is $\underline{5.47\times 10^{-5}\ \text{mol dm}^{-3}}$.

(b) The saturated aqueous solution of calcium oxalate contains 5.47×10^{-5} mol dm^{-3} of each Ca^{2+} and Ox^{2-} ions.

∴ apparent (concentration-based) solubility product $= [Ca^{2+}][Ox^{2-}]$

$$= (5.47\times 10^{-5})^2 = \underline{2.99\times 10^{-9}}$$

The thermodynamic (activity-based) solubility product

$$= a_{Ca^{2+}} \times a_{Ox^{2-}}$$

$$\text{and since } y_{\pm} = (y_{Ca^{2+}} \times y_{Ox^{2-}})^{1/2}$$

$$\text{thermodynamic } K_{sp} = y_{\pm}{}^2 \times [Ca^{2+}][Ox^{2-}]$$

† The silver chloride concentration does not in fact decrease to 1.7×10^{-7} mol dm^{-3} because of complex ion formation.

The value of the mean ionic activity coefficient ($y_{\pm}$) is given by the Debye–Hückel equation,

$$\log y_{\pm} = -Kz_{+}z_{-}\sqrt{\mu}$$

The ionic strength of the solution $\mu = \frac{1}{2}\sum mz^2$, and in this dilute solution, m = molality = mol dm^{-3} (approximately) = 5.47×10^{-5}, while z = valence of ion = 2.

$$\therefore \mu = \tfrac{1}{2}[(5.47 \times 10^{-5} \times 2^2) + (5.47 \times 10^{-5} \times 2^2)]$$
$$= \underline{2.134 \times 10^{-4}}$$

Since both Ca^{2+} and Ox^{2-} ions have a valence of 2,

$$\log y_{\pm} = -K \times 2 \times 2\sqrt{2.134 \times 10^{-4}}$$

Assuming K to equal 0.51 (as at 298 K for an aqueous solution, p. 83),

$$\log y_{\pm} = -0.51 \times 4 \times 1.46 \times 10^{-2}$$
$$= -0.0298 = \bar{1}.9702$$
$$\therefore y_{\pm} = \text{antilog } \bar{1}.9702 = 0.934$$

Substituting this value for the mean ion activity coefficient in the equation that gives the value of the thermodynamic solubility product constant of calcium oxalate, we obtain the following result:

$$\text{thermodynamic solubility product} = y_{\pm}^2 \times [Ca^{2+}][Ox^{2-}]$$
$$= (0.934)^2 \times [Ca^{2+}][Ox^{2-}]$$

But $[Ca^{2+}][Ox^{2-}]$ equals the apparent solubility product = 2.99×10^{-9}

$$\therefore \text{thermodynamic solubility product} = (0.934)^2 \times 2.99 \times 10^{-9}$$
$$= \underline{2.61 \times 10^{-9}}$$

Note that although calcium oxalate is only slightly soluble in water, because it is a bi-bivalent salt the ionic strength of its saturated solution is sufficiently high to ensure that it behaves non-ideally (apparent K_{sp} is appreciably larger than the thermodynamic K_{sp}).

(c) Even in the solution that contains 10^{-3} mol dm^{-3} calcium chloride, the true solubility product of calcium oxalate equals

$$2.61 \times 10^{-9} = a_{Ca^{2+}} \times a_{Ox^{2-}}$$

but,

(i) the concentration of oxalate ions in the saturated solution will be less than the concentration of these ions in the saturated solution of calcium oxalate in pure water, due to the high concentration of calcium ions provided by the calcium chloride (common ion effect);

(ii) the activity coefficients of both calcium and oxalate ions will be smaller in the saturated solution in calcium chloride (10^{-3} mol dm^{-3}) than in the saturated solution in water, due to the greater ionic strength of the former solution (the salt effect).

The ionic strength of the calcium chloride-containing solution is virtually entirely due to its content of calcium chloride (particularly as the solubility of the calcium oxalate is reduced in this solution due to the common ion effect).

In its 10^{-3} mol dm^{-3} solution calcium chloride is completely ionized, i.e. $Ca^{2+}Cl_2^- \rightarrow Ca^{2+} + 2Cl^-$, and $[Ca^{2+}] = 10^{-3}$ mol dm^{-3} while $[Cl^-] = 2 \times 10^{-3}$ mol dm^{-3}

$$\therefore \text{ ionic strength} = \mu = \tfrac{1}{2}\sum mz^2 = \tfrac{1}{2}[(0.001 \times 2^2) + (0.002 \times 1^2)]$$
$$= \tfrac{1}{2}[0.004 + 0.002]$$
$$= 0.003$$

Thus, in this solution, according to the Debye–Hückel equation

$$y_{Ca^{2+}} = y_{Ox^{2-}} \text{ where } \log y = -0.51 \times 2^2 \times \sqrt{0.003}$$
$$= -0.1118$$
$$= \bar{1}.8882$$

$$\therefore y_{Ca^{2+}} = y_{Ox^{2-}} = \text{antilog } \bar{1}.8882 = 0.773$$

(Note that the addition of only 10^{-3} mol dm^{-3} $CaCl_2$ reduces the value of $y_{\pm}$ for calcium oxalate in its saturated solution from 0.934 to 0.773.)

Since $[Ca^{2+}] = 0.001$ mol dm^{-3}, $a_{Ca^{2+}} = 0.773 \times 0.001$
$$= 7.73 \times 10^{-4}$$

But $a_{Ca^{2+}} \times a_{Ox^{2-}} = 2.61 \times 10^{-9}$
(the thermodynamic solubility product of calcium oxalate)

$$\therefore \quad a_{Ox^{2-}} = \frac{2.61 \times 10^{-9}}{7.73 \times 10^{-4}}$$

But, $a_{Ox^{2-}} = y_{Ox^{2-}} \times [Ox^{2-}] = 0.773[Ox^{2-}]$

$$\therefore \quad [Ox^{2-}] = \frac{2.61 \times 10^{-9}}{0.773 \times 7.73 \times 10^{-4}} = \underline{4.37 \times 10^{-6} \text{ mol dm}^{-3}}$$

Thus, in the saturated solution of calcium oxalate in a 10^{-3} mol dm^{-3} aqueous solution of calcium chloride, assuming that no complexes are formed, the concentration of oxalate ions is 4.37×10^{-6} mol dm^{-3} (compared with 5.47×10^{-5} mol dm^{-3} in the absence of calcium chloride).

In concluding, we can take some comfort from the fact that if we consider only dilute solutions of non-electrolytes ($< 10^{-1}$ mol dm^{-3}), we can assume for most purposes that they behave ideally. To be able to assume activity coefficients of unity for the ions of an electrolyte, we must work with even more dilute solutions (especially in the case of multi-

multivalent electrolytes). For example, $y_{\pm}$ has the following values in 10^{-3} mol dm^{-3} pure aqueous solutions of salts at 298 K: NaCl, 0.966; $CaCl_2$, 0.89; $CuSO_4$, 0.74.

PROBLEMS

(*Assume that the molal depression constant of water, $K_f = 1.86$ K, and that the gas constant, $R = 8.314$ J K^{-1} mol^{-1}.*)

1. Two beakers, A and B, containing solutions of sucrose in water, are placed under a bell jar at 293 K. Initially, A contains 250 g of 0.1 molal sucrose, and B contains 250 g of 0.5 molal sucrose. Approximately what will be the weights of the contents of these beakers when equilibrium is achieved?

2. The sugar stachyose is found in the seeds of several leguminous plants. It is a polysaccharide which on hydrolysis gives rise to galactose, glucose and fructose.

A solution of 100 mg of stachyose in 10 cm^3 water exhibited an osmotic pressure of 35.55 kPa at 285 K (i.e. 12°C). What is the molecular weight of stachyose? Is it a tri-, tetra-, or pentasaccharide?

3. A solution that is to be infused into the human blood stream must be made isotonic with the blood plasma. It is generally assumed that aqueous solutions with the same freezing point as blood serum are not only isosmotic, but are also isotonic with the serum.

Calculate the concentration (g/100 cm^3) of the saline solution that is isotonic with blood serum at 37°C, if the serum freezes at −0.56°C. (Molecular weight of Na^+Cl^- is 58.5; assume that the mean ion activity coefficient of sodium chloride in the isotonic saline is 0.75.)

4. Calculate the activity coefficient of water in an aqueous solution of a non-electrolyte at 298 K, given that the mole fraction of water in the solution is 0.95, and that the vapour pressure of water above the solution is 2.56 kPa (Vapour pressure of pure water at 298 K is 3.167 kPa.)

5. What is the freezing point of an aqueous solution of glycerol which has an osmotic pressure of 1215.6 kPa at 278 K (i.e. 5°C)?

6. The following osmotic pressure data were obtained for a protein dissolved at its isoelectric pH and at 278 K (i.e. 5°C) in 0.18 mol dm^{-3} buffer.

Protein concentration in g dm^{-3}	7.3	18.4	27.6	42.1	57.4
Osmotic pressure attributable to the protein (in kPa)	0.211	0.533	0.804	1.236	1.701

Calculate the molecular weight of the protein.

7. Calculate the ionic strengths of the following aqueous solutions at 298 K:

(i) 0.01 mol dm^{-3} $CaCl_2$;
(ii) 0.10 mol dm^{-3} $MgSO_4$;
(iii) a mixture of 0.5 mol dm^{-3} $(NH_4)_2SO_4$ with 0.5 mol dm^{-3} urea.

8. List the following aqueous solutions in order of increasing ionic strength:
0.1 mol dm^{-3} NaCl; 0.05 mol dm^{-3} $CaCl_2$; 0.25 mol dm^{-3} ethanol; 0.03 mol dm^{-3} $MgSO_4$; 0.03 mol dm^{-3} $FeCl_3$.

9. Calculate the mean ion activity coefficient at 298 K of magnesium chloride in its 10^{-4} mol dm^{-3} aqueous solution. (Assume a value of 0.51 for the constant in the Debye–Hückel equation.)

10. Potassium chloride has a mean ion activity coefficient of 0.77 in its 0.1 mol dm^{-3} aqueous solution at 298 K. Calculate the osmotic pressure of this solution.

11. Calculate the activities of K^+ and Cl^- ions in an aqueous solution which contains 10^{-5} mol dm^{-3} each of potassium chloride, magnesium sulphate and sodium nitrate.

12. List the following aqueous solutions in the probable order of increasing osmotic pressure at 298 K:

(i) 12.5 g protein in 100 g water (molecular weight of protein = 60 000);
(ii) 0.10 mol dm^{-3} sucrose;
(iii) 0.05 mol dm^{-3} sodium chloride (mean ion activity coefficient = 0.823);
(iv) 0.05 mol dm^{-3} calcium chloride (mean ion activity coefficient = 0.570).

13. Phosphoric acid is a weak electrolyte that ionizes as shown below in aqueous solution,

$$H_3PO_4 \rightleftharpoons H^+ + H_2PO_4^-$$

The degree of ionization of phosphoric acid in 10^{-3} mol dm^{-3} aqueous solution at 298 K, as determined from conductance studies, was found to be 0.93. Calculate approximate values for:

(i) the apparent (primary) ionization constant at 298 K;
(ii) the van't Hoff factor and osmotic pressure of its 10^{-3} mol dm^{-3} solution at 298 K.

14. To assay the radioactivity of a sample of $^{14}CO_2$, the radioactive gas is often absorbed in a solution of sodium hydroxide, and then precipitated as $Ba^{14}CO_3$ by the addition of excess barium chloride. The radioactivity of the precipitate is then assayed.

If 5 cm^3 of 10^{-2} mol dm^{-3} $BaCl_2$ is added to 5 cm^3 of 2×10^{-4} mol dm^{-3} sodium [^{14}C]-carbonate at 298 K, what percentage of the total radioactivity remains unprecipitated? (Assume that the apparent solubility product of $BaCO_3$ *in the final solution* is 8×10^{-9} at 298 K.)

15. Calculate (a) the activity, and (b) the concentration of Ba^{2+} ions present at 298 K in a saturated solution of barium sulphate in:

(i) water;
(ii) 0.1 mol dm^{-3} sodium chloride;
(iii) 0.1 mol dm^{-3} copper sulphate.

(Thermodynamic solubility product of barium sulphate in water at 298 K is 1.08×10^{-10}. Mean ion activity coefficients of the other electrolytes are, NaCl in 0.1 mol dm^{-3} aqueous solution at 298 K = 0.78; $CuSO_4$ in its 0.1 mol dm^{-3} aqueous solution at 298 K = 0.16.)

5

Acids, Bases and Buffers in Aqueous Solution

In this chapter we shall consider the hydrogen ion concentrations of aqueous solutions and the practice of reporting them as pH values. We shall discuss the nature of acids and bases, and the composition of buffer mixtures, which by acting as 'reservoirs' of acid and base, can stabilize the hydrogen ion concentration of an aqueous solution. The biological relevance of this study will then be considered in Chapter 6.

The ion product of water

All aqueous solutions contain positively charged hydrogen ions (or protons, H^+), and negatively charged hydroxyl ions (OH^-). In pure water these are entirely derived from the ionization of water molecules,

$$H_2O \rightleftharpoons H^+ + OH^-$$

a process that is also referred to as the *dissociation* of water (into its component ions), or as the *protolysis* of water (to emphasize the liberation of H^+ ions). In fact, it is not strictly true to speak of the liberation of H^+ ions in an aqueous medium, for in this situation the H^+ ion is always hydrated and is predominantly present as H_3O^+, the hydronium ion. Thus the protolysis of water might better be represented by the following equation,

$$H_2O + H_2O \rightleftharpoons H_3O^+ + OH^-$$

Yet, except when it is necessary to emphasize the part played by water in another protolytic reaction, we can overlook the fact that H^+ actually exists in an aqueous medium in the form of H_3O^+ ions, for this need not affect our appraisal of the origin and ultimate fate of these protons; we may therefore continue to think of water dissociating to yield H^+ plus OH^- ions.

If $H_2O \rightleftharpoons H^+ + OH^-$, then at equilibrium, $(H^+)(OH^-)/(H_2O) = K_a$, where K_a is the temperature-dependent, ***acid dissociation constant*** of water and the bracketed terms represent the equilibrium activities of H^+, OH^- and H_2O. At 298.15 K (i.e. 25°C), K_a of water $= 1.8 \times 10^{-16}$, and since the water is so little dissociated, (H_2O) in the denominator of the equation (above) that defines the values of K_a of water can be assumed to be insignificantly less than the total activity of pure water at 298 K which is a constant. Thus,

$$(H^+)(OH^-) = K_a(H_2O) = K_w$$

where K_w is the ***ion product*** (or ion product constant) of water.

In pure water, and in the dilute aqueous solutions that we shall consider in this chapter, we may assume that the activities of H^+ and OH^- ions will equal their concentrations. Therefore, $K_w = [H^+][OH^-]$, where $[H^+]$ and $[OH^-]$ are the equilibrium concentrations of these ions in any aqueous solution expressed in mol dm^{-3}. In pure water at 298 K, $K_w = 10^{-14}$, $[H^+]$ equals $[OH^-]$, and so both equal 10^{-7} mol dm^{-3} (the value of K_w increases by about 8% for every K rise in temperature, until at 310 K it is 2.4×10^{-14}).

The equation derived above to define K_w states that the product of the concentrations of hydrogen and hydroxyl ions in an aqueous solution is constant at a given temperature, and that any change in its hydrogen ion concentration will be reflected as a converse change in its hydroxyl ion concentration. Thus in any aqueous solution at 298 K, if the hydrogen ion concentration is 10^{-3} mol dm^{-3}, the hydroxyl ion concentration must be 10^{-11} mol dm^{-3}. Since values involving such negative powers of 10 are inconvenient to write and are troublesome to use in calculations, Sørensen suggested that the hydrogen ion concentrations of dilute aqueous solutions would be better expressed as pH values.

Meaning of the term pH

The relationship between a pH value and the hydrogen ion concentration (in mol dm^{-3}) that it represents, can be explained in various ways; e.g.

(i) The pH of a solution is the negative of the logarithm to the base 10 of its hydrogen ion concentration.

$$\text{pH} = -\log [H^+]$$

(ii) The pH of a solution equals the logarithm to the base 10 of the reciprocal of its hydrogen ion concentration.

$$\text{pH} = \log 1/[H^+]$$

(iii) If the $[H^+]$ is written as a power of 10, then the corresponding pH value is the index of this exponential term without its negative sign; e.g. a $[H^+]$ of $10^{-3.72}$ mol dm^{-3} is equivalent to a pH value of 3.72.

(To reassure yourself that these are indeed alternative definitions, you should consult Chapter 1, p. 4, where exponential terms and logarithms are discussed.)

The equation that is most useful in calculating pH values is that supplied by definition (i), i.e. $pH = -\log [H^+]$.

EXAMPLE

Calculate (a) the pH of a solution whose hydrogen ion concentration is 2.3×10^{-9} mol dm^{-3}; (b) the hydrogen ion concentration in a solution of pH 4.31.

(a) Since $[H^+] = 2.3 \times 10^{-9}$ mol dm^{-3} and $pH = -\log [H^+]$

$$pH = -\log (2.3 \times 10^{-9}) = -(\bar{9}.36) = -(-9+0.36)$$
$$= -(-8.64)$$
$$pH = \underline{8.64}$$

(Note: If in doubt concerning the use of logarithms, consult p. 7. There you will find that the logarithm of the product of two terms is equal to the sum of the logarithms of each term taken separately, i.e. $\log ab = \log a + \log b$.

Thus, $\log (2.3 \times 10^{-9}) = (\log 2.3 + \log 10^{-9}) = (0.36 + \bar{9}) = \bar{9}.36$.)

(b) Since $pH = 4.31$ and $pH = -\log [H^+]$, then $\log [H^+] = -4.31 = \bar{5}.69$

$$\log [H^+] = \bar{5}.69 \text{ and } [H^+] = \text{antilog } \bar{5}.69 = 4.9 \times 10^{-5}$$

$\therefore$ in a solution of pH 4.31 the $[H^+] = \underline{4.9 \times 10^{-5} \text{ mol dm}^{-3}}$

(Note: To convert a logarithm whose value is -4.31 into the conventional form having a positive mantissa (p. 5), proceed as follows:

$$-4.31 = (-5+(1-0.31)) = (-5+0.69) = \bar{5}.69)$$

The expression of hydrogen ion concentrations as pH values is therefore a means of representing a wide range of these concentrations (measured on an arithmetic scale) on a conveniently 'compressed' exponential scale. Although pH values less than 0 (negative values), and greater than 14, are theoretically possible, the range of pH values from 0 to 14 covers all hydrogen ion concentrations found in dilute aqueous solutions (and biological media).† In pure water at 298 K, $[H^+] = 10^{-7}$ mol dm^{-3} and the pH is 7 (often called 'neutral pH').

† Always remember when using the pH notation, that (a) the pH of a solution decreases as its $[H^+]$ increases, and vice versa, and (b) a tenfold change in $[H^+]$ is represented by a pH difference of 1 unit.

In the same way that hydrogen ion concentrations are converted into pH values, so too may hydroxyl ion concentrations be represented as pOH values, where $pOH = -\log(OH^-)$. Thus, in aqueous solutions when $[H^+][OH^-] = 10^{-14}$, $\log[H^+] + \log[OH^-] = \log 10^{-14} = -14$. Reversing the signs throughout, $-\log[H^+] - \log[OH^-] = 14$

so that, $$pH + pOH = 14$$

ACIDS AND BASES

Solution of an acid in water raises the hydrogen ion concentration above 10^{-7} mol dm^{-3} and correspondingly lowers the pH. Solution of a base in water raises the hydroxyl ion concentration above 10^{-7} mol dm^{-3}, and correspondingly decreases the hydrogen ion concentration (i.e. pH rises). This behaviour suggests that acids and bases may be defined as compounds which dissociate in water to yield H^+ and OH^- ions respectively:

$$\text{Acid: } HA \rightleftharpoons H^+ + A^-$$

$$\text{Base: } BOH \rightleftharpoons OH^- + B^+$$

A compound which can yield both H^+ and OH^- ions by dissociation, and which may therefore act both as an acid and a base, is termed an *amphoteric substance* or *ampholyte*.

This dissociation theory is delightfully simple, but it gives rise to several anomalies in its appraisal of bases. According to the theory, sodium hydroxide whose OH^- anion is liberated in aqueous solution, is obviously a base. Yet ammonia (NH_3), which cannot possibly liberate OH^- ions in this manner, also gives aqueous solutions that are strongly alkaline. The dissociation theory explains this by supposing that ammonia is hydrated in water to form a dissociable base, ammonium hydroxide (NH_4OH), and proposes that ammonia is an 'anhydro-base'. In fact, there is no good evidence for the existence of such a molecule as NH_4OH (see p. 110).

An alternative view of the nature of acids and bases and of their interaction was proposed by Brönsted and by Lowry in 1923. In their view an acid is characterized by its tendency to lose H^+, whilst a base is characterized by its tendency to associate with H^+. In other words, *acids are potential proton donors and bases are potential proton acceptors*. Amphoteric compounds might, according to this theory, be termed *amphiprotic compounds* capable of acting both as proton donors and as proton acceptors.

Thus water is amphiprotic since it can both (a) behave as an acid by undergoing protolysis, $H_2O \rightleftharpoons H^+ + OH^-$, and (b) behave as a base by combining with protons to form hydronium ions, $H_2O + H^+ \rightleftharpoons H_3O^+$.

This means that one molecule of water acting as an acid can donate a proton to another molecule of water acting as a base, $H_2O + H_2O \rightleftharpoons H_3O^+ + OH^-$. The behaviour of ammonia in aqueous solution is readily explained by this theory as being that of a base which accepts protons donated by the water,

$$NH_3 + H_2O \rightleftharpoons NH_4^+ + OH^-$$

It follows, from the Brönsted and Lowry definition of acid and base, that the product other than H^+ formed by the dissociation of an acid must be a base. For example, when the acid HA dissociates, $HA \rightleftharpoons H^+ + A^-$, its anion A^- is manifestly a base since it serves as a proton acceptor in the reverse reaction. The special relationship of base A^- to the parent acid HA is acknowledged by calling it the *conjugate base* of that acid, so that the acid HA and the base A^- comprise a *conjugate pair*. This means, for example, that the cyanide ion CN^- is the conjugate base of hydrocyanic acid HCN, and that the acetate ion CH_3COO^- is the conjugate base of acetic acid CH_3COOH.

Analogously, a base and its protonated derivative form a conjugate pair, so that if $B + H^+ \rightleftharpoons BH^+$, then BH^+ is the *conjugate acid* of base B. It follows that acetic acid is the conjugate acid of the acetate ion, and that the ammonium ion, NH_4^+, is the conjugate acid of ammonia, NH_3.

The strength of acids and bases

One must be careful not to confuse the terms 'concentration' and 'strength' as applied to aqueous solutions of acids and bases, for these terms are not synonymous and cannot be used interchangeably. By *concentration* is meant the quantity of acid or base dissolved in a certain volume of water and usually expressed in mol dm^{-3}. The *strength* of an acid or base is an indication of the degree to which the compound demonstrates the properties of an acid or base relative to other compounds with similar properties. The acidic or basic strength of a compound is therefore a measure of the effectiveness with which that compound behaves as an acid or base.

The strength of an acid

The strength of an acid is determined by the efficiency with which it acts as a proton donor. This in turn will be affected by the proton-accepting or proton-donating properties of the medium in which the acid is dissolved. Here we are concerned only with aqueous solutions and can assess the relative acidic strengths of compounds in terms of the extent to which each increases the hydrogen ion concentration when it is dissolved in water in known concentration and at a standard temperature. This can be recorded in any one of the following ways:

1. as the hydrogen ion concentration (or pH) of a dilute solution of stated concentration;
2. as the degree of dissociation of the acid, i.e. % of acid molecules dissociated at equilibrium;
3. as the dissociation constant, K_a, of the acid.

The last of these methods is undoubtedly the most convenient. According to the dissociation theory, the acid HA dissociates in water as follows,

$$HA \rightleftharpoons H^+ + A^-, \quad \text{whence at equilibrium (p. 212),} \quad \frac{[H^+][A^-]}{[HA]} = K_a$$

Being a ratio of equilibrium concentrations, K_a is independent of the total concentration of acid in any aqueous solution in which its strength is determined (so long as this is dilute and possesses a low ionic strength, p. 217).

Although K_a is a *dissociation constant*, even according to the Brönsted and Lowry theory its value is a valid measure of the acid strength of a substance. In the Brönsted and Lowry view, a compound evidences its acidic nature in aqueous solution by donating protons to the water (which here acts as a base). Thus,

$$HA + H_2O \rightleftharpoons H_3O^+ + A^-, \quad \text{and at equilibrium} \quad \frac{[H_3O^+][A^-]}{[HA][H_2O]} = K_{eq}$$

But the water (as solvent) is present in such excess that its concentration is not measurably affected by the occurrence of this reaction, so that $[H_2O]$ is a constant, and at equilibrium, $[H_3O^+][A^-]/[HA] = K_{eq}[H_2O]$ = a constant. Since the hydronium ion concentration in an aqueous solution equals its 'H^+ ion concentration',

$$\frac{[H_3O^+][A^-]}{[HA]} = \frac{[H^+][A^-]}{[HA]}$$

whence,

$$K_{eq}[H_2O] = K_a$$

Thus the constant ($K_{eq}[H_2O]$), which in the Brönsted and Lowry view measures the strength of the acid in aqueous solution, is identical with K_a, the acid dissociation constant. Consequently, even if we insisted upon recognizing the part played by water in accepting the protons donated by a substance, its acid strength is still quite properly defined by the value of its dissociation constant K_a. It is therefore often simpler just to assume that an acid dissociates in water to yield H^+ ions in the manner suggested by the dissociation theory. Then, the stronger the acid the more fully does it dissociate in water, and the greater is the value of its acid dissociation constant K_a.

The pH of an aqueous solution of a strong acid

Strong acids give the appearance of being virtually completely dissociated in *dilute* aqueous solution. Consequently their values of K_a approach infinity, and they are all of approximately equal strength. The pH of a dilute aqueous solution of such an acid is very easily calculated, as shown in the following example.

EXAMPLE

Calculate the pH of a 0.025 M solution of an ideal, monobasic strong acid at 298 K.
[The symbol M has formerly been used in place of mol dm^{-3}, but its use is now deprecated save perhaps for rough concentration values in aqueous solution.]

A 1 mol dm^{-3} solution of an ideal, monobasic strong acid will contain 1 mol dm^{-3} of H^+ (see p. 18).

$\therefore$ a 0.025 mol dm^{-3} solution will contain 2.5×10^{-2} mol dm^{-3} of H^+

Since pH $= -\log [H^+]$, the pH of the 0.025 mol dm^{-3} solution is $-\log 2.5 \times 10^{-2}$

$$\therefore \text{pH} = -(\bar{2}.4) = -(-1.6) = 1.6$$

The pH of an aqueous solution of a weak acid

Amongst weak acids there exists a gradation of strengths evidenced by the wide spectrum covered by their values of K_a at a given temperature. All these values of K_a, like H^+ ion concentrations, involve negative powers of 10 and they are therefore frequently expressed as pK_a values, where $pK_a = -\log K_a$. For example, the statement that 'monochloroacetic acid ($K_a = 1.4 \times 10^{-3}$) is a stronger acid than acetic acid ($K_a = 1.82 \times 10^{-5}$)' can be restated as, 'monochloroacetic acid ($pK_a = 2.86$) is a stronger acid than acetic acid ($pK_a = 4.74$)'. The stronger the acid, the larger is the value of its K_a but the smaller is its pK_a value (if no temperature is mentioned, the values of K_a are assumed to have been determined at 298 K, i.e. 25°C).

In the special case of a *very* weak acid, it is possible to calculate the pH of its dilute solution from the values of K_a and of c (its concentration in mol dm^{-3}). This calculation assumes that the extent of dissociation is so small that the concentration of undissociated acid at equilibrium is negligibly less than the total concentration of acid present. If we consider the dissociation of the very weak acid HX in an aqueous solution in which it is present in a total concentration of c mol dm^{-3}, then,

$$\text{HX} \rightleftharpoons \text{H}^+ + \text{X}^-, \quad \text{and at equilibrium,} \quad \frac{[\text{H}^+][\text{X}^-]}{[\text{HX}]} = K_a$$

$$\text{or,} \quad [\text{H}^+][\text{X}^-] = K_a[\text{HX}]$$

But, $[X^-]$ equals $[H^+]$ at equilibrium, and if we now assume that [HX] approximately equals c,

$$[H^+][X^-] = K_a[HX] \quad \text{becomes} \quad [H^+]^2 = K_a c \quad \textit{or} \quad [H^+] = \sqrt{K_a c}$$

Taking logarithms of both sides of this last expression, we obtain the new equation,

$$\log [H^+] = \frac{\log K_a}{2} + \frac{\log c}{2}$$

or,
$$-\log [H^+] = -\tfrac{1}{2}\log K_a - \tfrac{1}{2}\log c$$

$$\therefore \quad pH = \tfrac{1}{2}pK_a - \tfrac{1}{2}\log c$$

EXAMPLE

Calculate the pH of a 0.01 mol dm^{-3} solution of the very weak acid HY (K_a of HY is 3.2×10^{-7}).

Assuming that [HY] at equilibrium is negligibly less than the total concentration of HY in the solution (0.01 mol dm^{-3}), then the approximate pH of this solution will be given by the equation,

$$pH = \tfrac{1}{2}pK_a - \tfrac{1}{2}\log c \qquad \text{where} \begin{cases} pK_a = -\log 3.2 \times 10^{-7} \\ c = 0.01 \text{ mol dm}^{-3} \end{cases}$$

$$\begin{aligned} \therefore \quad pH &= \tfrac{1}{2}(-\log 3.2 \times 10^{-7}) - \tfrac{1}{2}\log 0.01 \\ &= \tfrac{1}{2}(-\bar{7}.505) - \tfrac{1}{2}\bar{2}.0 \\ &= \tfrac{1}{2}(+6.495) - \tfrac{1}{2}(-2) \\ &= 3.247 + 1 \\ \therefore \quad pH &= \underline{4.25} \end{aligned}$$

The pH of a dilute aqueous solution of a more extensively dissociated weak acid may equally simply be calculated from its degree of dissociation (α) in the aqueous solution in which it is present at a concentration of c mol dm^{-3}.

EXAMPLE

Calculate the pH of (a) a 0.1 mol dm^{-3} solution of hydrochloric acid which is 83% dissociated at 298 K and (b) a 0.1 mol dm^{-3} solution of acetic acid which is 1.35% dissociated at the same temperature.

Were the acids fully dissociated, the $[H^+]$ in a 0.1 mol dm^{-3} solution would be 0.1 mol dm^{-3}.

(a) Since the HCl is 83% dissociated, the $[H^+]$ in its 0.1 mol dm^{-3} solution $= 83/100 \times 0.1 = 8.3 \times 10^{-2}$ mol dm^{-3}.

$$\begin{aligned} \therefore \text{ since } pH = -\log [H^+], \text{ its } pH &= -\log (8.3 \times 10^{-2}) \\ &= -(\bar{2}.92) = -(-2 + 0.92) \\ &= -(-1.08) \\ &= \underline{1.08} \end{aligned}$$

(Note: If it was assumed that HCl was an ideal, strong acid and was 100% dissociated in its 0.1 mol dm^{-3} solution, the pH would be assumed to be 1.00, i.e. not very different from the actual pH of 1.08.)

(b) Since the acetic acid is 1.35% dissociated, the $[H^+]$ in its 0.1 mol dm^{-3} solution $= 1.35/100 \times 0.1 = 1.35 \times 10^{-3}$ mol dm^{-3}.

$$\therefore \text{ its pH} = -\log(1.35 \times 10^{-3}) = -(\bar{3}.13) = -(-3+0.13)$$
$$= -(-2.87)$$
$$= \underline{2.87}$$

Evidently the degree of dissociation (α) of an acid in aqueous solution at a concentration c mol dm^{-3} will be related to its dissociation constant (K_a).

For example, if $HA \rightleftharpoons H^+ + A^-$, then at equilibrium, $[HA] = (1-\alpha)c$,

$$[H^+] = [A^-] = \alpha c, \text{ whence } K_a = \frac{[H^+][A^-]}{[HA]} = \frac{(\alpha c)^2}{(1-\alpha)c} = \frac{\alpha^2 c}{1-\alpha}$$

This simple equation, relating the values of α and K_a, is one expression of ***Ostwald's Dilution Law*** (p. 86).

Strength of a base

According to the dissociation theory, the strength of a base is determined by the extent to which it dissociates in aqueous solution to yield hydroxyl ions. If base B dissociates as follows,

$$B \rightleftharpoons B^+ + OH^-$$

then at equilibrium, $[B^+][OH^-]/[B] = K_b$, where the basic dissociation constant K_b measures the extent of dissociation and hence the basic strength of B.

In the Brönsted and Lowry view of the same base B, its basic strength is determined by the efficiency with which it accepts protons. In aqueous solution, the protons are donated by the water (here acting as an acid) so that,

$$B + H_2O \rightleftharpoons BH^+ + OH^-$$

The value of the equilibrium constant is therefore given by the equation

$$K_{eq} = \frac{[BH^+][OH^-]}{[B][H_2O]}$$

Once again, the concentration of water is so great that it can be assumed to remain constant. Therefore

$$\frac{[BH^+][OH^-]}{[B]} = K_{eq} \times [H_2O] = \text{Constant}$$

Since $[BH^+]$ in this equation is the same as $[B^+]$ in the equilibrium expression obtained by using the dissociation theory, the constant $K_{eq} \times [H_2O]$ derived from the Brönsted and Lowry theory is identical with K_b, the basic dissociation constant. This means that while the Brönsted and Lowry theory may provide the more satisfactory explanation of the true nature of a base, the strength of the base in aqueous solution is still defined by its value of K_b (even though the hydroxyl ions which appear in the solution need not have originated in its own molecules).

We have noted that the species formed by the association of a base with a proton is called its conjugate acid,

$$\underset{\text{(base)}}{B} + H^+ \rightleftharpoons \underset{\text{(conjugate acid)}}{BH^+}$$

When B is a strong base having a high affinity for protons, BH^+ will show little tendency to dissociate; that is to say, it will be a weak acid. Conversely, when B is a weak base it will form a strong conjugate acid. This inverse relationship between the strength of a base and the strength of its conjugate acid makes it possible, and valid, to define the strength of a base in terms of the K_a of its conjugate acid.

There are thus two methods of expressing the strength of a base in aqueous solution:

(i) as its basic dissociation constant K_b (or pK_b)

$$K_b = \frac{[BH^+][OH^-]}{[B]}$$

(ii) as the acid dissociation constant of its conjugate acid, K_a (or pK_a)

$$K_a = \frac{[B][H^+]}{[BH^+]}$$

The two dissociation constants are related in a simple manner, for K_a is inversely proportional to K_b and the proportionality constant is K_w, the ion product of water, i.e.:

$$K_a \times K_b = \frac{\cancel{[B]}[H^+]}{\cancel{[BH^+]}} \times \frac{\cancel{[BH^+]}[OH^-]}{\cancel{[B]}}$$

$$= [H^+][OH^-] = K_w$$

$$\therefore \qquad K_a \times K_b = K_w$$

$$\text{or} \qquad pK_a + pK_b = pK_w$$

The stronger the base, the larger is its K_b and the smaller is the K_a of its conjugate acid. In other words, the stronger the base, the smaller is its pK_b value and the larger is the pK_a value of its conjugate acid.†

The pH of an aqueous solution of strong base (or alkali)

A strong base can be defined *either* as a base which yields an infinitely weak, conjugate acid, *or* as a base with a K_b value which approaches infinity. Since $K_b = [BH^+][OH^-]/[B]$, this means in effect that a strong base, when dissolved in water, promotes the formation of an equivalent concentration of hydroxyl ions.

This raises the question of whether sodium hydroxide should be considered to be a base (as it is by the dissociation theory), or as a 'reservoir' of the true base OH^-. Such hydroxides of the alkali metals (e.g. sodium, potassium) are extremely soluble, and are fully ionized, both in the solid state and in water. Their metal cations (e.g. Na^+, K^+), are 'neutral' in the sense that they possess nil acidic or basic strength, and it is therefore legitimate to consider the basicity of these hydroxides to be due entirely to their content of OH^-, which is present independently of the presence of water. Since their solutions in water contain concentrations of hydroxyl ions equivalent to their molarity, even dilute solutions are very alkaline in pH, and in recognition of this fact, plus their somewhat special status, these hydroxides are called ***alkalis***. Use of this term avoids the necessity of calling the Na^+OH^- molecule a base, and thus satisfies those who would prefer to consider it as a salt possessing a 'neutral' cation and the strongly basic OH^- anion (see p. 134).

The pH of a dilute aqueous solution of alkali is easily calculated if the normality or molarity of the solution is known.

EXAMPLE

Calculate the pH of a 0.56 mol dm^{-3} aqueous solution of potassium hydroxide (assuming ideal behaviour of the alkali).

A 0.56 mol dm^{-3} solution of K^+OH^- will contain 0.56 mol dm^{-3} of OH^-

$$\therefore \quad [OH^-] = 5.6 \times 10^{-1} \text{ mol dm}^{-3}$$

$$\text{whence} \quad pOH = -\log [OH^-] = -\log (5.6 \times 10^{-1})$$
$$= -(\bar{1}.75) = -(-1 + 0.75)$$
$$= -(-0.25)$$
$$= 0.25$$

In any aqueous solution at 298 K, $pH = 14 - pOH$, $\therefore$ $pH = \underline{13.75}$

† K_a of a weak acid and K_b of a weak base are equilibrium constants and their values are temperature-dependent (see p. 220). Thus, although you can assume that the values of K_a and K_b quoted throughout this Chapter refer to 298 K (i.e. 25°C), in practice you must employ the values possessed by these constants at the actual temperature of the solution under consideration.

The composition of aqueous solutions of ammonia

Having just mentioned alkalis, it is appropriate that we now consider why it is wrong to speak of 'ammonium hydroxide solutions' when we are really referring to aqueous solutions of ammonia. Ammonia (NH_3) is a highly soluble, weak base which is capable of accepting protons to form its conjugate acid (NH_4^+). When dissolved in water, the protons are donated by the water (here acting as a weak acid), i.e.

$$NH_3 + H_2O \rightleftharpoons NH_4^+ + OH^-$$

The NH_4^+ and OH^- ions that are produced when ammonia is dissolved in water are therefore formed as the direct result of the action of ammonia as a base, and the extent to which they are produced is a function of the K_b of ammonia, where $K_b = [NH_4^+][OH^-]/[NH_3]$. It is misleading to call this an 'ammonium hydroxide solution', for this suggests quite wrongly that the only unique components are the NH_4^+ and OH^- ions (thus overlooking its content of NH_3), and furthermore infers the separate existence of the compound ammonium hydroxide (in the absence of NH_3 and H_2O). It is therefore much better to refer to these solutions as 'aqueous solutions of ammonia' for attention is then drawn to the actual, weakly basic component that determines their ionic composition.

The pH of an aqueous solution of a weak base

We have seen that a weak base is characterized by its relatively small affinity for protons; this is evident in aqueous solutions from the small value of its K_b. If $B + H_2O \rightleftharpoons BH^+ + OH^-$, then at equilibrium in an aqueous solution $[BH^+][OH^-]/[B] = K_b$. Assuming that the solution contains a total concentration of B equal to c mol dm^{-3}, then, since the extent to which it accepts protons and is transformed into its strong, conjugate acid BH^+, is relatively small, [B] at equilibrium will be insignificantly less than c. The BH^+ and OH^- ions are produced in equal concentration; therefore at equilibrium in the dilute aqueous solution of base B, $[B] = c$, $[BH^+] = [OH^-]$, and therefore,

$$K_b = \frac{[BH^+][OH^-]}{[B]} = \frac{[OH^-]^2}{c}$$

But in any aqueous solution $[OH^-] = K_w/[H^+]$, and substituting this term for $[OH^-]$ in the above equation, we obtain the expression,

$$K_b = \frac{K_w^2}{[H^+]^2 . c}, \text{ whence } [H^+]^2 = \frac{K_w^2}{K_b . c} \quad \textbf{or } [H^+] = \frac{K_w}{\sqrt{K_b . c}}$$

Taking logarithms,

$$\log [H^+] = \log K_w - \tfrac{1}{2} \log K_b - \tfrac{1}{2} \log c$$

Reversing the signs throughout,

$$-\log [H^+] = -\log K_w + \tfrac{1}{2}\log K_b + \tfrac{1}{2}\log c$$

so that,

$$pH = pK_w - \tfrac{1}{2}pK_b + \tfrac{1}{2}\log c$$

EXAMPLE

The pK_b values of ammonia and trimethylamine are, respectively, 4.74 and 4.21 at 298 K. Calculate:
(1) the pH of 0.05 mol dm^{-3} aqueous solutions of ammonia and trimethylamine,
(2) the acid dissociation constants of ammonium and trimethylammonium ions.

1 Ammonia is a weak base that in aqueous solution forms its conjugate acid NH_4^+, according to the equation: $NH_3 + H_2O \rightleftharpoons NH_4^+ + OH^-$. In a 0.05 mol dm^{-3} solution, the total concentration of 'ammonia' is sufficiently great, in comparison with its K_b value, for $[NH_3]$ at equilibrium to be assumed to be negligibly less than 0.05 mol dm^{-3}.
The pH of this solution is approximately given by the equation,

$$pH = pK_w - \tfrac{1}{2}pK_b + \tfrac{1}{2}\log c \qquad \text{where} \begin{cases} pK_w = 14 \\ pK_b = 4.74 \\ c = 0.05 \text{ mol dm}^{-3} \end{cases}$$

$$\therefore \quad pH = 14 - \tfrac{1}{2}(4.74) + \tfrac{1}{2}\log 0.05$$
$$= 14 - 2.37 + \tfrac{1}{2}(\bar{2}.7) = 14 - 2.37 + \tfrac{1}{2}(-1.3)$$
$$= 14 - 2.37 - 0.65 = 14 - 3.02 = \underline{10.98}$$

By similar reasoning, the pH in a 0.05 mol dm^{-3} solution of trimethylamine is given by the equation,

$$pH = 14 - \tfrac{1}{2}(4.21) + \tfrac{1}{2}\log 0.05$$
$$= 14 - 2.11 - 0.65$$
$$= \underline{11.24}$$

2 The pK_a of the conjugate acid of a base is related to its pK_b value according to the equation, $pK_a = pK_w - pK_b$;

$$\therefore \quad pK_a \text{ of } NH_4^+ = 14 - pK_b \text{ of } NH_3$$
$$= 14 - 4.74 = 9.26$$
$$\therefore \quad -\log K_a = 9.26, \text{ and } \log K_a = -9.26 = \overline{10}.74$$
$$\therefore \quad K_a \text{ of } NH_4^+ = \text{antilog } \overline{10}.74 = \underline{5.5 \times 10^{-10}}$$

Similarly, $\quad pK_a$ of $(CH_3)_3NH^+ = 14 - 4.21 = 9.79$

$$\therefore \quad \log K_a = -9.79 = \overline{10}.21$$
$$\therefore \quad K_a \text{ of } (CH_3)_3NH^+ = \text{antilog } \overline{10}.21 = \underline{1.62 \times 10^{-10}}$$

Note that ammonia is a weaker base than trimethylamine, and NH_4^+ is a stronger acid than $(CH_3)_3NH^+$.

THE INTERACTION OF AN ACID WITH A BASE

Accepting the definition of acids and bases that is proposed by the dissociation theory, the interaction of any acid with any base can be considered as the formation of H_2O by the association of H^+ ions liberated by the acid, with OH^- ions derived from the base,

$$HA \rightleftharpoons H^+ + A^-$$

$$\text{Base} \rightleftharpoons B^+ + OH^-$$

$$H^+ + OH^- \rightleftharpoons H_2O$$

$$\text{Net reaction:} \quad HA + \text{Base} \rightleftharpoons B^+ + A^- + H_2O$$

The ions (B^+, A^-) produced by this interaction of acid and base, are the ions of the salt BA, and in elementary textbooks a salt is still often defined as 'the product, other than water, that is formed by the interaction of an acid with a base'. Similarly the term 'neutralization' is generally used to describe the interaction of acid and base to yield a salt.

The Brönsted and Lowry conception of the interaction between acids and bases offers a much broader view of the process of neutralization. In this view, neutralization is the process of proton transference from an acid to a base; this need not involve water, and need not result in the formation of a recognizable salt.

$$\underset{\text{(acid)}}{HA} + \underset{\text{(base)}}{B} \rightleftharpoons \underset{\text{(conjugate acid)}}{BH^+} + \underset{\text{(conjugate base)}}{A^-}$$

Therefore, even though we shall only be considering the interaction between acids and bases in aqueous solution, we would be better employed in identifying the actual reactant acid and base in any neutralization process (and hence the conjugate acid and base produced) than in attempting to identify a salt among the products.

Neutralization of a strong acid with a strong base

The manner in which the pH changes when an aqueous solution of strong acid (HCl) is titrated with a strong base (NaOH) provides an insight into the nature of the neutralization process. We have noted that NaOH can be assumed to act as an equivalent concentration of OH^- ions. Therefore, according to the proton transfer view of neutralization, the interaction of HCl and NaOH can be represented by the equation,

$$\underset{\text{(acid)}}{HCl} + \underset{\text{(base)}}{OH^-} \rightleftharpoons \underset{\text{(conjugate acid)}}{H_2O} + \underset{\text{(conjugate base)}}{Cl^-}$$

The products of neutralization are the exceedingly weak acid H_2O and the infinitely weak base Cl^-. Thus neutralization will be complete, for the reverse reaction will not occur to any significant extent, and the neutralization equation can be written,

$$HCl + OH^- \longrightarrow H_2O + Cl^-$$

This means that:

1 When less than an equivalent quantity of NaOH has been added, the hydrogen ion concentration will be equal to the concentration of HCl that remains unneutralized, since HCl may be assumed to be completely dissociated in water, $HCl \rightarrow H^+ + Cl^-$.

2 At the equivalence point, the pH of the mixture will be 7 for it will contain, besides water, only Na^+ and Cl^- ions that are negligibly acidic or basic. (pH 7 is therefore the pH of an aqueous solution of sodium chloride.)

3 When more than an equivalent quantity of NaOH has been added, the residual hydrogen ion concentration is inversely proportional to the concentration of excess NaOH present (since this is a measure of the excess $[OH^-]$).

The relationship between added NaOH and the pH is shown in Fig. 5.1, which is generally called a *titration curve*.

The shape of this curve is explained by the logarithmic nature of the pH scale, which means that, compared with the quantity of alkali required to raise the pH from 1 to 2, it takes only one tenth this amount of alkali to raise the pH from 2 to 3, one hundredth the quantity to raise the pH from 3 to 4, and one hundred thousandth as much to change the pH by one unit from 6 to 7. Over 95% of the total quantity of alkali needed to raise the pH to 7 is consumed between pH 1 and 3. This zone, where relatively large additions of alkali produce comparatively small changes in pH, is called the *acid buffer zone*. Addition of extremely small quantities of alkali to a mixture containing almost equivalent concentrations of HCl and NaOH produces a very marked increase in pH. These mixtures of pH between 3 and 11, are insignificantly buffered compared with mixtures containing either a great excess of strong acid ($pH < 3$) or a great excess of alkali ($pH > 11$), which are said to possess *buffering ability* because of their relatively great resistance to pH change on the addition of a small quantity of alkali or strong acid.

The pH of any mixture of HCl and NaOH, represented on the titration curve of Fig. 5.1, can be easily calculated, as shown in the following example.

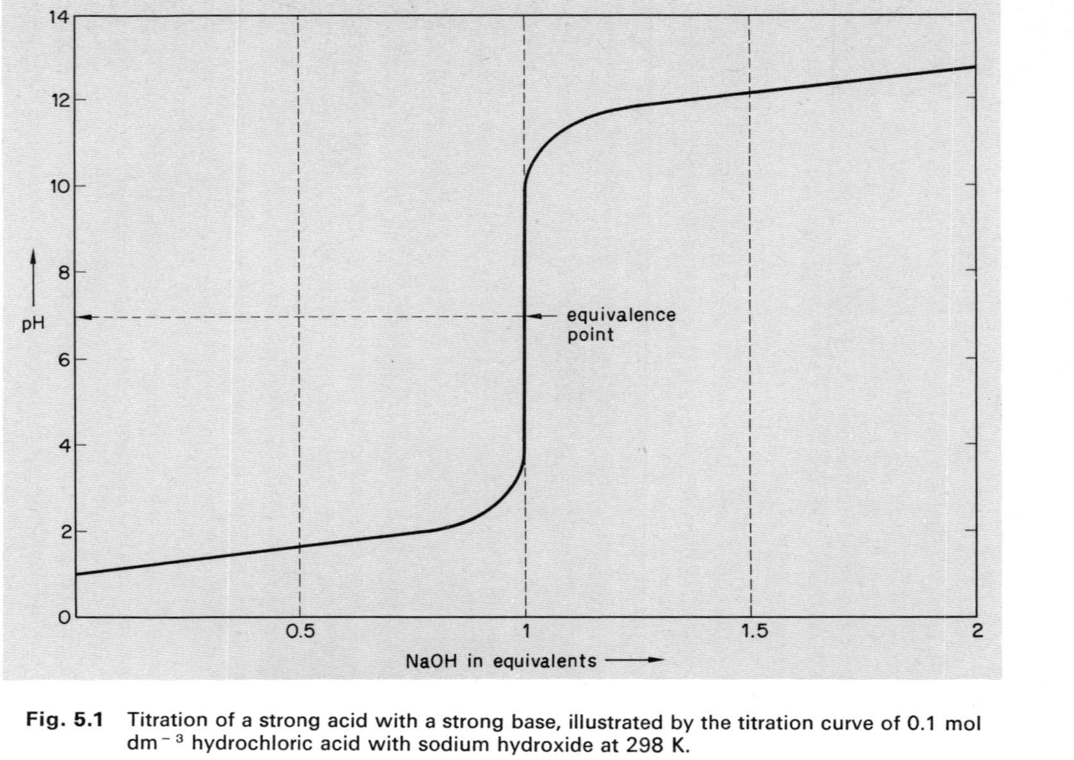

Fig. 5.1 Titration of a strong acid with a strong base, illustrated by the titration curve of 0.1 mol dm^{-3} hydrochloric acid with sodium hydroxide at 298 K.

EXAMPLE

To 10 cm³ of 0.1 mol dm⁻³ HCl was added 9.6 cm³ of 0.1 mol dm⁻³ NaOH. Calculate the pH of the final solution

9.6 cm^3 of 0.1 mol dm^{-3} NaOH will completely neutralize an equivalent quantity of the strong acid, HCl.

$$HCl + (Na^+)OH^- \longrightarrow H_2O + (Na^+)Cl^-$$

Thus if 10 cm^3 of 0.1 mol dm^{-3} HCl was initially present, a quantity equivalent to 0.4 cm^3 of 0.1 mol dm^{-3} HCl will remain unneutralized in the solution whose total volume is now 19.6 cm^3.

∴ concentration of HCl remaining unneutralized

$$= 0.4/19.6 \times 0.1 \text{ mol dm}^{-3}$$

$$= 2.04 \times 10^{-3} \text{ mol dm}^{-3}$$

$$[H^+] \text{ in } 1 \text{ mol dm}^{-3} \text{ HCl} = 1 \text{ mol dm}^{-3}$$

$$\therefore \quad [H^+] \text{ in } 2.04 \times 10^{-3} \text{ mol dm}^{-3} \text{ HCl} = 2.04 \times 10^{-3} \text{ mol dm}^{-3}$$

$$\therefore \text{ pH} = -\log [H^+] = -\log (2.04 \times 10^{-3}) = -(\bar{3}.31) = -(-3+0.31)$$

$$= -(-2.69) = \underline{2.69}$$

Neutralization of a weak acid with a strong base

When an aqueous solution of acetic acid is titrated with sodium hydroxide, the neutralization process can be written as

$$\underset{\text{(acid)}}{HAc} + OH^- \rightleftharpoons H_2O + \underset{\text{(conjugate base)}}{Ac^-}$$

The product of neutralization, other than water, is acetate ion Ac^-, which is the appreciably basic conjugate base of the weak acid HAc, i.e.

$$Ac^- + H^+ \rightleftharpoons HAc$$

This means that the pH of a solution of acetic acid, to which less than an equivalent of NaOH has been added, will depend not only on the quantity of acetic acid that remains unneutralized, but also upon the extent to which it is dissociated, which will be determined by the $[Ac^-]$ in the solution. Thus the course of the pH change during the titration will reflect:

1. the removal, by neutralization, of acetic acid, with the concurrent production of Ac^-;
2. the progressively increasing repression of the dissociation of the residual acetic acid in the face of the mounting $[Ac^-]$ produced by (1).

The quantitative effect of these events upon the pH during neutralization can be calculated from the equation that defines the acid dissociation constant of the weak acid, HA; i.e.

$$K_a = \frac{[H^+][A^-]}{[HA]}$$

whence, $$[H^+] = K_a \times \frac{[HA]}{[A^-]}$$

and $$\log [H^+] = \log K_a + \log \frac{[HA]}{[A^-]}$$

or, $$-\log [H^+] = -\log K_a - \log \frac{[HA]}{[A^-]}$$

But, $$-\log \frac{[HA]}{[A^-]} = +\log \frac{[A^-]}{[HA]}$$ (see p. 8)

∴ $$-\log [H^+] = -\log K_a + \log \frac{[A^-]}{[HA]}$$

i.e. $$pH = pK_a + \log \frac{[A^-]}{[HA]}$$

This means that the pH in any aqueous solution which contains a significant concentration of weak acid HA in the presence of its conjugate base A^-, will depend (1) on the pK_a value of the acid and (2) on the ratio of what concentrations of the acid and its conjugate base it contains, as described by the equation,

$$pH = pK_a + \log \frac{[\text{conjugate base}]}{[\text{acid}]}$$

This most important relationship is known as the ***Henderson–Hasselbalch Equation***. It means in practice that the weak acid-strong base titration curve will have the characteristic shape shown in Fig. 5.2 for the titration of acetic acid and NaOH.

It can be seen from this titration curve that the pH at the equivalence point is greater than 7. This is due to the basicity of the acetate ion, and the pH of the solution at equivalence can readily be calculated (see p. 135). Secondly, the zone of rapid pH change about the equivalence point is less extensive than that in the titration of a strong acid with NaOH (cf. Fig. 5.1). However, the most striking feature of this titration curve is that on the acid side of equivalence it has a characteristic sigmoidal shape with an inflection at half equivalence. At this point, half the acid initially present will have been converted into its conjugate base, and half will be unneutralized. Thus at this half-equivalence point [conjugate base] = [acid], and the pH equals ($pK_a + \log 1$) = pK_a (since log 1 equals 0). Therefore, by measuring the pH at the half-equivalence point when a weak acid is titrated with alkali, an experimental (apparent) value is obtained for the pK_a of the acid.† Figure 5.2 shows that the apparent pK_a of acetic acid in the titration mixture is about 4.7.

It further appears, from this titration curve, that for a considerable range on either side of half equivalence, the pH changes only slowly on addition of alkali (or of strong acid), though near the start, and again when approaching equivalence, the pH change is more rapid. In other words, mixtures in the middle range of the pre-equivalence curve possess considerable buffering ability, while those at both extremes are deficient in this respect. The sigmoidal shape of the titration curve is a consequence of the factors embodied in the Henderson–Hasselbalch equation. The change in pH brought about by the addition of a quantity of alkali will depend solely upon the change made by this addition in the value of the term log [conjugate base]/[acid], since pK_a is a constant in any titration. At the half-equivalence point, [conjugate base]/[acid] is 1, and the addition of, say, one-tenth of an equivalent of alkali to such a mixture would produce a relatively small change in the ratio, and consequently only a slight increase in the pH (the ratio would become 1.5, and log 1.5 = 0.176). On the other hand, addition of this amount of alkali to mixtures in which [conjugate base]/[acid] is initially 10 or 1/10 would produce a much larger change in the ratio and hence in the pH. In fact, the alteration in pH produced by a given small quantity of alkali (or of strong acid) is minimal at the half-equivalence point and increases the more the ratio [conjugate base]/[acid]

† The thermodynamic acid dissociation constant K_a, has a truly constant value at a given temperature, which is calculable from the *activities* of the components of the dissociation reaction at equilibrium. Apparent values of K_a determined, as described in this chapter, from the *concentrations* of the components at equilibrium, will be affected by the ionic strength of the solution (since the activity coefficients of ions in solution are affected by the total ionic strength of the solution, p. 82); thus, for the dissociation, $HA \rightleftharpoons H^+ + A^-$, thermodynamic $K_a = \frac{y_{H^+} \times y_{A^-}}{y_{HA}} \times$ apparent K_a. As the ionic strength of the solution is decreased, the ratio of the activity coefficients approaches unity, and the value of the apparent K_a approaches the value of the thermodynamic (true) K_a. Though in theory, the true K_a is the value of K_a at zero ionic strength, in practice, one can afford to ignore the difference between true and apparent values of K_a in dilute aqueous solutions of low ionic strengths. However, in aqueous solutions of appreciable ionic strength (e.g. buffer mixtures), the difference is too large to be overlooked.

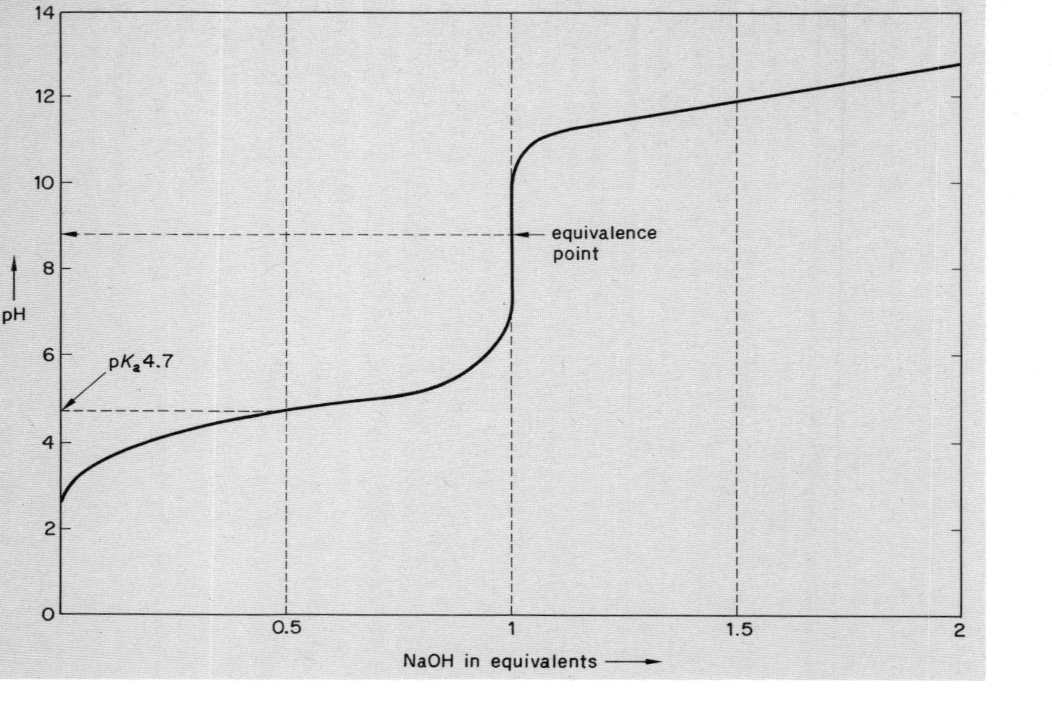

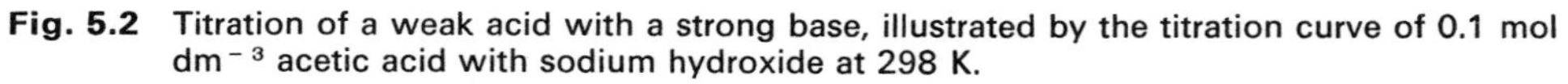
Fig. 5.2 Titration of a weak acid with a strong base, illustrated by the titration curve of 0.1 mol dm^{-3} acetic acid with sodium hydroxide at 298 K.

differs from 1; hence the sigmoidal shape of the titration curve. So, too, we can conclude that:

(a) a mixture with a [conjugate base]/[acid] ratio of 1, in which the pH is equal to the pK_a of the acid component, constitutes the most efficient buffer mixture,
(b) the buffering ability of the mixture diminishes the more its pH differs from the pK_a of the weak acid component.

In practice, it is considered that a mixture of a weak acid and its conjugate base makes a satisfactory buffer over the pH range of pH=(pK_a-1) to pH=(pK_a+1). Examination of the titration curve (Fig. 5.2) enables us to make rather more accurate predictions concerning the relative buffering abilities of mixtures containing equal total concentrations of the same weak acid and conjugate base, namely that:

1 The mixture whose pH equals the pK_a of the weak acid, in which the ratio [conjugate base]/[acid] is 1, is the optimal buffer mixture in the sense that it equally minimizes pH change on addition of equivalent, small quantities of either alkali or strong acid.
2 A mixture whose pH equals (pK_a-1), in which the ratio [conjugate base]/[acid] is 0·1, is an effective buffer 'against' alkali, but is much less effective 'against' strong acid.
3 The mixture at the other extreme of the buffer range, whose pH equals (pK_a+1) and in which the ratio [conjugate base]/[acid] is 10, is an effective buffer 'against' strong acid, but is much less effective as a buffer 'against' alkali.

However, the actual buffer capacity of a solution will be determined not only by the initial ratio of [conjugate base]/[acid], but also by the actual magnitudes of these concentrations (see p. 125). Note also that the same 'pre-equivalence' titration curve as in Fig. 5.2 would be obtained if a solution of sodium acetate (equivalence point mixture) were titrated with strong acid, though now the pH would decrease, moving 'back down the curve'.

We are now in a position to calculate the pH of a mixture prepared by partial neutralization of a weak acid with alkali. It should, however, be pointed out that values calculated by use of the Henderson–Hasselbalch equation are always approximate, and are extremely unreliable if the equation is applied to solutions in which the ratio [conjugate base]/[acid] is very large or small, or in which the strength of the acid component is too great or too small. In practice this means that it is generally unwise to apply the Henderson–Hasselbalch equation to aqueous solutions of pH less than 4 or greater than 10.

EXAMPLE

To 35 cm³ of 0.02 mol dm⁻³ acetic acid were added 10 cm³ of 0.05 mol dm⁻³ sodium hydroxide. What is the pH of the resulting solution? (pK_a of acetic acid = 4.74)

The neutralization of acetic acid (HAc) by sodium hydroxide can be represented as:

$$\underset{\text{(acid)}}{\text{HAc}} + (\text{Na}^+)\text{OH}^- \rightleftharpoons \text{H}_2\text{O} + \underset{\text{(conjugate base)}}{(\text{Na}^+)\text{Ac}^-}$$

Thus a quantity of NaOH will neutralize an equivalent quantity of HAc to form Ac^-. In the initial [HAc] = a mol dm^{-3}, and [NaOH] = x mol dm^{-3} was added with no volume change, then,

Initially	*Finally*
$[Ac^-]$ = negligible	$[Ac^-] = x$
$[HAc] = a$	$[HAc] = (a-x)$

The pH of the final solution is given by the Henderson–Hasselbalch equation,

$$\text{pH} = \text{p}K_a + \log \frac{[\text{conjugate base}]}{[\text{acid}]}$$

$$= 4.74 + \log \frac{[\text{Ac}^-]}{[\text{HAc}]} = 4.74 + \log \frac{x}{(a-x)}$$

The problem set by the present example is a little more complicated in that the addition of alkali in solution brought about a significant change in the volume of the solution. Two methods of calculation are possible:

(a) Calculating actual final concentrations of the conjugate base and acid.
(b) Recognizing that since what is used in the Henderson–Hasselbalch equation is the *ratio* of concentrations of conjugate base and acid, this will be identical with the ratio of their amounts in the final solution, whatever its volume might be.

We can compare the merits of these methods by applying both to the present example.

Method (a)

Initially: Volume = 35 cm^3
$[Ac^-]$ = negligible
$[HAc]$ = 0.02 mol dm^{-3}

Finally: Volume = 45 cm^3

$$[Ac^-] = [NaOH] = \frac{10}{45} \times 0.05 = 1.11 \times 10^{-2} \text{ mol dm}^{-3}$$

$$[HAc] = [\text{HAc if original soln. was diluted to } 45 \text{ cm}^3] - [Ac^-]$$

$$= (\frac{35}{45} \times 0.02) - (1.11 \times 10^{-2}) \text{ mol dm}^{-3}$$

$$= (1.55 \times 10^{-2} - 1.11 \times 10^{-2}) \text{ mol dm}^{-3}$$

$$= 0.44 \times 10^{-2} \text{ mol dm}^{-3}$$

The pH of the final solution is given by the equation

$$pH = 4.74 + \log \frac{[Ac^-]}{[HAc]} = 4.74 + \log \frac{1.11 \times 10^{-2}}{0.44 \times 10^{-2}}$$

$$= 4.74 + \log 2.52$$

$$= 4.74 + 0.4$$

$$= \underline{5.14}$$

Method (b)

Let us express all quantities as their equivalent in cm^3 of a 0.02 mol dm^{-3} solution.

Initially	*Finally*
Ac^- = negligible	$Ac^- \equiv$ NaOH added = 10 cm^3 of 0.05 mol dm^{-3}
	$\equiv$ 25 cm^3 of 0.02 mol dm^{-3}
	= 25
HA = 35	HAc = Initial HAc − Ac^- produced
	= 35 − 25 = 10

Thus in the final solution

$[Ac^-] \equiv$ 25 cm^3 of 0.02 mol dm^{-3} solution diluted to 45 cm^3

$[HAc] \equiv$ 10 cm^3 of 0.02 mol dm^{-3} solution diluted to 45 cm^3

$$\therefore \quad \frac{[Ac^-]}{[HAc]} = \frac{25}{10} = 2.5$$

Since in the final solution, $pH = 4.74 + \log \frac{[Ac^-]}{[HAc]}$,

$$pH = 4.74 + \log 2.5$$

$$= \underline{5.14}$$

Method (b), when carefully used, is simpler than method (a), at least when it is applied to elementary problems.

Neutralization of a weak base with a strong acid

In Fig. 5.3 is shown the titration curve that is obtained when an aqueous solution of ammonia is titrated with HCl. The pH at equivalence is now considerably less than 7 and the whole curve appears to be almost the mirror image of the weak acid-alkali titration curve of Fig. 5.2.

The neutralization reaction occurring during the titration can be represented by the general equation,

$$\underset{\text{(base)}}{\mathrm{B}} + \mathrm{HCl} \rightleftharpoons \underset{\text{(conjugate acid)}}{\mathrm{BH^+}} + \mathrm{Cl^-}$$

(in this instance, B is NH_3). The conjugate acid BH^+, formed by the neutralization of B, is a moderately strong acid and will appreciably dissociate to re-form B. Equilibrium at any stage in the titration is established with B and BH^+ present in concentrations that conform to the equation $\frac{[BH^+]}{[B][H^+]} = \frac{1}{K_a}$, where K_a is the acid dissociation constant of BH^+. Thus, so long as both B and BH^+ are present in significantly large concentrations, the $[H^+]$ in the solution will be given by the equation, $[H^+] = K_a \times \frac{[BH^+]}{[B]}$,

whence,
$$\mathrm{pH} = \mathrm{p}K_a + \log \frac{[B]}{[BH^+]}$$

(cf. derivation of the similar equation, p. 116)

This means that the Henderson–Hasselbalch equation is also applicable to solutions containing a weak base and its conjugate acid. To emphasize one's primary concern with the weak base in such a mixture, the equation may be written in the form,

$$\mathrm{pH} = \mathrm{p}K_a + \log \frac{[\text{base}]}{[\text{conjugate acid}]}$$

Note that the pH of the mixture at the half-equivalence point in the titration of a weak base with a strong acid will equal the apparent pK_a value of *the conjugate acid of the weak base*. Thus from Fig. 5.3, the pK_a of NH_4^+ is approximately 9.3.

Such mixtures of a weak base with its conjugate acid are just as effective pH buffers as the weak acid-conjugate base mixtures formed during the titration of a weak acid with alkali (p. 115). For identical reasons substantial buffering ability will be confined to mixtures whose pH's are within approximately 1 unit of the pK_a of the acid component. Since the pK_a values of the conjugate acids of weak bases are usually greater than 7, this means that these weak base-conjugate acid mixtures are useful buffers at alkaline pH's.

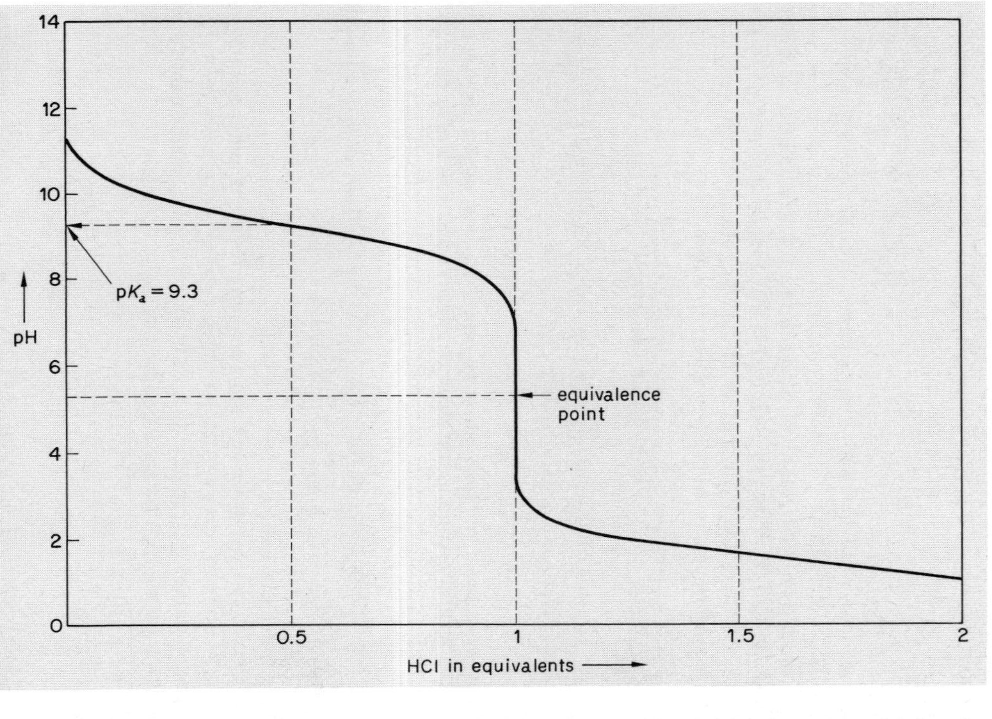

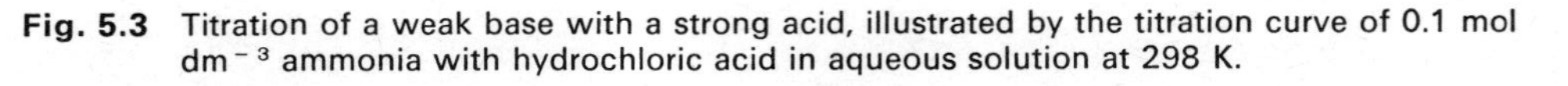

Fig. 5.3 Titration of a weak base with a strong acid, illustrated by the titration curve of 0.1 mol dm^{-3} ammonia with hydrochloric acid in aqueous solution at 298 K.

Neutralization of a weak acid with a weak base

In the titration of a weak acid with a weak base in aqueous solution, the pH at the equivalence point may be acid, alkaline or neutral depending on the relative strengths of the acid and base being titrated (p. 137). Furthermore, it is a feature of this type of titration that the equivalence point may be difficult to determine since the rate of change in pH as it is approached is much less striking than in any titration that involves either a strong acid or a strong base.

Mixtures of a weak acid with a weak base will buffer either at acid or at alkaline pH's, depending on which component is present in excess. This can be illustrated with mixtures of acetic acid and ammonia in water. When acetic acid is present in such excess that the pH is about 3.7 to 5.7, the mixture buffers due to its content of the components of the acetic acid-acetate conjugate pair in suitable proportion, when

$$pH = 4.74 + \log \frac{[Ac^-]}{[HAc]}$$

(The $[Ac^-]$ in this mixture is provided by the neutralization of HAc with NH_3;

$$HAc + NH_3 \rightleftharpoons NH_4^+ + Ac^-)$$

When equivalent amounts of HAc and NH_3 are mixed together in water, the resulting ammonium acetate solution has a pH of 7 (p. 138). Addition of excess ammonia to produce a mixture of pH 8.3 to 10.3 (approx.) also confers pH buffering ability on the solution which now contains the components of the $NH_4^+ - NH_3$ conjugate pair in suitable proportion $\left(\text{since } pH = 9.26 + \log \frac{[NH_3]}{[NH_4^+]}\right)$.

This means that although an ammonium acetate solution is not itself a good buffer, it can be converted into two quite different buffer mixtures: (a) by the addition of a very small quantity of a strong acid, or of a somewhat larger amount of HAc, (b) by the addition of a very small quantity of alkali, or of a somewhat larger amount of NH_3.

BUFFER MIXTURES AND THEIR BUFFER CAPACITY

From the appearance of the various titration curves, we have concluded that aqueous solutions of various types demonstrate buffering ability. Even solutions of strong acids or of strong bases exhibit such behaviour at their respective extremes of the normal pH range, viz. <3 and >11. Yet the solutions that make the best buffers in the pH range 4 to 10 are those that contain a weak acid with its conjugate base, or a weak base with its conjugate acid. Amongst the properties of such solutions that we have already mentioned, the following are particularly noteworthy (N.B. *Here 'acid' and 'base' refer to the members of the conjugate pair present in the mixture whether this is weak acid-conjugate base or weak base-conjugate acid*):

(a) that mixture with a [base]/[acid] ratio of 1, is optimally buffered against both strong acid and strong base, and its pH equals the pK_a of the acid component;
(b) mixtures with [base]/[acid] ratios between 0·1 and 10 are significantly buffering and their pH will fall within 1 unit of the pK_a value of their acid component;
(c) the pH of any mixture of this type can be calculated by applying the Henderson–Hasselbalch equation, $pH = pK_a + \log\frac{[base]}{[acid]}$, where pK_a is the pK_a value of its acid component.

The *buffer capacity* of a solution is an indication of its effectiveness in minimizing the pH change that results from the addition of a standard quantity of strong acid or strong base. A quantitative measure of this is given by the *buffer value*, a unit introduced by Van Slyke. If the addition of 1 mol of monoacidic strong base (e.g. OH^-) to 1 dm^3 of a solution causes its pH to increase by 1 pH unit, then that solution is said to have a buffer value of 1 unit, so that buffer value $= \beta = db/dpH$, where db is a quantity of monoacidic strong base in mol added to 1 dm^3 of buffer solution, and dpH is the resultant increase in pH. A quantity of monobasic strong acid would, in its effect, be equivalent to the addition of an equal but negative quantity of monoacidic strong base, i.e. $-db$, and would cause a decrease in pH equal to $-dpH$. Therefore, whether the buffer capacity is tested against strong acid or strong base, the buffer values, though they may be numerically different, will always be positive.

The actual choice of a buffer to stabilize the pH in a reaction mixture is always restricted by the requirement that its components should not interfere with the reaction in any other way. From an assortment of harmless buffer mixtures, it would be wise to choose that mixture whose acid component possessed at the working temperature a pK_a value very close to the desired pH. The actual composition of the required buffer could be approximately calculated using the Henderson–Hasselbalch equation, and the concentration of its components chosen to yield a mixture of sufficiently large buffer capacity to be able to maintain a constant pH during the course of the experiment.

To prepare a buffer mixture of this type, instead of partially neutralizing a weak acid with alkali, it is frequently more convenient to mix together a quantity of the weak acid in solution, with a quantity of its salt whose anion is its conjugate base and whose cation is 'neutral'. Thus an acetic acid-acetate buffer can be prepared by mixing together acetic acid and sodium acetate solutions. Similarly a weak base can be mixed with a solution of its salt whose cation is its conjugate acid and whose anion is 'neutral'. In this way, an ammonia-ammonium ion buffer can be made by mixing solutions of ammonia and ammonium chloride in calculated proportions. The salts that are used to prepare these buffer mixtures, e.g. salts of weak acids with alkalis (e.g. sodium acetate), or of weak bases with strong acids (e.g. ammonium chloride), can be considered to act as sources of an equimolar concentration of 'conjugate acid' or 'conjugate base'. Because of this, and because of the widespread use of such buffer mixtures, you will frequently find the Henderson–Hasselbalch equation written in the following forms:

(a) when applied to mixtures of a weak acid and its alkali metal salt, e.g. acetic acid-sodium acetate,

$$\text{pH} = \text{p}K_a + \log\frac{[\text{salt}]}{[\text{acid}]}$$

(b) when applied to mixtures of a weak base and its salt with a strong acid, e.g. ammonia-ammonium chloride

$$\text{pH} = \text{p}K_a + \log\frac{[\text{base}]}{[\text{salt}]}$$

Although Henderson in 1908 derived the equation in this form, since the Brönsted and Lowry theory of 1923 directed attention to the primary determinants of the pH of the mixture and suggested that the equation might be written as $\text{pH} = \text{p}K_a + \log\frac{[\text{conjugate base}]}{[\text{Brönsted acid}]}$, the use of a [salt] term in this equation has become redundant and, indeed, rather confusing.

EXAMPLE

Calculate the change in pH caused by dissolving 1.025 g of anhydrous sodium acetate in 100 cm³ of 0.25 mol dm⁻³ acetic acid (M. Wt. of anhydrous sodium acetate=82; assume the apparent $\text{p}K_a$ of acetic acid to be 4.74)

The initial concentration of acetic acid $= c = 0.25$ mol dm^{-3}. Since acetic acid is a reasonably weak acid, the pH of its aqueous solution will be given (approximately) by the equation, $\text{pH} = \frac{1}{2}\text{p}K_a - \frac{1}{2}\log c$ (see p. 106)

$$\therefore \quad \text{pH} = \tfrac{1}{2}(4.74) - \tfrac{1}{2}\log 0.25$$

Now, $$\log 0.25 = \bar{1}.4 = (-1+0.6) = -0.6$$

$$\therefore \quad \text{pH} = \tfrac{1}{2}(4.74) - \tfrac{1}{2}(-0.6) = 2.37 + 0.3 = \underline{2.67}$$

Sodium acetate is completely ionized, and 1.025 g dissolved in 100 cm^3 yields a solution containing $1.025/82 \times 100/1000$ mol dm^{-3} of acetate$^-$ ions = 0.125 mol dm^{-3}.

∴ in the final solution,

$$[Ac^-] = 0.125 \text{ mol dm}^{-3}$$
$$[HAc] = [\text{initial HAc}] = 0.25 \text{ mol dm}^{-3}$$

According to the Henderson–Hasselbalch equation,

$$pH = pK_a + \log\frac{[\text{conjugate base}]}{[\text{acid}]}$$

$$= 4.74 + \log\frac{[Ac^-]}{[HAc]} = 4.74 + \log\frac{0.125}{0.25}$$

$$= 4.74 + \log 0.5$$

Now, $$\log 0.5 = \bar{1}.7 \text{ (approx.)} = (-1+0.7) = -0.3$$

$$\therefore \quad pH = 4.74 - 0.3 = \underline{4.44}$$

∴ increase in pH due to the addition of the salt = $(4.44 - 2.67) = \underline{1.77}$

Although the Henderson–Hasselbalch equation is of considerable use in preparing buffer mixtures, it is always advisable to measure the accurate pH of the final mixture *at the temperature at which it is to be used.* Fine adjustment of its pH can then be made by the addition of small quantities of strong acid or alkali.†

THE DISSOCIATION OF POLYPROTIC WEAK ACIDS

So far, we have considered only monoprotic, weak acids whose molecules 'contain' only one proton that is available for the neutralization of a strong base. In biological media we frequently encounter polyprotic, weak acids (e.g. carbonic, phosphoric, citric acids), whose complete neutralization requires the addition of two or more equivalents of sodium hydroxide. Figure 5.4 shows the titration curve obtained when the triprotic, weak acid, phosphoric acid, is titrated with alkali.

Three distinct 'steps' are observable, each accomplished by the addition of one equivalent of alkali, and each evidencing a buffer zone centred on its

† In biochemical work, it is always best to make up a buffer mixture at the temperature at which it is subsequently to be used. Too many persons forget that buffer mixtures prepared at room temperature will have a different pH in the cold room, and again at 30°C or 37°C (for reasons explained in the Footnote to p. 96 and in Chapter 8, p. 220).

'half-equivalence' point. The first dissociation of phosphoric acid (pK_a about 2) must yield an amphiprotic conjugate base. This in turn dissociates as an acid (pK_a about 6.8) to form yet another amphiprotic conjugate base. This may in its turn dissociate (pK_a about 12) to form its conjugate

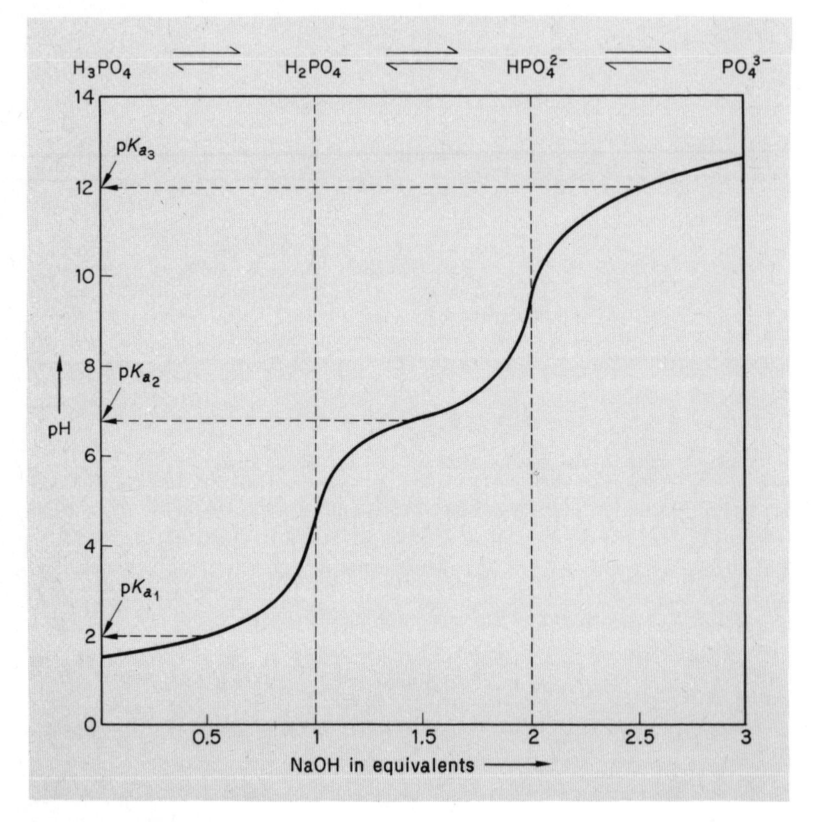

Fig. 5.4 Titration curve of 0.1 mol dm^{-3} phosphoric acid with sodium hydroxide at 298 K.

base. Ascribing formulae and apparent pK_a values† to these three dissociation 'steps', we obtain the following sequence for the dissociation of phosphoric acid (H_3PO_4),

† The discrepancy between apparent and thermodynamic (true) values of K_a is particularly marked in the case of phosphoric acid in reasonably dilute phosphate buffer mixtures (because of the comparatively high ionic strength of these solutions, arising from their content of multivalent anions, p. 91). Thus at 298 K (i.e. 25°C) the true values of the pK_a's of phosphoric acid (at zero ionic strength) are 2.12, 7.21 and ~12.4.

1st dissociation $H_3PO_4 \rightleftharpoons H^+ + H_2PO_4^-$ ... $pK_a = 1.96$

2nd dissociation $H_2PO_4^- \rightleftharpoons H^+ + HPO_4^{2-}$ ... $pK_a = 6.80$

3rd dissociation $HPO_4^{2-} \rightleftharpoons H^+ + PO_4^{3-}$ ... $pK_a = 12$ (approx.)

Therefore, when we talk of the 2nd dissociation of phosphoric acid (or of its 2nd acid dissociation constant), we are specifically referring to the dissociation of $H_2PO_4^-$ to yield its conjugate base HPO_4^{2-}. The multiple dissociation of the polyprotic, weak acid is thus seen to be a sequence of monoprotic dissociations yielding products of diminishing acid strength. The components of these reactions can be classified as follows: H_3PO_4 is a polyprotic (polybasic) acid; $H_2PO_4^-$ and HPO_4^{2-} are amphiprotic; and PO_4^{3-} is a polyacidic base.

The titration curve (Fig. 5.4) illustrates another important feature of the polyprotic, weak acid; namely that if its apparent pK_a values are sufficiently different, then the dissociation of any one acid intermediate is completed before its amphiprotic conjugate base is substantially dissociated (otherwise the 'steps' in the curve would tend to merge). At any one pH therefore, only the members of a single conjugate pair will be present in significant concentrations. At each of the 'equivalence points' shown in Fig. 5.4, one intermediate will greatly predominate over all others. At the first equivalence point when only the amphiprotic $H_2PO_4^-$ species is present, the pH will be 'half way' between the previous and subsequent 'half-equivalence' points; the pH will equal $\frac{pK_{a_1}+pK_{a_2}}{2} = \frac{1.96+6.80}{2} = 4.38$ (see p. 137). Similarly, at the second equivalence point, when the solution contains only HPO_4^{2-}, its pH will be $\frac{pK_{a_2}+pK_{a_3}}{2} = \frac{6.8+12}{2} = 9.4$

In a solution whose pH equals the pK_a value of one of the contributory weak acid intermediates, that weak acid and its conjugate base will be present in equal concentrations. Therefore, since

$$H_3PO_4 \underset{pK_{a_1}}{\rightleftharpoons} H^+ + H_2PO_4^- \underset{pK_{a_2}}{\rightleftharpoons} H^+ + HPO_4^{2-} \underset{pK_{a_3}}{\rightleftharpoons} H^+ + PO_4^{3-},$$

a solution whose pH is equal to pK_{a_1} will contain equal concentrations of H_3PO_4 and $H_2PO_4^-$, while a solution of pH equal to pK_{a_2} will contain equal concentrations of $H_2PO_4^-$ and HPO_4^{2-}, etc.

EXAMPLE

To 100 cm³ of 0.1 mol dm⁻³ phosphate buffer pH 7.1 were added:

(a) 1 cm³ of 1.0 mol dm⁻³ NaOH
(b) 5 cm³ of 1.0 mol dm⁻³ NaOH
(c) 5 cm³ of 1.0 mol dm⁻³ HCl

Calculate the new pH in each case. (Apparent pK_a *values of phosphoric acid are 1.96, 6.8 and 12)*

(Note: It is often helpful, in any calculation concerning a polyprotic weak acid, to draw a probable titration curve of the acid with alkali, positioning the 'steps' and equivalence points in relation to its apparent pK_a values (which will be the pH values at the medial points of the buffer zones). The ionic species that predominate at the 'equivalence' points should be indicated (as is done in Fig. 5.4). This diagram now aids one in deciding which species are present in the solution at any given pH, and also helps one to identify the pK_a value to be employed in calculating the effect on the pH of the addition of strong acid, alkali or salts.)

In phosphate buffer of pH 7.1, the predominant conjugate pair will be $H_2PO_4^-$ and HPO_4^{2-} (check with Fig. 5.4), for,

$$\underset{\text{(acid)}}{H_2PO_4^-} \underset{pK_{a2}=6.8}{\rightleftharpoons} H^+ + \underset{\text{(conjugate base)}}{HPO_4^{2-}}$$

According to the Henderson–Hasselbalch equation, the pH of this solution is related to pK_{a_2} by the equation, $pH = pK_{a_2} + \log \frac{[HPO_4^{2-}]}{[H_2PO_4^-]}$. Therefore, since pH = 7.1, and $pK_{a_2} = 6.8$,

$$\log \frac{[HPO_4^{2-}]}{[H_2PO_4^-]} = 0.3, \quad \text{and} \quad \frac{[HPO_4^{2-}]}{[H_2PO_4^-]} = \text{antilog } 0.3 = 2.0$$

Thus the initial 100 cm³ of 0.1 mol dm⁻³ phosphate buffer pH 7.1 is, in effect, a mixture of 66.7 cm³ of 0.1 mol dm⁻³ HPO_4^{2-} with 33.3 cm³ of 0.1 mol dm⁻³ $H_2PO_4^-$ (from $x+y=100$ and $x/y=2.0$).

Now,

1 Strong acid (HCl) would neutralize the conjugate base HPO_4^{2-} according to the equation,

$$HPO_4^{2-} + HCl \longrightarrow H_2PO_4^- + Cl^-$$

The complete neutralization of 66.7 cm³ of a 0.1 mol dm⁻³ solution of HPO_4^{2-} would consume 6.67 cm³ of 1.0 mol dm⁻³ HCl.

2 Alkali (NaOH) would neutralize the acid component of this buffer mixture ($H_2PO_4^-$) according to the equation,

$$H_2PO_4^- + (Na^+)OH^- \longrightarrow H_2O + (Na^+)HPO_4^-$$

The complete neutralization of 33.3 cm^3 of a 0.1 mol dm^{-3} solution of $H_2PO_4^-$ would consume 3.33 cm^3 of 1.0 mol dm^{-3} NaOH.

(a) *Addition of 1 cm^3 of 1 mol dm^{-3} NaOH*

Let us express all quantities in terms of cm^3 of a solution containing 0.1 mol dm^{-3}.

$\therefore$ NaOH added $\equiv$ 10 cm^3 of 0.1 mol dm^{-3} NaOH which will convert an equivalent quantity of $H_2PO_4^-$ into HPO_4^{2-}.

Initially		*Finally*	
Volume	= 100 cm^3	Volume	= 101 cm^3
HPO_4^{2-}	= 66.7	HPO_4^{2-}	$\equiv (66.7+10) = 76.7$
$H_2PO_4^-$	= 33.3	$H_2PO_4^-$	$\equiv (33.3-10) = 23.3$

The pH in the final solution is given by the equation

$$pH = 6.8+\log\frac{[HPO_4^{2-}]}{[H_2PO_4^-]}$$

where,

$[HPO_4^{2-}] \equiv$ 76.7 cm^3 of a 0.1 mol dm^{-3} solution made up to 101 cm^3
$[H_2PO_4^-] \equiv$ 23.3 cm^3 of a 0.1 mol dm^{-3} solution made up to 101 cm^3

$$\therefore \qquad \log\frac{[HPO^2{}_4{}^-]}{[H_2PO_4^-]} = \log\frac{76.7}{23.3} = \log 3.3 = 0.52$$

$$\therefore \qquad pH = 6.8+0.52 = \underline{7.32}$$

(b) *Addition of 5 cm^3 of 1 mol dm^{-3} NaOH*

As noted above, addition to the buffer mixture of 3.33 cm^3 of 1 mol dm^{-3} NaOH is sufficient to convert it into the equivalent of 100 cm^3 of a 0.1 mol dm^{-3} solution of HPO_4^{2-}. Thus, when 5 cm^3 of 1 mol dm^{-3} NaOH is added, 1.67 cm^3 of 1 mol dm^{-3} NaOH remains to be neutralized by the HPO_4^{2-} acting as an acid (check with Fig. 5.4), when

$$\underset{\text{(acid)}}{HPO_4^{2-}}+(Na^+)OH^- \rightleftharpoons \underset{\text{(conjugate base)}}{PO_4^{3-}}+Na^++H_2O$$

The pH of the resulting solution will therefore be determined by the 3rd dissociation constant of phosphoric acid,

$$HPO_4^{2-} \underset{pK_{a3}=12}{\rightleftharpoons} H^++PO_4^{3-}$$

and in a solution containing the conjugate pair HPO_4^{2-} and PO_4^{3-},

$$pH = 12{\cdot}0+\log\frac{[PO_4^{3-}]}{[HPO_4^{2-}]}$$

Let us express all quantities in terms of cm^3 of a solution containing 0.1 mol dm^{-3}.

Then, NaOH added $\equiv$ 50, of which 33.3 are consumed in converting all the $H_2PO_4^-$ initially present, into HPO_4^{2-}. The remaining 16.7 of NaOH will convert an equivalent quantity of HPO_4^{2-} into PO_4^{3-}.

Initially	*After addition of 33.3 NaOH*	*After further addition of 16.7 NaOH*
PO_4^{3-} = negligible	PO_4^{3-} = negligible	PO_4^{3-} = 16.7
HPO_4^{2-} = 66.7	HPO_4^{2-} = 100	HPO_4^{2-} = (100 − 16.7) = 83.3
$H_2PO_4^-$ = 33.3	$H_2PO_4^-$ = negligible	$H_2PO_4^-$ = negligible

$\therefore$ in the final solution formed by the addition of 5 cm^3 of 1 mol dm^{-3} NaOH,

$[PO_4^{3-}] \equiv$ 16.7 cm^3 of a 0.1 mol dm^{-3} solution made up to 105 cm^3

$[HPO_4^{2-}] \equiv$ 83.3 cm^3 of a 0.1 mol dm^{-3} solution made up to 105 cm^3

$$\therefore \qquad \log\frac{[PO_4^{3-}]}{[HPO_4^{2-}]} = \log\frac{16.7}{83.3} = \log 0.2 = \bar{1}.3 = -0.7$$

$\therefore$ in this solution, $\quad pH = 12.0+(-0.7) = \underline{11.3}$

(c) *Addition of 5 cm^3 of 1 mol dm^{-3} HCl*

As noted above, it would take 6.67 cm^3 of 1 mol dm^{-3} HCl to neutralize completely the HPO_4^{2-} (conjugate base) present in 100 cm^3 of 0.1 mol dm^{-3} phosphate buffer pH 7.1. Addition of only 5 cm^3 of 1 mol dm^{-3} HCl would produce a solution in which the conjugate pair is still $H_2PO_4^-$ and HPO_4^{2-}, where

$$H_2PO_4^- \underset{pK_{a2} = 6\cdot 8}{\rightleftharpoons} HPO_4^{2-} + H^+, \quad \text{and the pH} = 6.8+\log\frac{[HPO_4^{2-}]}{[H_2PO_4^-]}.$$

Let us express all quantities in terms of cm^3 of a solution containing 0.1 mol dm^{-3}.

Then HCl added $\equiv$ 50, which is entirely consumed in converting an equivalent quantity of HPO_4^{2-} into $H_2PO_4^-$ according to the equation,

$$\underset{\text{(base)}}{HPO_4^{2-}} + HCl \longrightarrow \underset{\text{(conjugate acid)}}{H_2PO_4^-} + Cl^-$$

Initially	*Finally*
HPO_4^{2-} = 66.7	HPO_4^{2-} = (66.7 − 50) = 16.7
$H_2PO_4^-$ = 33.3	$H_2PO_4^-$ = (33.3 + 50) = 83.3

$\therefore$ in the final solution formed by the addition of 5 cm^3 of 1 mol dm^{-3} HCl,

$[HPO_4^{2-}] \equiv$ 16.7 cm³ of a 0.1 mol dm⁻³ solution made up to 105 cm³
$[H_2PO_4^-] \equiv$ 83.3 cm³ of a 0.1 mol dm⁻³ solution made up to 105 cm³

$$\therefore \quad \log \frac{[HPO_4^{2-}]}{[H_2PO_4^-]} = \log \frac{16.7}{83.3} = -0.7 \qquad \text{(see above)}$$

$$\therefore \text{ in this solution,} \quad pH = 6.8 + (-0.7) = \underline{6.1}$$

pH INDICATORS

A number of substances are available, known as 'acid-base indicators' or 'pH indicators', whose colour in solutions more acid than a characteristic pH differs markedly from their colour in solutions more alkaline than this. Very conveniently, this colour change is not abrupt but is gradual over a certain range of pH (usually of about 2 pH units), which is called the 'colour change interval' of the indicator. The pH range over which an indicator responds in this fashion is characteristic, evidently being determined by its chemical constitution.

In general, pH indicators are very weak organic acids or bases, and their behaviour is explained by the fact that the Brönsted acid form of the indicator is strikingly different in colour from its conjugate base, i.e.

$$\underset{\text{(colour 1)}}{HIn} \rightleftharpoons H^+ + \underset{\text{(colour 2)}}{In^-}$$

The Henderson–Hasselbalch equation is applicable to their aqueous solutions, and if these were titrated with strong acid or alkali as required, the pH at the half-equivalence point would be the pK_a value of the indicator (often termed the pK_{In}). At the half-equivalence point (known as the 'change-over point' in the case of an indicator), the concentrations of the indicator weak acid and of its conjugate base, or of the indicator weak base and of its conjugate acid, will be equal. Therefore the colour adopted by the indicator at its change-over point is that of an equal mixture of the acid and alkaline colours of that indicator. These colours are so intense that a pH indicator may be used in such low concentration that it confers negligible buffer capacity on the solution to which it is added. It should be possible to assay spectrophotometrically the concentrations of HIn and In^- present in a solution whose pH falls within the colour change interval of the indicator. The pH of this solution could then be calculated from the equation, $pH = pK_{a\,(In)} + \log\,[In^-]/[HIn]$. Much more often, pH indicators are used to denote visually the approximate pH of a solution.

The above explanation applies to those indicators that are monoprotic, weak acids or monoacidic weak bases. In fact, some indicators are polyprotic with more than one pK_a value, others exist in tautomeric forms of differing colour, and still others adopt colours that depend in part on the ionic strength of the solution in which they are employed. Such complications need not concern us here.

Indicators are frequently used in volumetric analyses when the concentration of an acid solution is to be determined by titration against a standard concentration of base (or vice versa). To obtain an accurate 'end point' in such a titration, an indicator should be used whose pK_a is very nearly equal to the pH at the equivalence point. The more marked the pH change as equivalence is approached, the sharper will be the visual colour change of the indicator. Furthermore, a greater latitude is allowable in the choice of indicators for a titration in which the pH rapidly changes as the end point is approached. Thus, in the strong acid-alkali titration (Fig. 5.1), although an indicator of $pK_a = 7$ would theoretically be best, others whose pK_a values fall within the range 5 to 9 would be acceptable. In the weak acid-alkali titration (Fig. 5.2), the pH change about the equivalence point is not so extensive, and the choice of indicators is restricted to those whose pK_a values fall within a smaller range—perhaps within 1 unit on either side of the pH value at the true end point. The anticipated pH at the end point of an acid-base titration can be calculated as described below.

THE pH's OF DILUTE, AQUEOUS SOLUTIONS OF SALTS

Many salts form acid or alkaline aqueous solutions. The dissociation theory explains this as being due to hydrolysis of the salt by interaction with one or other of the ions of water, which then creates an imbalance in the residual concentrations of H^+ and OH^- ions. The extent of this hydrolysis is expressed as a hydrolysis constant whose value is correlated with the strengths of the parent acid and base by whose interaction the salt was formed.

The Brönsted and Lowry concept offers a much clearer explanation, attributing the effect of the salt on the pH of its aqueous medium to the relative acid and basic strengths of its component ions. This view is greatly to be preferred, if only because it avoids the needlessly confusing use of the term 'hydrolysis' to describe what is, after all, a neutralization process.

1. *A salt both of whose ions are 'neutral'*, e.g. sodium chloride

This type of salt is created by the interaction of a strong acid with a strong base (generally OH^- added as an alkali). Its ions are negligibly acidic and basic and yield an aqueous solution of neutral pH (7.0).

2. *A salt, one of whose ions is basic, the other 'neutral'*, e.g. sodium acetate

This type of salt is formed by the neutralization of a weak acid by a strong base (or alkali). One of its ions will be the significantly strong conjugate base of the parent weak acid, the other will be negligibly acidic or basic. The pH of its aqueous solution will be alkaline due to proton-acceptance from water by the basic ion; e.g. in the case of an aqueous solution of sodium acetate,

$$(Na^+)Ac^- + H_2O \rightleftharpoons HAc + OH^-$$

Thus the pH of a dilute aqueous solution of this type of salt will be given by the equation applicable to the solution of any weak base (p. 111); i.e.

$$pH = pK_w - \tfrac{1}{2}pK_b + \tfrac{1}{2}\log c$$

where K_b is the basic dissociation constant of the salt's basic ion, c is its concentration in mol dm^{-3}, and K_w is the ion product constant of water.

EXAMPLE

What, approximately, is the pH of a 0.02 mol dm^{-3} aqueous solution of sodium propionate at 298 K? (pK_a of propionic acid=4.85.)

Of the component ions of sodium propionate, Na^+ is 'neutral' and propionate$^-$ being the conjugate base of the relatively weak acid, propionic acid, is significantly basic. Thus the pH of an aqueous solution of sodium propionate will be given by the equation,

$$pH = pK_w - \tfrac{1}{2}pK_b + \tfrac{1}{2}\log c$$

where K_b is the basic dissociation constant of the propionate$^-$ ion present in c mol dm^{-3}.

The pK_b of propionate$^-$ is related to the pK_a of its conjugate acid (propionic acid) according to the equation,

$$pK_b = pK_w - pK_a \quad \text{(see p. 108)}$$

$$\therefore \quad pK_b \text{ of propionate}^- \text{ anion} = (14 - 4.85) = 9.15$$

Since 1 mol of sodium propionate dissolved in 1 dm^3 of water yields a solution containing 1 mol dm^{-3} of propionate$^-$ anions, the concentration of propionate$^-$ in a 0.02 mol dm^{-3} solution of sodium propionate = 0.02 mol dm^{-3} = c.

$$\therefore \quad \text{pH of 0.02 mol dm}^{-3} \text{ sodium propionate at 298 K} = pK_w - \tfrac{1}{2}(9.15) + \tfrac{1}{2}\log 0.02$$

Now, $\log 0.02 = \bar{2}.3 \text{ (approx.)} = (-2+0.3) = -1.7$

$$\therefore \quad pH = 14 - \tfrac{1}{2}(9.15) + \tfrac{1}{2}(-1.7)$$

$$= 14 - 4.58 - 0.85$$

$$= \underline{8.57}$$

3. *A salt, one of whose ions is acidic and the other 'neutral'*, e.g. ammonium chloride

This type of salt is formed by the interaction of a weak base with a strong acid. One of its ions will be the significantly strong conjugate acid of the parent weak base, the other will be the infinitely weak conjugate base of the parent strong acid. Aqueous solutions of these salts are acid, their actual pH depending on the concentration of salt present, as well as on the pK_a value of the acid ion, i.e. $pH = \frac{1}{2}pK_a - \frac{1}{2}\log c$, where pK_a is the pK_a of the acid ion and c is its concentration in mol dm^{-3} (see p. 106).

EXAMPLE

Calculate the pH of a 0.05 mol dm^{-3} solution of 'tris'hydrochloride at 298 K. ('Tris' is the weak base, tris(hydroxymethyl)-aminomethane, $(CH_2OH)_3.C.NH_2$, $pK_b = 5.92$ at 298 K.)

The salt 'tris'hydrochloride contains the two ions, $(CH_2OH)_3.C.NH_3^+$ and Cl^-. The former is the significantly strong conjugate acid of the weak base 'tris', the latter is the infinitely weak conjugate base of the strong acid HCl and is therefore 'neutral'. Thus the pH in an aqueous solution of 'tris'hydrochloride of concentration c mol dm^{-3}, will be given by the equation,

$$pH = \tfrac{1}{2}pK_a - \tfrac{1}{2}\log c$$

where K_a is the acid dissociation constant of the acid cation $(CH_2OH)_3.C.NH_3^+$.

The pK_b of 'tris' and the pK_a of its conjugate acid cation are related according to the equation, $pK_a = pK_w - pK_b$. Therefore the pK_a of the acid cation $= (14 - 5.92) = 8.08$. In a 0.05 mol dm^{-3} solution of 'tris' hydrochloride the [cation] $= 0.05$ mol dm^{-3}.

$$\therefore \tfrac{1}{2}\log c = \tfrac{1}{2}\log 0.05 = \tfrac{1}{2}(\bar{2}.7) = \tfrac{1}{2}(-1.3) = -0.65$$

$\therefore$ pH of the 0.05 mol dm^{-3} solution of 'tris' hydrochloride

$$= \tfrac{1}{2}(8.08) - (-0.65) = 4.04 + 0.65 = \underline{4.69}$$

4. *A salt, one of whose ions is acidic and the other basic*, e.g. ammonium acetate

This type of salt is formed by the interaction of a weak acid with a weak base. Its solution in water may be acid, neutral or alkaline in pH, depending on the relative strengths of its acid and basic ions. In the example mentioned, i.e. ammonium acetate, the acid strength of the NH_4^+ ion ($pK_a = 9.26$) is exactly matched by the basic strength of the acetate$^-$ ion ($pK_b = 9.26$), and the mutual neutralization of these ions means that a dilute aqueous solution of ammonium acetate, whatever its actual concentration may be, will possess a pH of 7 (cf. p. 124).

5. *A salt, one of whose ions is amphiprotic, the other 'neutral'*, e.g. sodium bicarbonate, sodium dihydrogen phosphate, or a diamine monohydrochloride

This type of salt is formed when one 'acidic group' of a weak, diprotic acid is neutralized with alkali, or when one 'basic group' of a diacidic base is neutralized with strong acid. The pH of its aqueous solution will depend on the relative acid and basic strengths of its amphiprotic ion. These may be assessed as follows:

Basic strength: as the acid dissociation constant of the conjugate acid formed from the amphiprotic ion acting as a base $= K_{a_1}$

Acid strength: as the acid dissociation constant of the amphiprotic ion $= K_{a_2}$

It can be shown that when the values of K_{a_1} and K_{a_2} are small in comparison with the concentration of the amphiprotic ion, the pH of its dilute aqueous solution will be given by the equation

$$\mathrm{pH} = \frac{\mathrm{p}K_{a_1} + \mathrm{p}K_{a_2}}{2}$$

(See also p. 145).

EXAMPLE

What is the pH of a 0.03 mol dm^{-3} solution of sodium bicarbonate at 298 K? (pK_{a_1} and pK_{a_2} of carbonic acid = 6.37 and 10.25.)

A sodium bicarbonate solution contains 'neutral' Na^+ ions and amphiprotic HCO_3^- ions.

As an acid, HCO_3^- dissociates as follows,

$$HCO_3^- \rightleftharpoons H^+ + CO_3^{2-} \quad \ldots \quad K_a \text{ of } HCO_3^- = K_{a_2} \text{ of } H_2CO_3$$

As a base, HCO_3^- associates with protons as follows,

$$HCO_3^- + H^+ \rightleftharpoons H_2CO_3 \quad \ldots \quad K_a \text{ of conjugate acid} = K_{a_1} \text{ of } H_2CO_3$$

Since pK_{a_1} of carbonic acid = 6.37, K_{a_1} must be between 10^{-6} and 10^{-7} and is therefore negligibly small in comparison with the concentration of HCO_3^- ions in the solution (0.03 mol dm^{-3}). Thus the pH of the sodium bicarbonate solution will be given approximately by the equation,

$$\mathrm{pH} = \frac{\mathrm{p}K_{a_1} + \mathrm{p}K_{a_2}}{2}$$

$$\therefore \qquad pH = \frac{6.37+10.25}{2} = \frac{16.62}{2} = 8.31$$

We will consider 'multidissociable' compounds and their amphiprotic ions in more detail when we discuss the acidic and basic properties of amino acids and proteins in Chapter 6.

We are now in a position to calculate the pH at the equivalence point in the titration of an acid with a base, and hence to select indicators which will demonstrate the correct end point.

EXAMPLE

You are provided with the following pH indicators: methyl orange (pK_a= 3.7), methyl red (pK_a=5.1), bromothymol blue (pK_a=7.0), and α-naphthylphthalein (pK_a=8.4). Which of these would you select to demonstrate the end point in the following titrations, in which the reagents are used in 0.2 mol dm^{-3} concentration?

(a) hydrochloric acid with sodium hydroxide
(b) formic acid with sodium hydroxide
(c) pyridine with hydrochloric acid

(Assume pK_a of formic acid=3.75; pK_b of pyridine=8.85.)

(a) *HCl with NaOH*

At the end point of the titration, the solution contains the 'neutral' ions Na^+ and Cl^- and neutralization would be complete,

$$HCl+(Na^+)OH^- \longrightarrow H_2O+(Na^+)Cl^-$$

The pH at the equivalence point would be 7, as in a solution of sodium chloride (p. 134), and, of the indicators listed, bromothymol blue (pK_a=7) should be chosen. (Yet, since the pH change is very great as the equivalence point is approached (Fig. 5.1), methyl red (pK_a=5.1) or α-naphthylphthalein (pK_a=8.4) might also be used.)

(b) *Formic acid (HCOOH) with NaOH*

At the end point of this titration, the solution contains the 'neutral' ion Na^+ and the appreciably basic ion $HCOO^-$ (i.e. the conjugate base of the weak acid HCOOH), for

$$HCOOH+(Na^+)OH^- \rightleftharpoons H_2O+Na^++HCOO^-$$

Thus the pH of the solution at the end point will be a function of the pK_b of the $HCOO^-$ ion and of its concentration, i.e.

$$pH = pK_w - \tfrac{1}{2}pK_b + \tfrac{1}{2}\log c$$

Now

$$pK_b \text{ of } HCOO^- = pK_w - pK_a \text{ of HCOOH} = 14 - 3.75 = 10.25$$

In the final solution produced by mixing the 0.2 mol dm^{-3} formic acid with an equal volume of 0.2 mol dm^{-3} NaOH (necessary to attain the equivalence point), the $[HCOO^-] = 0.1$ mol dm^{-3} = [HCOOH] had this been merely diluted to twice its original volume.

$$\therefore \quad pH = 14 - \tfrac{1}{2}(10.25) + \tfrac{1}{2}\log 0.1$$

$$\tfrac{1}{2}\log 0.1 = \tfrac{1}{2}(-1) = -0.5$$

$$\therefore \quad pH = 14 - 5.13 - 0.5 = \underline{8.35}$$

Therefore, α-naphthylphthalein ($pK_a = 8.4$) would best indicate this end point.

(c) *Pyridine with HCl*

The neutralization process can be written as,

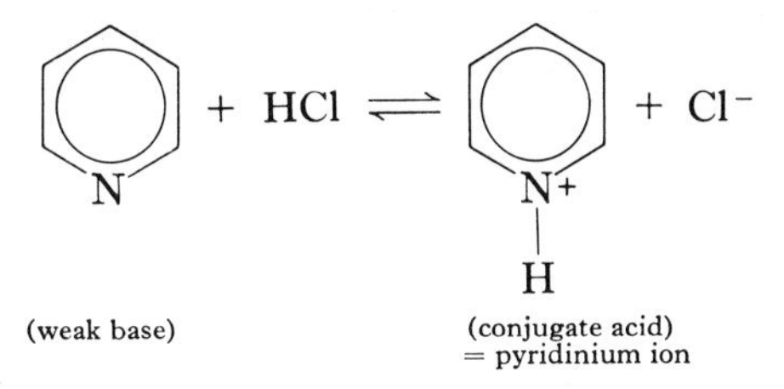

(weak base) (conjugate acid) = pyridinium ion

The solution at the end point in this titration would contain the 'neutral' Cl^- ion and the quite strongly acidic pyridinium ion,

$$pK_a \text{ of the pyridinium ion} = pK_w - pK_b \text{ of its conjugate base, pyridine.}$$

Thus the pH of the solution at the end point will be a function of the pK_a of the pyridinium ion and its concentration; i.e.

$$pH = \tfrac{1}{2}pK_a - \tfrac{1}{2}\log c$$

where $pK_a = 14 - 8.85 = 5.15$, and $c = [\text{pyridinium}^+] = 0.1$ mol dm^{-3} (see $[HCOO^-]$ above).

$$\therefore \quad pH = \tfrac{1}{2}(5.15) - \tfrac{1}{2}\log 0.1 = 2.58 - (-0.5)$$
$$= 2.58 + 0.5 = \underline{3.08}$$

Thus methyl orange ($pK_a = 3.7$) would, of those indicators given, be the best able to demonstrate this end point.

The equations which relate the pH of a solution of salt to the pK_a and pK_b values of its component ions, and to their concentration, have been recorded chiefly for reference purposes. You might wish to use them in calculations of the approximate pH of dilute salt solutions, or of the probable pH at the end point of an acid-base titration, as in the worked examples. Yet in practice this is rarely necessary since the glass electrode-pH meter (p. 363) can so easily be used to give a more accurate experimental value. Therefore you should not necessarily memorize these equations. It is much more important that you understand why a certain salt should yield an acid or alkaline solution, and why this pH should be temperature and perhaps concentration dependent.

PROBLEMS

(*All solutions are assumed to be at 298 K, and all dissociation constants refer to this temperature.*)

1. (a) Convert the following hydrogen ion concentrations (in mol dm^{-3}) into pH values: 10^{-3}; 1.5×10^{-4}; 3.1×10^{-6}; 1.1×10^{-10}
   (b) Convert the following pH values into hydrogen ion concentrations: 3.2; 7.7; 10.6; 13.5

2. Calculate the pH of a 0.1 mol dm^{-3} solution of butyric acid, $CH_3(CH_2)_2COOH$. (Assume K_a of butyric acid $= 1.5 \times 10^{-5}$.)

3. What is the pH of a 0.05 mol dm^{-3} solution of aniline, $C_6H_5NH_2$? (Assume K_b of aniline $= 4 \times 10^{-10}$.)

4. Given 100 cm^3 of 0.05 mol dm^{-3} formic acid, what volume of 0.05 mol dm^{-3} sodium hydroxide would you need to add to obtain a buffer of pH 4.23? (Assume K_a of formic acid $= 1.77 \times 10^{-4}$.)

5. What are the resultant pH values when the following volumes of 0.1 mol dm^{-3} sodium hydroxide are added to 100 cm^3 of 0.1 mol dm^{-3} acetic acid: (a) 10 cm^3, (b) 25 cm^3, (c) 50 cm^3, (d) 75 cm^3, (e) 90 cm^3? (Assume K_a of acetic acid $= 1.82 \times 10^{-5}$.)

6. What are the pH values of the resulting solutions when 100 cm^3 of 0.1 mol dm^{-3} sodium acetate are mixed with (a) 100 cm^3 of 0.05 mol dm^{-3} hydrochloric acid, (b) 100 cm^3 of 0.05 mol dm^{-3} acetic acid? What are the hydroxyl ion concentrations in the final mixtures? (Assume K_a of acetic acid $= 1.82 \times 10^{-5}$.)

7. The following results were obtained when 100 cm^3 of 0.1 mol dm^{-3} monoacidic base was titrated against 0.1 mol dm^{-3} hydrochloric acid. Interpret these results and determine the apparent basic dissociation constant (K_b) of the base.

0.1 mol dm^{-3} HCl added (in cm^3)	0	10	25	50	90	99	99.8
pH	11.1	10.2	9.8	9.3	8.3	7.3	6.6

8. A bacterial suspension in 0.1 mol dm^{-3} phosphate buffer pH 7.1 containing 0.3% glucose completely converts the sugar, by fermentation, into lactic acid. What is the final pH of the suspension? (1 glucose $\xrightarrow{\text{fermentation}}$ 2 lactic acid; molecular weight of glucose 180; apparent pK_{a2} of phosphoric acid in buffer at this ionic strength is 6.80.)

9. Three solutions were prepared by mixing together the following:

(a) 30 cm^3 of 0.02 mol dm^{-3} hydrochloric acid and 90 cm^3 of 0.02 mol dm^{-3} trimethylamine,
(b) 30 cm^3 of 0.02 mol dm^{-3} hydrochloric acid and 20 cm^3 of 0.04 mol dm^{-3} trimethylamine,
(c) 100 cm^3 of 0.05 mol dm^{-3} Na_3PO_4 and 150 cm^3 of 0.05 mol dm^{-3} hydrochloric acid.

What would be the pH values of these solutions? (Assume the following acid dissociation constants: trimethylammonium ion $K_a = 1.74 \times 10^{-10}$; phosphoric acid, $K_{a_1} = 1.1 \times 10^{-2}$; $K_{a_2} = 1.6 \times 10^{-7}$; $K_{a_3} = 10^{-12}$.)

10. Over what range of hydrogen ion concentrations may an ammonium chloride buffer be expected to show reasonable buffering ability? (Assume pK_b ammonia = 4.74.)

11. How would you prepare 1 dm^3 of a 0.02 mol dm^{-3} pyridine buffer pH 5.0 given a 0.1 mol dm^{-3} solution of pyridine and 2.0 mol dm^{-3} hydrochloric acid? (Assume pK_b of pyridine = 8.64.)

12. How would you prepare 500 cm^3 of 0.1 mol dm^{-3} sodium phosphate buffer pH 7.1 given $Na_2HPO_4.2H_2O$ (M. Wt. = 178) and $NaH_2PO_4.H_2O$ (M. Wt. = 138)? (Assume apparent pK_a values for phosphoric acid of 1.96, 6.8 and 12.)

13. If 5.2 cm^3 of 0.2 mol dm^{-3} hydrochloric acid are added to 100 cm^3 of a solution containing 820 mg of the monosodium salt of veronal (i.e. sodium diethylbarbiturate), a buffer solution of pH 8.4 is obtained. What is the molecular weight of this monosodium salt of veronal? (Apparent pK_a of veronal = 7.95.)

14. Calculate the pH values of 0.02 mol dm^{-3} solutions of the following: (a) potassium cyanide, (b) 'tris' hydrochloride. (Assume the following values: K_a of hydrocyanic acid = 7.24×10^{-10}; K_a of 'tris' = 8.32×10^{-9}.)

15. Which of the following indicators would you use when titrating an approximately 0.2 mol dm^{-3} solution of triethanolamine against 0.2 mol dm^{-3} hydrochloric acid?

(a) bromocresol green (pK_{In} = 4.7),
(b) methyl orange (pK_{In} = 3.7),
(c) methyl red (pK_{In} = 5.1),
(d) phenol red (pK_{In} = 7.9).

(Assume pK_a of triethanolamine = 7.76.)

6

Biochemical Relevance of pH

Many biologists would state with little hesitation that the pH of the protoplasm of a living cell must be maintained virtually constant, and close to pH 7. They would support this contention by citing the death of tissue cultures and of many micro-organisms in markedly acid or alkaline media, the existence of elaborate pH-buffering mechanisms which maintain the pH of the circulating and extracellular fluids of higher animals near neutrality, and the loss of metabolic activity suffered by cell extracts in inadequately or wrongly buffered media. Others, pointing to the structural nonhomogeneity and minute size of a living cell, would question whether the term 'intracellular pH' can have any readily understandable meaning when applied to a volume so small that the production of a very few 'free' H^+ ions would cause a dramatic rise in the hydrogen ion concentration. However, what is not disputed is that many of the most important components of a living cell are acidic, or basic, or amphiprotic, and that any alteration in the pH of their environment could profoundly affect their state of ionization, and hence their molecular conformation and biological activity.

In this chapter we are not able to discuss all types of biochemical compounds that are likely to be pH-sensitive, nor to consider all metabolic processes in which protons are reactants or products. Instead we shall in the main consider only one group of compounds, namely proteins; these are perhaps the most characteristic of all cytoplasmic constituents having structural and catalytic functions and contributing to the pH-buffering ability of the cellular contents. As the acidic and basic properties of proteins are attributable to their component amino acids, we shall examine these before venturing to consider what effects pH changes might have on the structure and properties of proteins themselves. Remember, though, that while we shall concentrate our attention on amino acids and proteins, much that will be said about their pH-dependence and its analytical usefulness might also have been said of other multidissociable constituents of protoplasm that receive only brief mention at the close of this chapter.

THE pH-DEPENDENT IONIZATION OF AMINO ACIDS

All amino acids are amphiprotic compounds and the majority of those present in protein hydrolysates conform to the general formula R.CH.(NH_2)COOH. The amino group is known to be weakly basic, associating with protons to form a positively charged ion, which in aliphatic amines has a pK_a of about 10.6 ($R—NH_2+H^+ \rightleftharpoons R—NH_3^+$). The carboxylic group, on the other hand, is weakly acidic, and when present as the chain terminating group in a fatty acid it dissociates, with a pK_a of about 4.5, to yield a negatively charged ion ($R—COOH \rightleftharpoons H^+ + RCOO^-$). In amino acids, the attachment of the amino and carboxylic groups to the same (α) carbon atom enhances their acidity, the pK_a of the α-carboxyl group now being about 1.7 to 2.4, whilst the pK_a of the α-NH_3^+ group is about 9 to 10.5. Thus all amino acids will be ionized in aqueous solution in a manner determined by the prevailing pH.

Those amino acids whose side chains (R) bear no dissociable groups may exist in three ionic forms as shown:

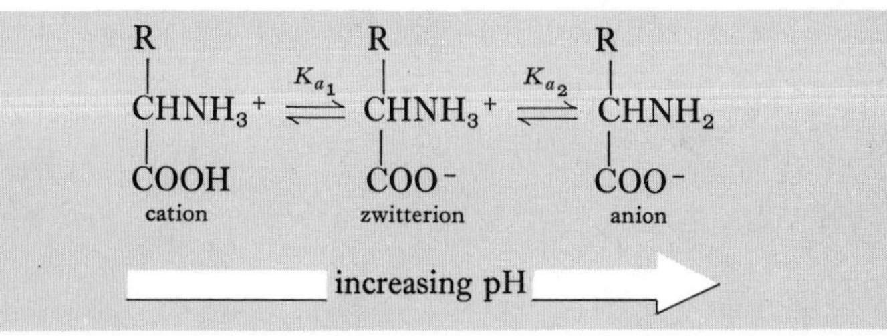

At low pH's the amino acid will exist as a cation in which only the α-amino group is ionized. As this solution is made more alkaline, so will the α-carboxyl group dissociate and become ionized. The product is the dipolar form of the amino acid which bears no net electrical charge. This is called the ***zwitterion*** (i.e. hybrid ion). Raising the pH of the solution still further, causes the α-NH_3^+ group to dissociate and thereby lose its charge. The product which predominates at very alkaline pH's is therefore the anion in which only the α-carboxyl group is ionized. At the pH equal to the pK_a value of the carboxyl group (pK_{a_1} of this amino acid), cation and zwitterion will coexist in equal concentration. Similarly, at the pH equal to the pK_a of the $-NH_3^+$ group (pK_{a_2} of this amino acid), zwitterion and anion will be present in equal concentration. In its crystalline state, or in solution in pure water, the amino acid will be predominantly in its zwitterionic form.

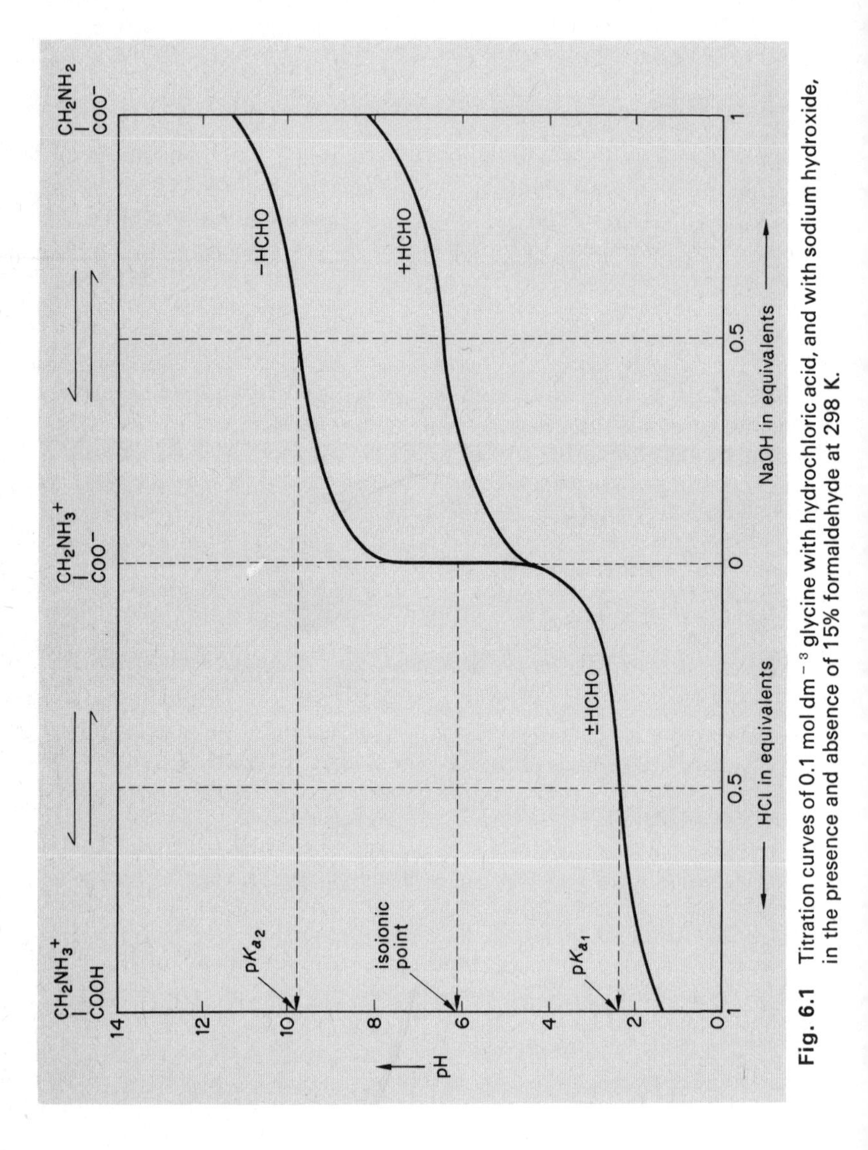

Fig. 6.1 Titration curves of 0.1 mol dm^{-3} glycine with hydrochloric acid, and with sodium hydroxide, in the presence and absence of 15% formaldehyde at 298 K.

Take for example, the simplest amino acid, glycine ($H_2N.CH_2.COOH$). In solution in pure water it exists as the zwitterion, $^+H_3N.CH_2.COO^-$, and will behave as a base when titrated with hydrochloric acid (Fig. 6.1). The neutralization process can be represented by the equation,

$$^+H_3N.CH_2.COO^- + HCl \rightleftharpoons {}^+H_3N.CH_2.COOH + Cl^-$$

and the titration curve is that of a weak base; the pH at half equivalence is the apparent pK_a value of the carboxylic group. When titrated with alkali (Fig. 6.1), the zwitterion of glycine behaves as a weak acid, the neutralization equation being,

$$^+H_3N.CH_2.COO^- + (Na^+)OH^- \rightleftharpoons H_2N.CH_2.COO^- + Na^+ + H_2O$$

Now the pH at the half-equivalence point equals the apparent pK_a value of the α-NH_3^+ group. Thus the acid group of the zwitterion of glycine is its α-NH_3^+ group, whilst its basic group is its α-COO^- group. The pH of a dilute aqueous solution of this zwitterion, as of any other amphiprotic substance (p. 137), is determined by its basic and acidic strengths, and is given by the equation, $pH = \frac{pK_{a_1} + pK_{a_2}}{2}$, where K_{a_1} and K_{a_2} are the acid dissociation constants of its basic and acidic groups respectively. For glycine, pK_{a_1} equals 2.4, and pK_{a_2} is 9.8, so that the pH of a dilute aqueous solution of glycine is $(2.4+9.8)/2=6.1$. In this solution the zwitterion predominates, and pH 6.1 is the ***isoionic point*** of glycine where the number of negative charges on the molecule produced by protolysis equals the number of positive charges acquired by proton gain. It is also the ***isoelectric point*** of glycine, for at pH 6.1 its molecules carry no net charge and are electrophoretically immobile (see p. 156).

The ionization of glycine can therefore be summarized as follows,

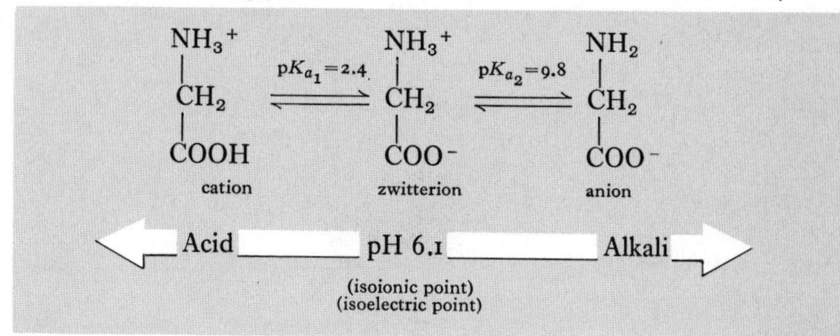

EXAMPLE

1.068 g of a crystalline α-amino acid with pK_{a_1} and pK_{a_2} values of 2.4 and 9.7 gave a solution of pH 10.4 when dissolved in 100 cm^3 of 0.1 mol dm^{-3} sodium hydroxide. Calculate the molecular weight of the amino acid, and suggest a possible molecular formula. (Atomic weights: C=12, N=14, O=16, and H=1.)

Let the amino acid be represented by the formula R.CH(NH_2)COOH. Then it will dissociate in aqueous solution as follows:

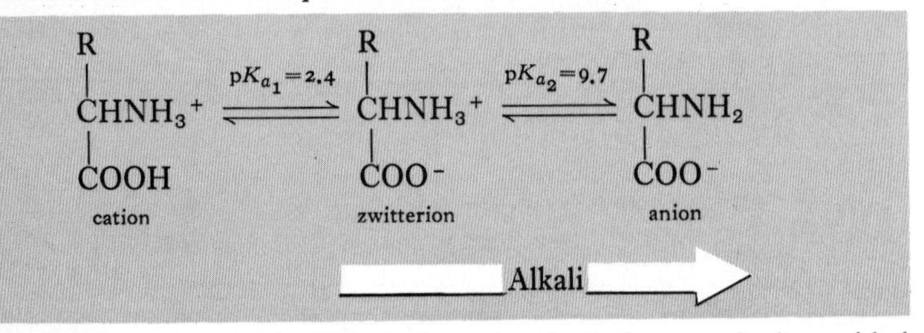

The crystalline amino acid will be in its zwitterionic form, and when added to alkali it will behave as a weak acid, dissociating according to the equation:

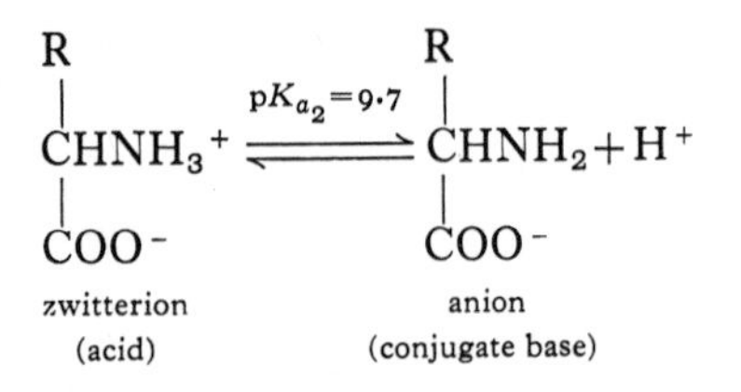

The quantity of alkali used was evidently insufficient to neutralize the zwitterion completely, since the pH of the final solution is within 1 pH unit of the pK_a value for this dissociation. Therefore the pH of the final solution defines its composition, for, by the Henderson-Hasselbalch equation,

$$\text{pH} = pK_a + \log \frac{[\text{conjugate base}]}{[\text{acid}]}$$

so that,

$$10.4 = 9.7 + \log \frac{[\text{anion}]}{[\text{zwitterion}]}$$

whence, $\log \frac{[\text{anion}]}{[\text{zwitterion}]} = 0.7$, and $\frac{[\text{anion}]}{[\text{zwitterion}]} = \text{antilog } 0.7 = 5$

Since $\quad \text{zwitterion}^{\pm} + (Na^+)OH^- \longrightarrow \text{anion}^- + Na^+ + H_2O$

the [anion] in the final solution

$$\equiv [Na^+OH^-] \text{ 'consumed'} = 0.1 \text{ mol dm}^{-3}$$

$$\therefore \qquad [\text{anion}] = 0.1 \text{ mol dm}^{-3}$$

$$\therefore \qquad [\text{zwitterion}] \text{ being equal to } \tfrac{1}{5}[\text{anion}] = 0.02 \text{ mol dm}^{-3}$$

$$\therefore \qquad [\text{total amino acid}] = [\text{anion}] + [\text{zwitterion}] = 0.12 \text{ mol dm}^{-3}$$

$$\therefore \qquad [\text{total amino acid}] = 0.12 \text{ mol dm}^{-3}$$

Thus 100 cm³ of 0.12 mol dm⁻³ amino acid solution contains 1.068 g of amino acid

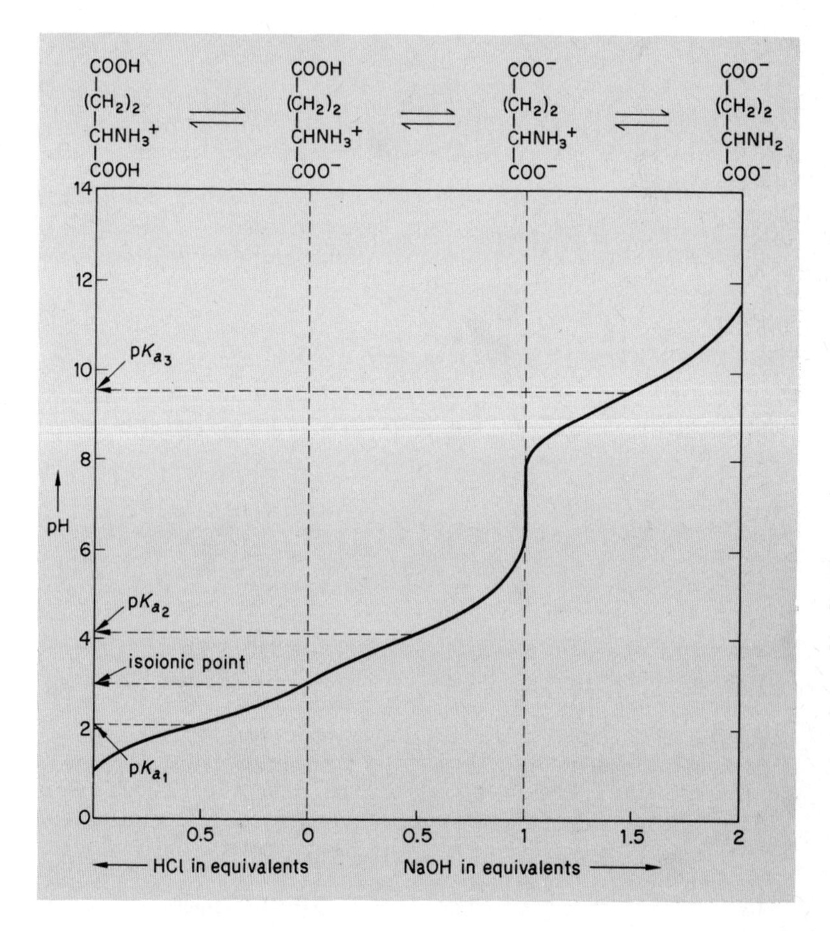

Fig. 6.2 Titration curves of 0.1 mol dm^{-3} glutamic acid with hydrochloric acid, and with sodium hydroxide, at 298 K.

$\therefore$ 1 dm^3 of 1 mol dm^{-3} solution contains

$$\left(1.068\times\frac{1000}{100}\times\frac{1}{0.12}\right)\text{g} = 89\text{ g}$$

$\therefore$ molecular weight of the amino acid = 89

If the amino acid is R.CH(NH_2)COOH with a molecular weight of 89, then, since the formula weight of —CH(NH_2)COOH is 74, the side chain R must contribute 15 units and must surely be —CH_3.

Therefore it is likely that the amino acid is *alanine*, whose molecular formula is CH_3.CH(NH_2).COOH.

The ionization of amino acids which carry supplementary dissociable groups

The naturally-occurring amino acids have distinctive side chains (R–), some of which carry additional basic or acidic groups that give the molecule an overall basic or acidic character; e.g.

valine ('neutral')	glutamic acid ('acidic')	lysine ('basic')
$(CH_3)_2$CH–CHNH$_2$–COOH	COOH–$(CH_2)_2$–CHNH$_2$–COOH	NH$_2$–$(CH_2)_4$–CHNH$_2$–COOH

The presence of supplementary, dissociable groups complicates the ionic behaviour of an amino acid. This is evident in the titration curves that are obtained when glutamic acid is titrated with (a) HCl and (b) NaOH (Fig. 6.2).

The aqueous solution of glutamic acid again contains the zwitterionic form of the amino acid but its pH is 3.1, which makes it considerably more acid than the solution of the 'neutral' amino acid, glycine. Comparison of the titration curves of glutamic acid with those of glycine, shows that two equivalents of alkali are required to neutralize glutamic acid. The new dissociation step apparent in the titration curve of glutamic acid with NaOH demonstrates the existence of a buffer zone centred on the half-equivalence point of pH 4.1. This is the apparent pK_a of the γ-carboxyl group so that glutamic acid ionizes according to the following dissociation sequence,

$$\begin{array}{ccccccc}
\mathrm{COOH} & & \mathrm{COOH} & & \mathrm{COO^-} & & \mathrm{COO^-} \\
| & & | & & | & & | \\
\mathrm{(CH_2)_2} & \overset{pK_{a_1}=2.1}{\rightleftharpoons} & \mathrm{(CH_2)_2} & \overset{pK_{a_2}=4.1}{\rightleftharpoons} & \mathrm{(CH_2)_2} & \overset{pK_{a_3}=9.5}{\rightleftharpoons} & \mathrm{(CH_2)_2} \\
| & & | & & | & & | \\
\mathrm{CHNH_3^+} & & \mathrm{CHNH_3^+} & & \mathrm{CHNH_3^+} & & \mathrm{CHNH_2} \\
| & & | & & | & & | \\
\mathrm{COOH} & & \mathrm{COO^-} & & \mathrm{COO^-} & & \mathrm{COO^-} \\
 & & \text{zwitterion} & & & &
\end{array}$$

← Acid pH 3.1 Alkali →

When crystalline glutamic acid is dissolved in water, a proportion of its zwitterions will dissociate, and the pH will fall due to the ionization of their γ-carboxyl groups. Equilibrium is achieved when the pH has fallen to pH 3.1, with the bulk of the amino acid still present in its zwitterionic form. When glutamic acid is dissolved in buffers of pH 2.2 to 4.0, its zwitterionic form will again predominate, though in buffers of pH about pH 7 the amino acid will be present almost entirely as the net negatively charged species, $^-OOC.(CH_2)_2.CH(NH_3^+).COO^-$.

The 'basic' amino acid, lysine, behaves quite differently when it is titrated with HCl and NaOH (Fig. 6.3).

Lysine is a diamino-monocarboxylic acid which yields an alkaline solution in pure water (isoionic point of lysine is pH 10). Two equivalents of HCl are consumed in the titration of lysine which commences at this isoionic point, and the values of pK_a (apparent) supplied by this titration (9.2 and 2.2) are the values to be expected of its α-NH_3 group and α-carboxyl group respectively. On the other hand, one equivalent of NaOH is consumed in the titration of lysine from its isoionic point, the pH at half equivalence being 10.8. It follows that 10.8 must be the apparent pK_{a_3} of lysine,

$$\begin{array}{ccccccc}
\mathrm{NH_3^+} & & \mathrm{NH_3^+} & & \mathrm{NH_3^+} & & \mathrm{NH_2} \\
| & & | & & | & & | \\
\mathrm{(CH_2)_4} & \overset{pK_{a_1}=2.2}{\rightleftharpoons} & \mathrm{(CH_2)_4} & \overset{pK_{a_2}=9.2}{\rightleftharpoons} & \mathrm{(CH_2)_4} & \overset{pK_{a_3}=10.8}{\rightleftharpoons} & \mathrm{(CH_2)_4} \\
| & & | & & | & & | \\
\mathrm{CHNH_3^+} & & \mathrm{CHNH_3^+} & & \mathrm{CHNH_2} & & \mathrm{CHNH_2} \\
| & & | & & | & & | \\
\mathrm{COOH} & & \mathrm{COO^-} & & \mathrm{COO^-} & & \mathrm{COO^-} \\
 & & & & \text{zwitterion} & &
\end{array}$$

← Acid pH 10 Alkali →

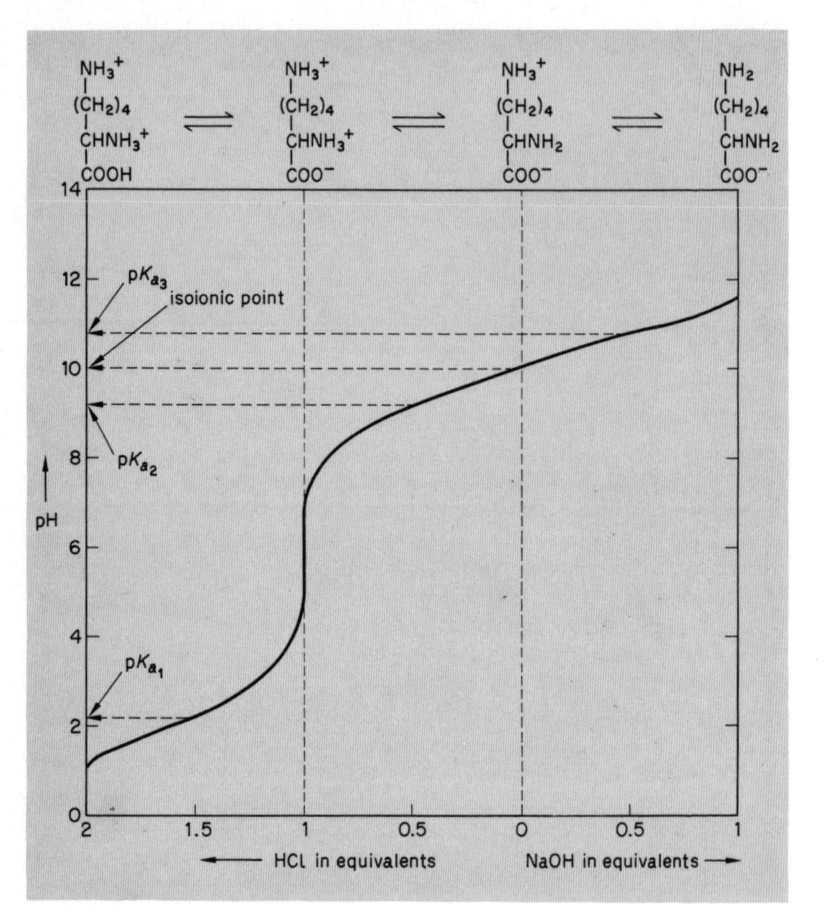

Fig. 6.3 Titration curves of 0.1 mol dm^{-3} lysine with hydrochloric acid, and with sodium hydroxide, at 298 K.

characteristic of its ε-NH_3^+ group. It further follows that it will be this ε-NH_3^+ group that will be ionized in the zwitterion of lysine, for the amino acid must dissociate as shown at the foot of p. 149.

In buffer at pH 7, at which pH glycine carries no net charge and glutamic acid is negatively charged overall, the 'basic' amino acid, lysine, will carry a net positive charge since predominant in its solution will be the cation $^+H_3N.(CH_2)_4.CH(NH_3^+).COO^-$.

Note that the isoionic point is always determined by the pK_a values that define the basic and acidic strengths of the *zwitterion*. Thus, for glutamic acid the isoionic point is $\frac{pK_{a_1}+pK_{a2}}{2}$, but for lysine the isoionic point is $\frac{pK_{a_2}+pK_{a3}}{2}$. Yet, although we calculate an isoionic *point* in this way, the fact is, that when the contributory pK_a values are very different, there exists a considerable range of intermediate pH values in which the molecule will be virtually entirely in its zwitterionic form.

EXAMPLE

Calculate the isoionic points of the following amino acids:

(*a*) *aspartic acid* ($pK_{a_1}=1.99$; $pK_{a_2}=3.90$; $pK_{a_3}=9.90$);
(*b*) *arginine* ($pK_{a_1}=1.82$; $pK_{a_2}=8.99$; $pK_{a_3}=12.48$);
(*c*) *histidine* ($pK_{a_1}=1.80$; $pK_{a_2}=6.04$; $pK_{a_3}=9.33$);
(*d*) *tyrosine* ($pK_{a_1}=2.20$; $pK_{a_2}=9.11$; $pK_{a_3}=10.13$).

(Note: It is always advisable to obtain the formula of the compound and then to write out equations for the sequential dissociations that it undergoes, using your knowledge of the probable pK_a values of its dissociable groups (p. 154) to place these in the correct order of diminishing acid strength. Then the zwitterion, and the pK_a values that define its acidic and basic strengths, are readily identified.)

(*a*) *Aspartic acid.* Formula: $HOOC.CH_2.CH(NH_2).COOH$
This dicarboxylic, monoamino acid will dissociate as follows:

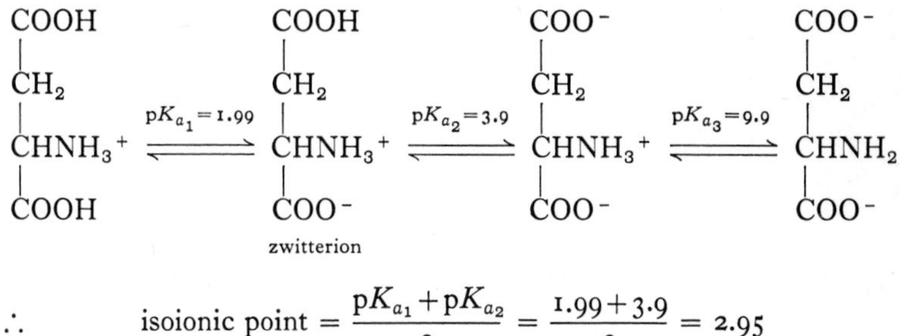

$$\therefore \quad \text{isoionic point} = \frac{pK_{a_1}+pK_{a_2}}{2} = \frac{1.99+3.9}{2} = \underline{2.95}$$

(b) *Arginine.* Formula: $H_2N.C(=NH).NH.(CH_2)_3.CH(NH_2).COOH$

Because of its guanidyl group, this is a 'basic' amino acid that will dissociate as follows:

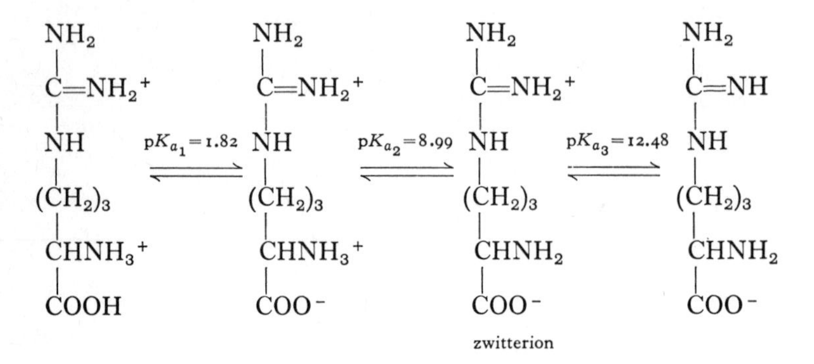

$\therefore$ isoionic point $= \dfrac{pK_{a_2}+pK_{a_3}}{2} = \dfrac{8.99+12.48}{2} = \underline{10.74}$

(c) *Histidine.* Formula: $CH_2.CH(NH_2).COOH$ attached to the imidazole ring (N, NH)

Because of its imidazolyl ring, this is a 'basic' amino acid that dissociates as follows:

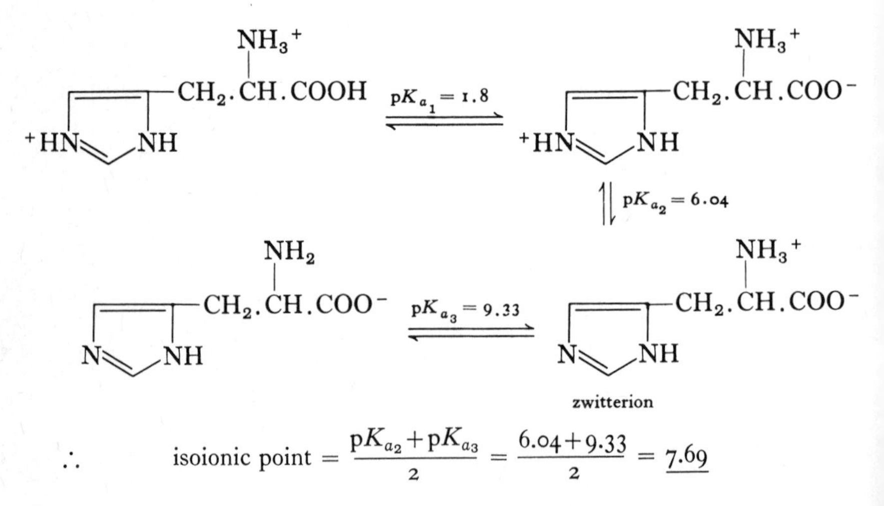

$\therefore$ isoionic point $= \dfrac{pK_{a_2}+pK_{a_3}}{2} = \dfrac{6.04+9.33}{2} = \underline{7.69}$

(d) *Tyrosine.* Formula: HO–C_6H_4–$CH_2.CH(NH_2).COOH$

The phenolic hydroxyl group confers upon tyrosine a net 'acidic' character. It dissociates as follows:

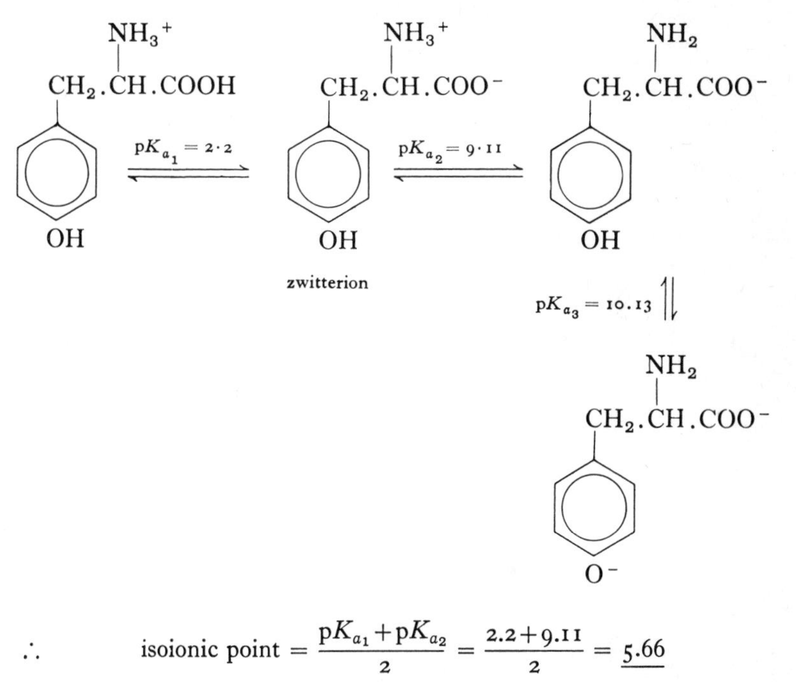

$$\therefore \quad \text{isoionic point} = \frac{pK_{a_1} + pK_{a_2}}{2} = \frac{2.2 + 9.11}{2} = \underline{5.66}$$

Assignment of apparent pK_a values to dissociable groups in amino acids

It is generally not difficult to correlate any pK_a value which might be manifested in the titration curve of an amino acid, with the dissociation of one specific group in its molecule. After all, any one amino acid contains only a few dissociable groups with 'well separated' pK_a values. It is, anyway, also possible to identify the contribution made by one class of dissociable group to a titration curve, by examining what change in the shape of the curve is made by a reagent which combines specifically with that dissociable group to alter its acidic or basic strength. As an illustration of this technique, compare the shapes of the titration curves of glycine obtained in the presence of 15% formaldehyde and in its absence (Fig.

6.1). Formaldehyde reacts with the amino group to yield a product of considerably diminished basic strength,

$$H_3N^+.CH_2.COO^- + HCHO \longrightarrow HOCH_2.NH.CH_2.COO^- + H^+$$

The titration curve of glycine with HCl (characteristic of neutralization of its carboxyl group) is not affected by the presence of formaldehyde. However, the titration curve of glycine with NaOH is considerably displaced, reflecting the more strongly acidic behaviour of the α-NH_3^+ group in the presence of formaldehyde. In this way, the contribution made by the α-NH_3^+ group to the normal titration curve is easily identified. From the formulae of glutamic acid and of lysine, we can predict that formaldehyde will cause the similar displacement of only one segment of the titration curve of glutamic acid, but of two segments of the titration curve of lysine.

By similar methods, and by other techniques which are based on quite different principles (e.g. microcalorimetry, spectrophotometry and polarography), it has proved possible to assign apparent pK_a values to all of the dissociable groups found in the common amino acids:

(a) *Uncharged groups that dissociate to leave a negatively charged, basic group on the molecule*

	Apparent pK_a *at 298 K*
α-carboxyl	1.7 to 2.4
β-carboxyl (aspartate)	3.9
γ-carboxyl (glutamate)	4.1
sulphydryl (cysteine)	8 to 9 (possibly)
phenolic hydroxyl (tyrosine)	10.1

(b) *Positively charged groups that dissociate to leave an uncharged basic group on the molecule*

	Apparent pK_a *at 298 K*
imidazolyl$\equiv NH^+$ (histidine)	6
α-NH_3^+	9 to 10.5
pyrrolidone$=NH_2^+$ (hydroxyproline)	9.7
pyrrolidone$=NH_2^+$ (proline)	10.6
ε-NH_3^+ (lysine)	10.8
δ-NH_3^+ (ornithine)	10.8
guanidyl$=NH_2^+$ (arginine)	12.5

pH-Dependent properties of amino acids

The properties of an amino acid which are most likely to be affected by changes in pH are those that are possessed in different measure by its variously ionized forms, since the pH determines which ionic forms are present, and in what proportions.

As one example of this, the solubility of an amino acid is often very evidently affected by the pH in a manner that suggests that the zwitterion is its least soluble form.† This is obviously true of cystine and tyrosine, for these amino acids are sparingly soluble in the pH range in which their zwitterions predominate (pH 4 to pH 9), and are much more soluble in both more acid or more alkaline media in which they exist as cations or anions.

Other properties which may be possessed by the various ionic forms of an amino acid in differing degrees, include optical rotation, ultraviolet absorbance at a given wavelength, metal chelating ability, even biological activity, so that these properties also can be expected to be affected by changes in environmental pH.

One compensation for the behavioural complications that stem from the way in which the pH determines the predominant ionic state of an amino acid is the practical use to which this effect can be put in separating and identifying amino acids by electrophoresis and ion-exchange chromatography.

Electrophoresis of amino acids

This technique exploits the migration of ions in an electrical field. When a direct current is passed between two electrodes in an ion-containing medium, the negative ions (anions) move towards the positive electrode (anode), while the positive ions (cations) move towards the negative electrode (cathode). The rate of migration of an ion will depend, among other things, on the size of the charge that it carries, and the magnitude of the current employed. In the version of the technique that is known as high voltage, paper electrophoresis, a mixture is applied in a small spot or transverse band in the centre of a long strip of filter paper which is then saturated with a buffer of suitable pH. The ends of the paper strip are placed in electrical contact with electrodes across which is applied a voltage (about 100 V/cm) sufficient to cause the passage of a current of about 40 milliamperes through the buffered medium. A compound that is homogeneously charged at the pH used will migrate at a characteristic rate towards the electrode of opposite charge, and will have travelled a characteristic distance relative to other compounds when the current is switched off. After the paper (electrophoretogram) has been dried, the new sites now

† Note, however, that since the charges on the amino acid molecule may be neutralized by becoming electrostatically bound to alien anions or cations, the ionic strength of the medium can also affect the solubility of an amino acid.

occupied by the charged components of the starting mixture can be revealed by spectrophotometric or staining techniques.

Paper electrophoresis is particularly suitable for the separation of ionized compounds of relatively small molecular weight such as amino acids. The method can also be used to study the way in which the net charge on the molecules varies with the pH of the medium, which may in turn indicate the number of dissociable groups in the molecule and their approximate pK_a values. The isoelectric point of a compound, at a known ionic strength, is particularly easy to determine by finding (quite empirically) at what pH it does not move from the starting line, i.e. the pH at which it is electrophoretically immobile. However, the most common use of paper electrophoresis is as a means of fractionating mixtures for analytical purposes. Here the pH of the electrophoresis medium must be most carefully chosen. For example, to separate glycine, glutamic acid and lysine by paper electrophoresis, one should use a buffer of about pH 6, for at this pH:

(a) glycine is in its zwitterionic form and will be immobile;
(b) glutamic acid will be negatively charged and will move towards the positive electrode (anode);
(c) lysine will be positively charged and will move towards the negative electrode (cathode).

A very clean separation of the differently charged components will therefore be obtained. Yet resolution of a mixture of similarly charged compounds can also be accomplished by this technique, since the rate of migration of an ion in a constant electrical field is determined not only by its net charge but also by its size—which in an aqueous medium generally means the size of the hydrated ion. Indeed, mixtures of amino acids are frequently separated into their components by paper electrophoresis at pH 1.8 to 2, at which pH all amino acids carry a net positive charge and all migrate towards the negative electrode. The rate of migration of any one amino acid at these low pH's will be determined by the size of its hydrated cation, and the degree to which it is charged (since values of pK_{a_1} differ, though all are in the range of 1.7 to 2.4).

Ion-exchange chromatography of amino acids

This procedure makes use of ion-exchange materials that consist of a water-insoluble polymeric matrix whose surface carries one type of acidic or basic group. The pH-controlled dissociation of these groups determines the number of *fixed charges* on the ion exchange material which are available to hold (by electrostatic attraction) an equivalent number of oppositely charged *mobile ions* supplied by the medium.

Ion-exchange materials are classified according to the charge of their *mobile ions*.

(a) *Cation exchangers*

These ionize to reveal fixed negative charges that 'bind' mobile cations. They are also called 'acidic' ion-exchange materials since their fixed negative charges are produced by the protolysis of acidic groups; they are in fact catalogued according to the strength of these acidic groups.

1. STRONGLY ACIDIC CATION EXCHANGERS For example, a resin lattice of the polystyrene type, bearing sulphonic acid groups (Dowex 50)

$$\text{ʃ}\!-SO_3H \rightleftharpoons \text{ʃ}\!-SO_3^- + H^+$$

These strongly acidic groups will be ionized at all except very low pH's.

2. 'INTERMEDIATE' AND WEAKLY ACIDIC CATION EXCHANGERS For example, a polyacrylic resin lattice bearing carboxylic acid groups, or a polysaccharide lattice bearing carboxymethyl groups (CM-Cellulose).

$$\text{ʃ}\!-O-CH_2-COOH \rightleftharpoons \text{ʃ}\!-O-CH_2-COO^- + H^+$$

(b) *Anion exchangers*

These bear positive fixed charges that bind mobile anions. They are also known as 'basic' ion-exchange materials since their fixed positive charges are generally the product of association of H^+ ions with fixed basic groups.

1. STRONGLY BASIC ANION EXCHANGERS For example, a polystyrene lattice bearing quaternary ammonium groups (Dowex 1). Such materials are fully ionized at all save very alkaline pH's. An 'intermediate' strength material is given by a polysaccharide lattice bearing diethyl-amino groups (DEAE-cellulose).

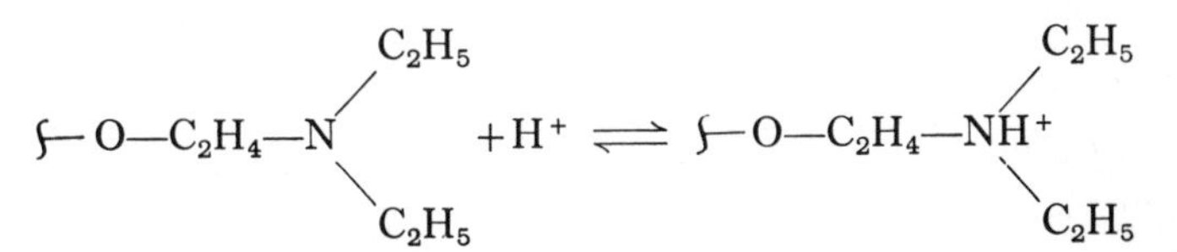

2. WEAKLY BASIC ANION EXCHANGERS For example, a polysaccharide lattice bearing *p*-aminobenzyl groups.

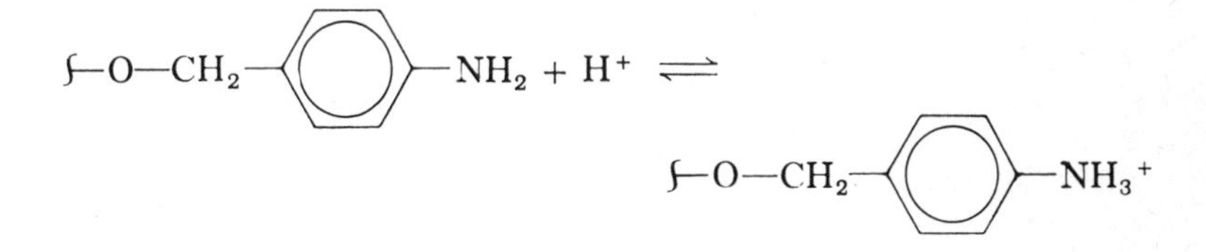

The usefulness of these materials can be illustrated by considering a cationic exchanger which, at a pH where it is completely ionized, is fully saturated with a mobile cation X^+; i.e.

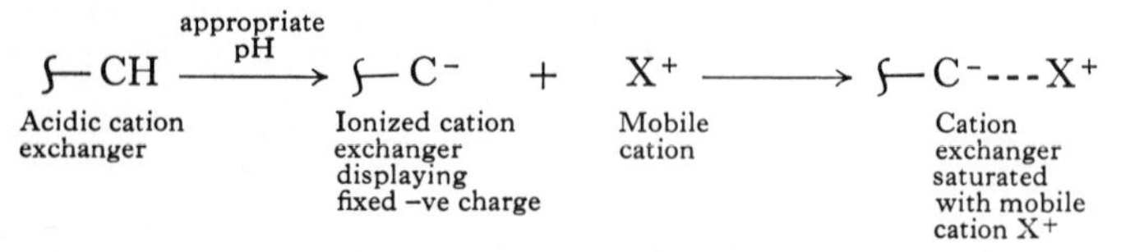

Cation exchange takes place when this ion-exchange material, with its bound cation, is added to a solution containing a different cation Y^+ at a pH at which the exchange material remains fully ionized. Some X^+ will be displaced by Y^+ until equilibrium is established,

$$\int\!\!-C^{-}\text{---}X^{+} + Y^{+} \rightleftharpoons \int\!\!-C^{-}\text{---}Y^{+} + X^{+}$$

The composition of the equilibrium mixture will depend on several factors, including the temperature, the initial concentrations of $\int\!\!-C^{-}\text{---}X^{+}$ and of Y^+, and the relative affinity of the ion-exchange material for X^+ and Y^+. It is because the exchange of cations proceeds to an equilibrium, and not to completion, that column treatment with ion-exchange materials is generally more effective than batch treatment. If instead of proceeding as in the batch treatment just described, the solution of Y^+ is allowed to percolate slowly through a column of $\int\!\!-C^{-}\text{---}X^{+}$, then progressive removal of Y^+ ensures that successive layers of fresh $\int\!\!-C^{-}\text{---}X^{+}$ are exposed to diminishing concentrations of Y^+, until Y^+ is virtually completely removed from solution. By the correct choice of ion-exchange material and its initial mobile ion, and with suitably controlled operating conditions of pH, ionic strength, temperature and flow rate, Y^+ can be 'bound' within a relatively narrow band on the column material and will remain bound while the column is washed with a buffer of low ionic strength, and of the same pH as the applied solution of Y^+. To recover Y^+, it can be displaced and eluted by passing a suitable buffer solution through the column. This eluting buffer should differ from the loading buffer by being:

(a) of lower pH—sufficiently low to cause the fixed negative charges to be lost by association with H^+ ions;

or (b) of identical pH but increased ionic strength;

or (c) of identical pH and ionic strength, but containing a cation for which the ion-exchange material demonstrates a much greater affinity than it does for Y^+.

When the net charge carried by the mobile ions is also pH-determined, as it is when these are amino acids, then the operating pH has to be even more carefully chosen. In fact, the ion-exchange material most frequently employed for the column chromatography of amino acids is a strongly acidic, cation-exchange resin whose fixed charges are sulphonate ions (i.e. Dowex 50). This material is completely ionized at all save very acid pH's and is generally used in its Na^+ form (i.e. saturated with Na^+ mobile cations). Since all amino acids will be positively charged in a solution buffered at pH 2 to pH 3, when such a solution is loaded onto a Dowex 50 (Na^+) column, the amino acid cations will exchange with Na^+ and will be bound to the resin. The amino acids may now be individually eluted from the column by applying buffer solutions of gradually increasing pH. As the pH increases, so will each amino acid progressively lose its net positive charge at a rate determined by the strengths of its distinctive acidic groups. As it loses its net positive charge, so too the amino acid becomes progressively less firmly bound to the negatively charged resin and will be washed down (and eventually out of) the column. Amongst the first amino acids to be eluted in this way from a strongly acidic resin, would be the most acidic amino acids, i.e. aspartic and glutamic acids (possibly at pH 3·5 to 4, depending on the ionic strength, etc.). The basic amino acids will retain a net positive charge in media of much higher pH, so that lysine and arginine would be the last to emerge in an eluate of perhaps pH 10 to 11.

Although one can in general predict the behaviour of an ion-exchange material and its mobile ions at various pH's and ionic strengths, complete separation of all the components of a protein hydrolysate on a single column of ion-exchange material, relies greatly on somewhat empirical observations; e.g. whether it is preferable at a certain stage in the elution to raise the pH, or to increase the ionic strength, or both, or to do neither but instead raise the temperature. Yet skilled use of ion-exchange methods requires above all else an understanding of the pH-dependence of the fixed and mobile charges that are involved.

EXAMPLE

Suggest how you might separate the components of the following pairs of amino acids by ion exchange chromatography:

(*a*) *aspartic acid and glycine;*
(*b*) *aspartic acid and glutamic acid;*
(*c*) *lysine and arginine.*

(a) *Aspartic acid and glycine*

To separate these amino acids, we can exploit the fact that in buffer of pH 6, glycine will bear no net charge ($^+H_3N.CH_2.COO^-$) whilst aspartic acid will be negatively charged ($^-OOC.CH_2.CH.(NH_3^+).COO^-$). We might use an anion-exchange material which is completely ionized (positive fixed charges) at pH 6. A suitable resin would be the strongly basic anion exchanger Dowex 1 in its chloride form.

A column of Dowex 1 (Cl^-) is equilibrated with a suitable buffer of low ionic strength and pH 6, and the amino acid mixture in solution in this buffer is loaded onto the column. The aspartate anions will be bound to the Dowex 1 by anion exchange, i.e.

$$\int\!-N^+(CH_3)_3\text{---}Cl^- + asp^- \rightleftharpoons \int\!-N^+(CH_3)_3\text{---}asp^- + Cl^-$$

The glycine, of nil net charge, will not be bound, and will appear in the effluent and in the first 'washings' obtained when the loading buffer is used to wash the column. The aspartate can then be eluted with a buffer of greater ionic strength, and/or lower pH (when the ionization of its β-carboxyl group is repressed by association with H^+).

(b) *Aspartic and glutamic acids*

Both of these amino acids are negatively charged at pH 6 and will be bound at this pH to Dowex 1 anion-exchange resin as described in (a). However, the γ-carboxyl group of glutamate (pK_a 4·1) is a little less strongly acidic than the β-carboxyl group of aspartate (pK_a 3·9), and glutamate will lose its net charge in advance of aspartate as the pH of the eluting buffer is gradually lowered. Together with the somewhat larger size of the hydrated glutamate ion, this means that glutamate can be completely eluted from the column in advance of aspartate if a long enough column of the ion-exchange material is used, together with a slow flow rate of eluting buffer, and a 'gentle' gradient of decreasing pH.

(c) *Lysine and arginine*

To separate these amino acids, we can exploit the fact that the ε-NH_3^+ group of lysine (pK_a 10.8) is more strongly acidic than the guanidino $=NH_2^+$ group of arginine (pK_a 12·5), and that therefore the isoionic point of lysine (pH 10) is lower than that of arginine (pH 10·74). In buffer at pH 7, both amino acids will be positively charged, and will be bound by a cation exchanger which is fully ionized (−ve fixed charges) at this pH. A suitable material would be the strongly acidic, cation-exchange resin Dowex 50, which could be used in its Na^+ form.

Thus, a column of Dowex 50 (Na^+) is equilibrated with a low ionic strength buffer of pH 7 and the amino acid mixture loaded in solution in this same buffer. Both amino acids will be bound by cation exchange.

$$\int\!\begin{matrix}-SO_3^- \;\ldots\; Na^+ \\ -SO_3^- \;\ldots\; Na^+\end{matrix} + \begin{matrix}lys^+ \\ arg^+\end{matrix} \rightleftharpoons \int\!\begin{matrix}-SO_3^- \;\ldots\; lys^+ \\ -SO_3^- \;\ldots\; arg^+\end{matrix} + 2Na^+$$

After washing the column with the loading buffer, the amino acids will be differentially eluted when the pH and ionic strength of the eluting buffers are gradually increased. As the pH rises, lysine will lose its net positive charge sufficiently ahead of arginine to ensure that with care, and the use of an adequately long column, the lysine could be fully eluted before any arginine appeared in the eluant.

Hybridization of dissociations

When we considered the titration curve of glutamic acid (Fig. 6.2), we attributed its apparent pK_a values of 2.1 and 4.1 to the dissociation of the α- and γ-carboxyl groups respectively. We concluded that, as the pH was raised, the glutamate cation would dissociate in the following steps, the α-carboxyl group ionizing before the γ-carboxyl group,

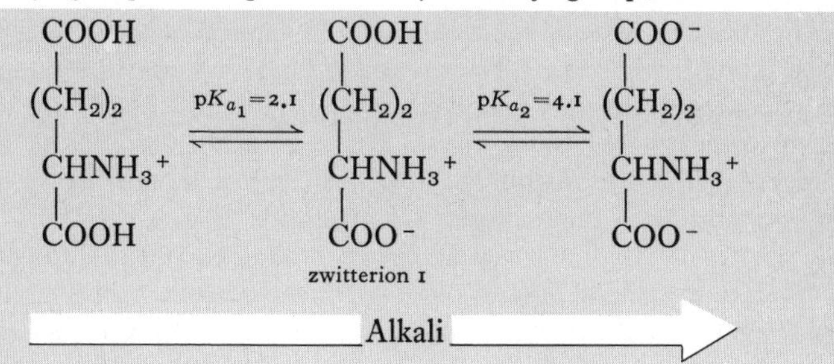

Yet it is conceivable that in a small proportion of the cations the γ-carboxyl group ionizes before the α-carboxyl group to produce a different zwitterion, as shown below,

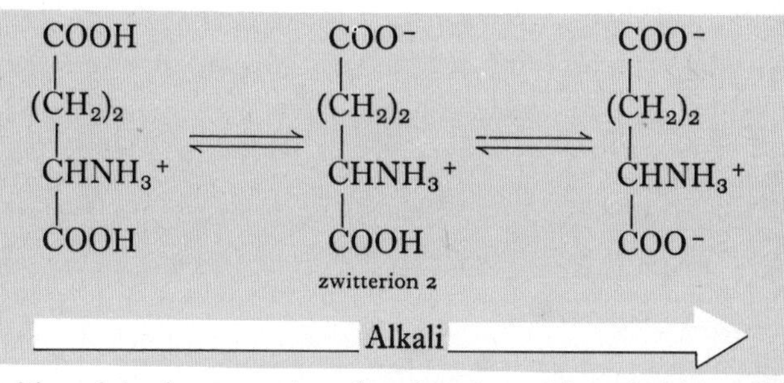

The acid and basic strengths of zwitterions (1) and (2) would not be identical, and the pK_a values for the reverse sequence of dissociation would not be the same as the pK_a values for the identical groups dissociating in their normal sequence. Fortunately, the α-carboxyl group of glutamic acid is so much more strongly acidic than the γ-carboxyl group, that the reverse order of dissociation will be very rare, and zwitterion (2) will be produced in negligibly small concentration. The phenomenon would be much more prominent in the case of a compound possessing two groups of more nearly equal pK_a values, and would be still more pronounced if more than two such groups were involved. The phenomenon of several groups dissociating simultaneously by alternative sequences is called 'hybridization of dissociation'. In this situation we no longer obtain a clear picture of one predominant dissociation sequence from the titration curve. Any apparent pK_a values read from the curve will now be *hybrid values*, containing contributions from several dissociation events, and cannot be attributed to the dissociation of any specific groups.

THE pH-DEPENDENT IONIZATION OF PROTEINS

The multidissociable polypeptide chains of large molecular weight that make up a protein molecule are composed of uniquely ordered sequences of amino acids linked together by peptide bonds. As these bonds are formed by the condensation of the α-amino group of one amino acid with the α-carboxyl group of its neighbour, the majority of these groups will no longer be ionizable in the completed polypeptide (Fig. 6.4).

Only the terminal, free α-amino group at one end of the polypeptide chain, and the terminal, α-carboxyl group at the other end, remain as potential sources of the ionized α-NH_3^+ and α-COO^- groups that were so characteristic of the component amino acids. The very large number of dissociable groups that are actually possessed by a polypeptide must mainly be those carried by the protruding side chains.

The final three-dimensional protein molecule results from the very specific coiling and aggregation of its polypeptide chains, and it can be very large, with a molecular weight of from ten thousand to several millions.

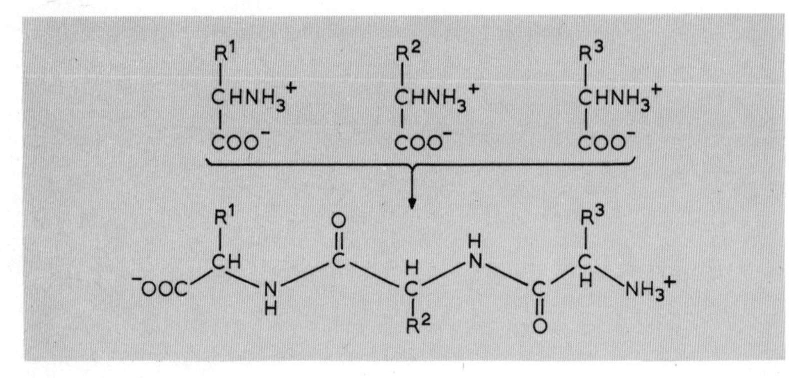

Fig. 6.4 How charged α-NH_3^+ and α-COO^- groups are 'lost' in the formation of a polypeptide.

Although other types of bond are involved in maintaining the stability of the final structure, an important contribution is made by electrostatic bonding between ionized groups of opposite charge. As the number of these is determined by the pH of the medium, the stability of all proteins is likely to be affected by pH. Furthermore, it is conceivable that in the three-dimensional protein molecule the polypeptide chains may be folded in such a manner that amino acids with polar side chains are predominant in one part of the molecule, while those with non-polar side chains are congregated in another area. This is the case in the molecule of myoglobin, wherein those amino acids with ionizable, or other polar side chains (e.g. glutamate, lysine, serine) are virtually entirely excluded from its centre.

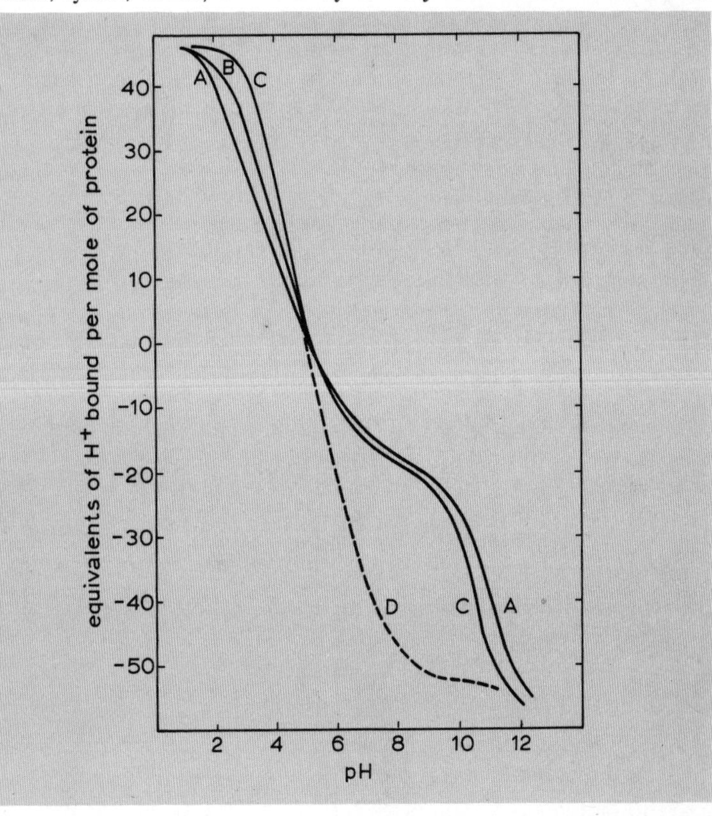

Fig. 6.5 Dissociation curves of β-lactoglobulin. Curve A, 0.019 mol dm^{-3} potassium chloride, 0.5% protein; curve B, 0.135 mol dm^{-3} potassium chloride, 0.5% protein; curve C, 0.67 mol dm^{-3} potassium chloride, 0.5% protein; curve D, 1 mol dm^{-3} formaldehyde, 2% protein. (From R. Keith Cannan, 1942, *Chemical Reviews*, **30**, 401.)

This means that a hydrophobic micro-environment is created within the molecule by the central congregation of those amino acids with non-ionizable, non-polar side chains (e.g. valine, tryptophan). The surface of the molecule will, on the other hand, be hydrophilic since it is here that the ionizable and polar side chains are located. As we find out more about the three-dimensional structure of protein molecules and its relation to their special properties and functions, it is likely that the importance of the congregation of polar and of non-polar side chains to create hydrophilic and hydrophobic areas will be emphasized. Indeed it might be one determinant of the manner in which the constituent polypeptide chains must fold to give the stable, biologically functional, three-dimensional structure.

The multidissociable character of a protein is reflected in its titration curve (Fig. 6.5), which demonstrates that a protein has great pH-buffering ability, especially in the pH ranges of 3 to 5, and 9 to 12. Since so many dissociable groups contribute to the shape of this curve (β-lactoglobulin of Fig. 6.5 contains about 110 dissociable groups per molecule), it is difficult to single out any segment of the curve as being characteristic of the dissociation of one type of group. The situation is further complicated by the fact that the actual pK_a value of any individual group on the protein molecule is likely to be affected to some extent by its proximity to other dissociable groups in the three-dimensional structure and by the hybridization that occurs. Because of this, special measures are taken to identify the contribution made to a protein's titration curve by various types of dissociable group. Thus one can examine the change in shape of the titration curve when the dissociation of one class of group has been specifically repressed (e.g. carboxyl groups in alcohol), or when the acid strength of others is selectively enhanced (e.g. $—NH_3^+$ groups in the presence of formaldehyde). In this way it has been possible to show that the quantitatively most important sources of the charges borne by proteins around neutral pH are the following groups, which ionize within the pH range indicated:

(a) Acidic groups that dissociate to create *negatively* charged sites on the protein molecule

β-carboxyl of aspartic acid } pH 4 to 4.8
γ-carboxyl of glutamic acid }

(b) Basic groups that associate with protons to create *positively* charged sites on the protein molecule

imidazolyl of histidine	pH 6.4 to 7.4
ε-amino of lysine	pH 9.5 to 10.5
guanidyl of arginine	pH > 12

A quantitatively smaller contribution is made by the chain-terminating α-carboxyl groups (pH 3.3 to 3.8) and α-amino groups (pH 7.5 to 8). The amide groups of glutamine and asparagine are extremely weakly basic and will not be ionized at usual pH's, whilst the weakly acidic sulphydryl group of cysteine and the hydroxyl group of tyrosine (both pH 9 to 10), will only ionize in moderately alkaline solution.

The isoionic and isoelectric points of a protein

The *isoionic point* of a protein can be defined as that pH at which the protein molecule bears equal numbers of negative charges formed by the dissociation of its acidic groups and of positive charges resulting from the association of its basic groups with protons. In other words it is the pH at which the protein molecule carries no net charge because of the complete self-neutralization of its 'fixed' positive charges by its own 'fixed' negative charges.

The *isoelectric point* is also a pH at which the protein molecule bears no net charge, for it is that pH at which the protein is electrophoretically immobile. Why is it then that this isoelectric point is generally not identical with the isoionic point of the same protein? The answer lies in the fact that, to determine the electrophoretic mobility of a protein, it is dissolved in a buffered medium containing small molecular weight cations and anions. The large multi-ionized protein molecule tends to bind these buffer ions after the manner of an ion-exchange material, and the equal balance of charges on its molecule at the isoelectric point could in part be due to there being more 'mobile' anions than cations bound at this pH (or vice versa); this could mask an actual imbalance of fixed charges on the molecule. Thus the isoelectric point of a protein is not constant, but is likely to depend on the nature and concentration of the ions present in the medium in which its electrophoretic mobility is tested. In turn, this means that the isoionic point of a protein can only be experimentally determined by measuring the pH of a solution of the protein in salt-free water, which in practice requires that the protein solution be exhaustively dialysed against 'de-ionized' water. But even at its isoionic point, thanks to extensive hybridization, a protein solution will not be homogeneous. For the relatively small molecule of insulin (51 component amino acids, less than a quarter of which carry ionizable side chains) it has been calculated that there are over 8000 possible structures of nil net charge that can coexist at the isoionic point.

Practical importance of the isoelectric point of a protein

As proteins are generally handled in buffered media, we are usually interested in their isoelectric points (pI). Fairly reliable values have been obtained for many proteins in media of comparatively low ionic strength and it has been discovered that the pI values of most proteins are generally less than 7, though they cover a very wide range, from 1.1 for pepsin to 12 for some protamines. The pI value of a protein is an indication of its net acidic or basic character which is also reflected in the net charge that it bears in solution at pH 7. At this pH, positive charges are contributed by the side chains of arginine and lysine and by approximately half of its histidine residues (since the apparent pK_a of the imidazole ring of histidine in protein is close to 7). Also at pH 7, negative charges are contributed by the side chain carboxyl groups of aspartate and glutamate, while although the chain-terminating α-amino and α-carboxyl groups will be fully ionized their charges will usually 'balance out'. Thus an 'acidic' protein with a low pI will be negatively charged at pH 7, whilst a 'basic' protein with a high pI will be positively charged at neutral pH. In fact, knowing the amino acid composition of a protein molecule, you can make an intelligent guess at its probable isoelectric point (though it is much easier to determine the pI experimentally). Calf thymus histone possesses 11 lysine, 15 arginine and 2 histidine residues in its molecule but only 1 aspartate and 1 glutamate residue. Thus it will be positively charged at pH 7 with each molecule bearing 27 positive charges and only 2 negative charges derived from these ionized side chains. One can predict from this that the histone is a 'basic' protein with a high pI value somewhere near the pK_{a_3} of arginine, i.e. 10 to 12.

The interaction between the basic proteins of a cell's nucleus and its nucleic acid (DNA) to form nucleoprotein complexes is well known, and might prove to be of importance in 'masking' certain portions of the DNA of higher organisms so as to control their genetic expression. It is, however, less widely recognized that proteins with greatly different pI values can also interact. Thus insulin (pI about 5) can bind with a protamine (pI about 12) to create a protamine insulinate that would bear a much smaller net charge at pH 7 than either of its components. Indeed the high pI value of protamine makes it very useful as a precipitant at near neutral pH's of acidic compounds of high molecular weight, such as nucleic acids, and proteins of low pI.

In general, the isoelectric point of a protein is also the pH at which it is least soluble and most easily precipitated. This is probably due to mutual attraction between oppositely charged groups of neighbouring molecules, for the number of opposing charges will be greatest at the isoelectric point. Factors other than the pH also affect a protein's solubility, e.g. the hydrophilic character of its ionized groups, the ionic strength of the medium. Even so, it is frequently possible to precipitate, and so purify, a protein by suitably adjusting the ionic strength of its solution at a pH somewhat removed from the isoelectric point, and then gradually altering the pH until it is identical with the pI, causing precipitation of the required protein but not of others with different pI values.

We can summarize such ionic behaviour of a protein by stating that, if it is in solution at a pH higher than its isoelectric point in that medium, it will carry a net negative charge, while at pH's lower than its pI value it will be positively charged; i.e.

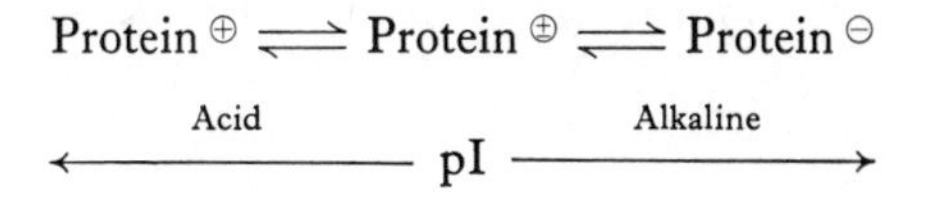

The ability to vary the charge carried by a protein molecule by changing the pH of its solution can be exploited to purify and characterize proteins by electrophoresis and ion-exchange chromatography.

Electrophoresis of proteins

The same principles are involved in the electrophoresis of proteins as in the electrophoretic separation of smaller molecules. By careful adjustment of the pH, ionic strength, and temperature of the electrophoresis medium, it has proved possible to separate proteins that differ only slightly in isoionic points and molecular size. The electrophoresis can be performed in rigid media in which the buffer solution is supported by a polymer matrix, e.g. starch and polyacrylamide gels, or cellulose acetate strips. Using these media, it is usual to complete the electrophoresis and then identify the separated bands either by staining, or by sectioning the medium, eluting the sections with buffer, and assaying each eluate for its content of protein or characteristic biological activity. However, it is also possible to follow the migration of 'bands' of protein during electrophoresis in unsupported buffered media, by making use of the refractivity of their boundaries or their characteristic ultraviolet absorbancy. In this form of electrophoresis, which is usually performed in a vertical vessel, among the factors other than the size of the current and net charge on the molecules that will affect their rate of migration will be the size and shape of the hydrated molecules and the viscosity of the medium. Tiselius (1937) demonstrated with this 'moving boundary' method that it was possible to derive a value for the molecular weight of a protein from its observed electrophoretic mobility if some idealized molecular shape was assumed. Yet, so many 'imponderables' are involved in the calculation that this has been superseded by more reliable methods for the determination of protein molecular weights (e.g. from their sedimentation velocities during ultracentrifugation).

Isoelectric focusing of proteins

Whereas in conventional electrophoresis the protein molecules retain their initial charge during their passage through a medium of constant pH, in the technique known as *isoelectric focusing* they progressively lose their charge as they migrate along a stable pH gradient, eventually coming to rest (and being concentrated) in that pH zone in which they have nil net charge (i.e. when the pH equals their isoelectric point). In practice, the stable pH gradient is produced and maintained by the electrolysis of an aqueous solution of suitably low molecular weight ampholytes (called carrier ampholytes), the pH within the vessel increasing progressively from anode to cathode. If a protein is then introduced at a pH lower than its pI, its molecules will carry a net positive charge, and as they migrate towards the cathode will move through zones of increasing pH. The positive charge on the molecules will in consequence progressively diminish until eventually (when pH = pI) the molecules will bear nil net charge and will cease their migration. By this means, separation of the components of a mixture of proteins can be achieved, since each will accumulate at that pH which is its isoelectric point. To counteract convective remixing of the separated components a suitable 'density gradient' (e.g. of glycerol or polyglycan) can be concurrently employed to aid stabilization of the focused bands of protein. In this way isoelectric focusing can be exploited as a useful analytical and preparative separation technique which has the additional advantage of allowing the pI values of the separated proteins to be quite accurately determined.

Ion exchange chromatography of proteins

The most useful ion-exchange materials for protein chromatography have proved to be weak cationic and 'intermediate strength' anionic exchangers whose fixed charges are carried on a polysaccharide lattice of cellulose or dextran. Suitably charged proteins are relatively weakly bound to these materials and can be eluted by employing buffers of slightly altered pH, or increased ionic strength. Again, no new principle is involved, though to maintain the native structure and biological activities of the proteins the chromatography is usually performed at low temperatures (0°–4°C) and no use is made of highly acid or alkaline buffers. The technique can be illustrated by considering the use of the anionic-exchange material DEAE-cellulose, and the weak cationic exchanger CM-cellulose (see p. 157).

(a) *DEAE-cellulose*—an intermediate strength anion exchanger

This material is very suitable for the chromatography of proteins with pI values less than 7. The DEAE-cellulose is pre-equilibrated with a buffer of comparatively low ionic strength and of pH about 8, whilst the protein is applied in solution in the same buffer. At this pH, DEAE-cellulose is appreciably ionized and bears positive fixed charges, while the protein, being 'on the alkaline side' of its pI, will bear a net negative charge and will be bound by anion exchange. Lowering the pH will decrease the net negative charge on the protein and it may be eluted by the use of buffers of decreasing pH. Alternatively, elution may be accomplished at the loading pH using buffers of increased ionic strength, or buffers containing a competing di- or trivalent anion, i.e.

$$\underbrace{\int\!\!-\text{DEAE}^{\oplus}+\text{Pr}^{\ominus}}_{\text{pH } 8} \xrightarrow{\text{'Bound'}} \int\!\!-\text{DEAE}^{\oplus}\text{---Pr}^{\ominus} \xrightarrow[\substack{\text{(a) decreasing pH}\\ \text{(b) increasing ionic strength}\\ \text{(c) competing anions}}]{\text{Eluted by}} \text{Protein}$$

(b) *CM-cellulose*—a weak cation exchanger

This material is generally most suitable for the chromatography of proteins with pI values greater than 7. The CM-cellulose is pre-equilibrated with a buffer of comparatively low ionic strength at a pH of about 5, and the protein is loaded onto the column in solution in the same buffer. At this pH the CM-cellulose bears negative fixed charges whilst the protein, being 'on the acid side' of its pI, will carry a net positive charge and will be bound by cation exchange. Increasing the pH will decrease the net positive charge on the protein and it may be eluted using buffers of increasing pH. Alternatively, it might be eluted at the loading pH by employing buffers of gradually increased ionic strength, or buffers to which a competing di- or trivalent cation has been added, i.e.

$$\underbrace{\int\!\!-\text{CM}^{\ominus}+\text{Pr}^{\oplus}}_{\text{pH } 5} \xrightarrow{\text{'Bound'}} \int\!\!-\text{CM}^{\ominus}\text{---Pr}^{\oplus} \xrightarrow[\substack{\text{(a) increasing pH}\\ \text{(b) increasing ionic strength}\\ \text{(c) competing cations}}]{\text{Eluted by}} \text{Protein}$$

Proteins with pI values close to 7 that are stable at moderately acid or alkaline pH's could be chromatographed on either anionic or cationic exchangers. More often, the choice of ion-exchange material is then dictated by the instability of the protein either (a) in mildly acid solution, when DEAE-cellulose would be used, or (b) in mildly alkaline solution, when CM-cellulose would be employed.

pH-dependence of the distribution of ions across a protein-restraining semi-permeable membrane

Any protein in solution at a pH other than its isoelectric point will carry a net electrical charge. In order to maintain electrical neutrality, such a solution of protein ions will also contain an equivalent concentration of ions of opposite charge. Thus at a pH higher than the pI of a protein, its negative charge may be nullified by an equivalent concentration of Na^+ ions and its solution will behave as if it were a solution of a protein salt, i.e. sodium proteinate. An interesting situation arises if such a solution of sodium proteinate (concentration C_1) is separated from a solution of sodium chloride (concentration C_2) by a membrane freely permeable to the small ions Na^+ and Cl^-, but completely impermeable to the large protein ion Pr^-.

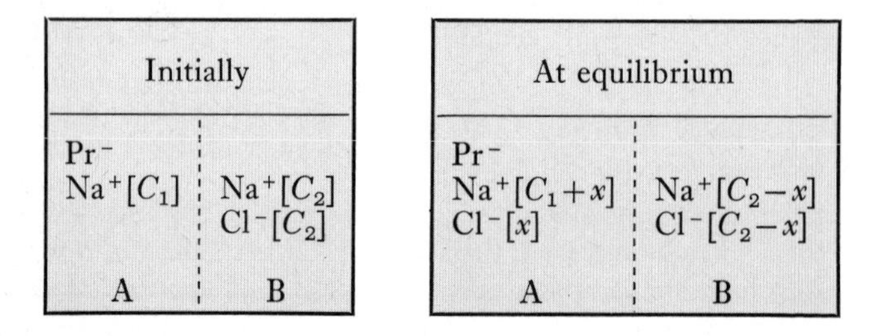

The restriction of the protein anions to compartment A, together with the need to maintain electrical neutrality in each compartment, ensures that equilibrium will be attained with this compartment containing a lower concentration of the diffusible anions (Cl^-) than remains in compartment B. The resultant imbalance between the concentrations of 'mobile' ions maintained at equilibrium in the two compartments is known as the ***Gibbs–Donnan effect.*** The extent of the imbalance can be calculated if we assume the two compartments to be of equal volume. In the above example, at equilibrium,

$$\frac{[Cl^-]_B}{[Cl^-]_A} = 1 + \frac{C_1}{C_2}$$

Thus the greater the concentration of protein anions the greater will be the discrepancy at equilibrium between the chloride (and NaCl) concentrations in the two compartments.

The pH may be involved in such events in two main ways:

1 The prevailing pH, by determining the net charge on the protein molecules, determines its effective equivalent concentration as assessed by the concentration of monovalent ions of opposite charge required for its electrical neutralization. At the isoelectric point of the protein, the Gibbs–Donnan effect will be minimal (see also the determination of the osmotic pressure of an aqueous solution of protein, p. 76.)
2 In a situation where the mobile cations are H^+ ions, then the Gibbs–Donnan effect would result in the establishment of a pH difference between the two compartments at equilibrium.

It is very difficult to assess the importance of the Gibbs–Donnan effect in any biological situation, for by expending energy a living cell can maintain an otherwise implausible distribution of ions between itself and its environment. This active transport and retention of ions against normal concentration gradients can very easily obscure the distribution that would have been achieved in the absence of these phenomena. One example of unequal ion distribution between the contents of a cell and its surroundings, that can be fairly attributed to the Gibbs–Donnan effect, is the lower concentration of chloride ions in erythrocytes (red blood cells) than in the plasma in which they are bathed. The erythrocyte membrane is freely permeable to chloride ions, yet the concentration of chloride within these cells is only 70% of that which is present in the external blood plasma. It is very likely that this is due to the much higher concentration of protein anions retained within the erythrocyte, chief amongst which would be the haemoglobin that accounts for a third of its dry weight. The fact that, at least in stationary phase cultures of the bacterium *Escherichia coli*, the intracellular Cl^- ion concentration is linearly related to the extracellular Cl^- ion concentration, suggests that the passive distribution of chloride ions between this organism and its culture medium may similarly be attributable to the operation of the Gibbs–Donnan effect.

pH and the biological activities of proteins

Since electrostatic bonds contribute to the maintenance of the 'native' three-dimensional structure of every protein molecule, it is not surprising that a change in pH which alters the proportions of positive and negative charges on the molecule can lead to unfolding and disorientation of the component polypeptide chains. Minor conformational changes caused by small pH changes may not markedly affect the physical properties of the protein, and restitution of a favourable pH may be followed by re-establishment of the usual protein structure. More drastic pH changes may cause more profound and irreversible structural changes. If the pH were sufficiently lowered, the negatively charged COO^- groups would be neutralized, leaving positively charged $—NH_3^+$ groups free to exercise electrostatic repulsion, thus aggravating distortion and unfolding of the molecule. Were the pH to be sufficiently raised, the COO^- groups, in the absence now of nullifying $—NH_3^+$ groups, might exert this same disruptive tendency. Thus a drastic change in pH can so alter the conformation of a protein molecule that it undergoes irreversible precipitation and coagulation (i.e. denaturation). Yet the effects of pH changes on proteins cannot be attributed solely to alterations in the number and distribution of their intramolecular electrostatic (ionic) bonds, for at least three other types of bond are of major importance in the maintenance of the secondary and tertiary structures of proteins, viz: apolar (hydrophobic) bonds, hydrogen bonds between peptide groups, and hydrogen bonds between side-chain groups. Since the pH also affects the extent of hydrogen bonding, some of the pH-induced denaturation of proteins can be ascribed to the 'weakening' of these structure-stabilizing hydrogen bonds.

A few proteins are unusual in maintaining their structure at extremely high or low pH's and it is quite often found that they also demonstrate optimal biological activity at these pH's; for example, the optimal pH for the enzymic activity of pepsin is about pH 1 to 2. Biological activity is generally an attribute of one specific conformation of the protein molecule so that pH changes that drastically affect the physical characteristics of a protein are likely to have at least an equivalently profound effect on its biological activity. Frequently, the biological activity of a protein is even more sensitive to pH changes than would be expected from physical measurements of the distortion of the molecule that these cause. Indeed, this is one of the many pieces of evidence which together suggest that the biological activity of a protein may primarily be the function of only part of its molecule, which can be called its biologically active site. Should a slight change in pH affect the ionization of a single dissociable group at this active site, then it could greatly change the biological activity without significantly altering the overall structure of the protein molecule. Examination of the effect of pH changes on the biological activity of a protein may therefore provide evidence of its possession of an active site, and might even suggest what dissociable groups this contains.

Proteins as pH buffers

All proteins, irrespective of their specialist function, must contribute to the general buffer capacity of the cellular contents by virtue of their high content of weakly acidic and basic groups. Haemoglobin provides a spectacular example of a protein of specialist function being called upon to undertake the role of an efficient pH buffer in an unusual manner.

The occurrence of O_2-consuming and CO_2-releasing cellular respiration in tissues of the body far distant from the lungs, requires that O_2 be transported in the arterial blood supplied to these tissues and that CO_2 be carried from the tissues to the lungs in the venous blood. In man, arterial O_2 transport is accomplished by the combination of haemoglobin with O_2 at the lungs to form oxyhaemoglobin. Arriving at the respiring tissue, the oxyhaemoglobin delivers up its O_2 and reverts to haemoglobin. At first sight the problem of CO_2 transport appears less formidable, since CO_2 is much more soluble in aqueous media than is O_2 and erythrocytes are known to contain an enzyme (carbonic anhydrase) which promotes the rapid reaction of CO_2 with water to form carbonic acid. At the pH of blood (about pH 7.4), carbonic acid (pK_{a_1} 6.1), will be 96% dissociated into H^+ plus bicarbonate ions, i.e.

$$CO_2(g) \longleftrightarrow CO_2 + H_2O \rightleftharpoons H_2CO_3 \rightleftharpoons H^+ + HCO_3^-$$

Respiring tissue: ⟶

⟵ Lungs

Thus the carriage of considerable amounts of CO_2 in venous blood would tend to decrease its pH. The scale of the problem becomes evident from the calculation that a quantity of CO_2 equivalent to between 20 and 40 dm^3 of 1 mol dm^{-3} monobasic acid is excreted via the lungs of man in the course of 1 day. The observation that the pH of the CO_2-carrying venous blood is only marginally lower than that of the CO_2-depleted arterial blood argues the presence in the blood of concentrations of bases (supplied by 'blood buffers') sufficient to associate with all the H^+ ions which are formed in equivalent concentration to the HCO_3^- ions.

Although a portion of the buffer base required for CO_2 transport at pH 7.4 is supplied by plasma phosphates and plasma proteins (as HPO_4^{2-} and Pr^-), more than three-quarters is provided by haemoglobin.

We have noted that a portion of the haemoglobin that arrives at the respiring tissue is in its oxygenated form ($HHbO_2$) and that it there yields up its O_2 and so re-forms haemoglobin (HHb). To represent these complex, multidissociated protein molecules as the model, monoprotic weak acids $HHbO_2$ and HHb is a gross simplification, but one that serves to emphasize the fact that at the pH of the blood they will be dissociated to different extents, since it is known that oxyhaemoglobin is a stronger acid than haemoglobin. This is reflected in their isoelectric points (pI oxyhaemoglobin = 6.7; pI haemoglobin = 6.81) but is best expressed in terms of the apparent pK_a values of the representative model compounds,

$$HHbO_2 \rightleftharpoons H^+ + HbO_2^-, \quad pK_a = 6.62$$

$$HHb \rightleftharpoons H^+ + Hb^-, \quad pK_a = 8.18$$

From these values we can calculate that, at the normal pH of blood (pH 7.4), only 14% of oxyhaemoglobin will be in its undissociated state ($HHbO_2$), but 85% of haemoglobin will be present in this condition (HHb). Therefore, when at pH 7.4 oxyhaemoglobin loses its O_2 and becomes converted into haemoglobin, a quantity of H^+ ions must be 'taken up'.

$$\underset{\text{(predominantly } HbO_2^-)}{\text{Oxyhaemoglobin}} \xrightarrow[\text{At pH 7.4}]{(H^+)} \underset{\text{(predominantly HHb)}}{O_2 + \text{Haemoglobin}}$$

While the O_2-consuming tissue by producing CO_2 causes H^+ ions to be liberated, by simultaneously obtaining its oxygen from oxyhaemoglobin it forms sufficient base (Hb^-) to associate with the majority of these H^+ ions. This briefly is the basis of the phenomenon known as ***isohydric exchange.*** You will notice that this differs from normal pH buffering which relies on the buffering capacity of a single conjugate pair (of one pK_a value).

Instead, in isohydric exchange, one buffer ($HHbO_2$—HbO_2^-) is converted, at the site of H^+ 'loading', into another buffer of higher pK_a (HHb—Hb^-).

EXAMPLE

What amount of CO_2 can be transported in blood as carbonic acid and bicarbonate ions without pH change at pH 7.4, as a consequence of the deoxygenation of 1 mole of oxyhaemoglobin? (Apparent pK_a values: oxyhaemoglobin = 6.62; haemoglobin = 8.18; carbonic acid = 6.1.)

The 'solution' of CO_2 in blood at pH 7.4 proceeds by the following reaction sequence:

$$CO_2 + H_2O \longrightarrow H_2CO_3 \longrightarrow H^+ + HCO_3^-$$

Since H^+ and HCO_3^- ions are formed in equivalent amounts, the quantity of CO_2 that can be carried at pH 7.4 as a consequence of the deoxygenation of 1 mole of oxyhaemoglobin will be determined by the quantity of H^+ ions that will be 'taken up' by this process. This will equal the difference between the quantities of H^+ 'bound' in 1 mole of oxyhaemoglobin and in 1 mole of haemoglobin at this pH. Throughout this calculation we will assume that haemoglobin and its oxygenated derivative can be represented by the model compounds HHb and $HHbO_2$ respectively.

(a) *H^+ 'bound' in 1 mole of oxyhaemoglobin at pH 7.4*

Assuming that $\underset{\text{(acid)}}{HHbO_2} \rightleftharpoons H^+ + \underset{\text{(conjugate base)}}{HbO_2^-} \quad \ldots \quad pK_a = 6.62$

then since, according to the Henderson–Hasselbalch equation,

$pH = pK_a + \log \dfrac{[\text{conjugate base}]}{[\text{acid}]}$, for oxyhaemoglobin at pH 7.4,

$$7.4 = 6.62 + \log \frac{[HbO_2^-]}{[HHbO_2]}$$

i.e. in any given volume, $\log \dfrac{HbO_2^-}{HHbO_2} = 0.78$, or $\dfrac{HbO_2^-}{HHbO_2} = \text{antilog } 0.78 = 6.03$

If 1 mol of oxyhaemoglobin contained x mol of $HHbO_2$, the HbO_2^- content $= (1-x)$

but
$$\frac{(1-x)}{x} = 6.03 \quad \text{or} \quad 1-x = 6.03x$$

$\therefore$
$$7.03x = 1 \quad \text{and} \quad x = \frac{1}{7.03} = 0.142$$

$\therefore$ 1 mol of oxyhaemoglobin at pH 7.4 contains 0.142 mol of $HHbO_2$; or, to state it somewhat differently, 1 mol of oxyhaemoglobin at pH 7.4

effectively has 'bound' to it 0.142 mol of H^+ ions (which have associated with HbO_2^- to produce $HHbO_2$).

(b) *H^+ 'bound' by 1 mole of haemoglobin at pH 7.4*

Assuming: $$\underset{\text{(acid)}}{HHb} \rightleftharpoons H^+ + \underset{\text{(conjugate base)}}{Hb^-} \quad \ldots \quad pK_a = 8.18$$

Then at pH 7.4, $$7.4 = 8.18 + \log \frac{[Hb^-]}{[HHb]}$$

and in any given volume,

$$\log \frac{Hb^-}{HHb} = -0.78 = \bar{1}.22, \text{ and } \frac{Hb^-}{HHb} = \text{antilog } \bar{1}.22 = 0.17$$

If 1 mol of haemoglobin contains x mol of HHb, then it contains $(1-x)$ mol of Hb^- and,

$$\frac{(1-x)}{x} = 0.17 \text{ whence } 1.17x = 1 \text{ and } x = \frac{1}{1.17} = 0.85$$

∴ 1 mol of haemoglobin at pH 7.4 contains 0.85 mol of HHb, i.e. it effectively has 'bound' to it 0.850 mol of H^+ (associated with Hb^- to form HHb).

Thus the quantity of H^+ ions that can be taken up at pH 7.4 as a consequence of the loss of O_2 from 1 mol of oxyhaemoglobin = (Quantity of H^+ bound by 1 mol of haemoglobin) minus (Quantity of H^+ bound by 1 mol of oxyhaemoglobin) = (0.850 − 0.142) mol = 0.708 mol.

By extrapolation, this means that 0.708 mol of bicarbonate ions can be carried by blood at pH 7.4 without any change of pH as a consequence of the H^+ uptake which accompanies the deoxygenation of 1 mol of oxygenated haemoglobin. In turn, this quantity of bicarbonate represents the uptake of 0.708 mol of CO_2.

(c) *CO_2 'carried' as H_2CO_3*

At pH 7.4 bicarbonate ions will be in equilibrium with H_2CO_3,

$$\underset{\text{(acid)}}{H_2CO_3} \rightleftharpoons H^+ + \underset{\text{(conjugate base)}}{HCO_3^-} \quad \ldots \quad pK_a = 6.1$$

Thus an extra load of CO_2 will be carried as undissociated H_2CO_3 as a consequence of O_2 loss from 1 mol of oxyhaemoglobin. Its quantity can be calculated by applying the Henderson–Hasselbalch equation to the dissociation of H_2CO_3, when at pH 7.4,

$$7.4 = 6.1 + \log \frac{[HCO_3^-]}{[H_2CO_3]}$$

and in a given volume,

$$\log \frac{HCO_3^-}{H_2CO_3} = 7{\cdot}4 - 6{\cdot}1 = 1{\cdot}3, \text{ and } \frac{HCO_3^-}{H_2CO_3} = \text{antilog } 1{\cdot}3 = 20$$

Thus at pH 7.4, in a volume of blood containing 0.708 mol of HCO_3^-, there must also be present 0.708/20=0.035 mol of H_2CO_3 derived from 0.035 mol CO_2. Therefore, total amount of CO_2 carried at pH 7.4 without pH change as a consequence of the liberation of 1 mol of O_2 from oxyhaemoglobin,

$$= (CO_2 \text{ as } HCO_3^-) + (CO_2 \text{ as } H_2CO_3)$$

$$= 0{\cdot}708 + 0{\cdot}035 \text{ mol}$$

$$= \underline{0{\cdot}743 \text{ mol}}$$

EFFECTS OF pH CHANGE ON NON-PROTEIN PROTOPLASMIC COMPONENTS

The distribution of charges on all multidissociable macromolecules in the living cell, e.g. nucleic acids, phospholipids, mucopolysaccharides, will be affected by the pH; like proteins these compounds could suffer structural changes at 'unfavourable' pH's that would diminish their ability to perform their normal functions. Metabolites of small molecular weight are likely to be similarly affected. In an essay entitled 'The Importance of being Ionised', B. D. Davis[19] remarked that it was surely more than coincidence that virtually all low molecular weight intermediates in biosynthetic pathways carry dissociable groups that are ionized at neutral pH. This could have something to do with the fact that cell membranes are usually less permeable to charged than to uncharged molecules, so that intracellular concentrations of metabolites are more easily maintained when these are ionized. Among the weak acid and basic groups present in these compounds, carboxyl and amino groups are very common. For example, carboxyl groups are found in fatty acids, keto acids and the di- and tricarboxylic acids of terminal respiratory cycles, whilst amino groups are present in amino sugars, nucleotides and amines. The phosphate group too is widely distributed, being present in cellular constituents as diverse as sugar phosphates, nucleic acids, phospholipids and many coenzymes. Indeed, the ubiquity of inorganic phosphate and organic phosphoesters makes them biologically important pH buffers.

The biological effectiveness of metabolites bearing such dissociable groups may be very dependent on their ionic condition within the cell for reasons other than selective impermeability. If a multidissociable compound is biologically active in only one of its several ionized forms (or solely in its undissociated state), then this activity must be affected by the prevailing pH since this dictates what proportion of its molecules will be in this active form. This relationship between pH and biological activity is frequently noted in the interaction of an enzyme with a dissociable substrate (p. 317). More dramatically it is demonstrated in the effect of pH on the efficacy of several dissociable drugs; for example, the cationic forms of the acridines are generally more potent bacteriostatic agents than the uncharged molecules.

pH AND METABOLIC REACTIONS INVOLVING PROTONS

Chemical reactions affected by pH change are chiefly of two types:

(a) reactions catalysed by H^+ ions
(b) reactions in which H^+ ions are reactants.

In metabolic systems, non-enzymic acid catalysis by H^+ ions is perhaps not as important as in non-biological systems (p. 273). However, metabolic reactions of type (b) are very common and the possible effect of pH change on these is best illustrated by a specific example.

Yeast in its fermentation of glucose reduces acetaldehyde to yield ethyl alcohol. The reaction is catalysed by alcohol dehydrogenase and utilizes reduced pyridine nucleotide as hydrogen donor:

$$CH_3CHO + \text{reduced NAD} \rightleftharpoons CH_3CH_2OH + \text{oxidized NAD}$$

At equilibrium, by the Law of Mass Action,

$$\text{Equilibrium constant} = \frac{[CH_3CH_2OH][NAD_{ox}]}{[CH_3CHO][NAD_{red}]}$$

But the equilibrium constant calculated from these equilibrium concentrations (K'_c) varies with the pH, being 10^4 at pH 7, but 10^2 at pH 9. This is explained by the participation of H^+ ions in this reaction, which should in fact be written as,

$$CH_3CHO + NADH + H^+ \rightleftharpoons CH_3CH_2OH + NAD^+$$

whence

$$\text{real } K_c = \frac{[CH_3CH_2OH][NAD^+]}{[CH_3CHO][NADH][H^+]}$$

Since the concentration of hydrogen ions at equilibrium contributes to the value of K_c, the environmental pH will affect the extent to which acetaldehyde will have been converted into alcohol at equilibrium. The more acid the environment, the more complete will be the reduction of acetaldehyde to alcohol; the more alkaline the medium, the more will oxidation of ethanol be favoured. We will see later that this effect of pH on the value of the *apparent* equilibrium constant of a reaction that involves protons is mirrored in certain other of its characteristics, e.g. the modified standard free energy change, $\Delta G^{\ominus\prime}$ (p. 199), or electrode potential, $\mathbf{E}'_{\ominus}$(p. 345), and that the extent of ionization of acidic and basic reactants (also affected by pH) can determine whether a reaction will proceed spontaneously (p. 225).

In the living cell the situation is complicated by the establishment of an overall dynamic equilibrium in which any one reaction is merely one member of an integrated sequence of metabolic events. In such a situation it is highly unlikely that the steady state concentrations attained by the substrates and products of any individual reaction will be identical with those concentrations achieved by the same reaction proceeding in isolation. Nevertheless, it should be obvious that a change in pH must disturb the overall dynamic equilibrium by virtue of its effect on those reactions that yield or utilize protons, even if the outcome of this disturbance is not entirely predictable in the metabolic situation.

PROBLEMS

1. If the acid dissociation constants of the amino acid serine ($CH_2OH.CHNH_2.COOH$) are 6.2×10^{-3} and 7.1×10^{-10}, which ionic form will predominate in solutions at the following pH's: (a) pH 2, (b) pH 5, (c) pH 7, and (d) pH 11?

To 100 cm³ portions of 0.01 mol dm^{-3} serine at its isoionic point are added (i) 50 cm³ of 0.01 mol dm^{-3} sodium hydroxide, and (ii) 66.7 cm³ of 0.01 mol dm^{-3} sodium hydroxide; what will be the pH of the resulting mixtures? Will these solutions be buffers?

2. The α-amino acid, methionine, has apparent pK_a values of 2.1 and 9.3 in its 0.1 mol dm^{-3} solution at 298 K. What will be the zwitterion concentration in this solution adjusted to (a) pH 3.1, (b) pH 5.7, and (c) 10.3?

3. Calculate the isoionic points of the following amino acids (whose pK_a values are appended):

(a) L-valine (2.29; 9.74)
(b) DL-phenylalanine (2.16; 9.18)
(c) L-ornithine, $CH_2NH_2.(CH_2)_2.CHNH_2.COOH$ (1.7; 8.69; 10.76 δ-NH_3^+)
(d) DL-cystine, $S.CH_2.CHNH_2.COOH$ (< 1; 2.1; 8.02; 8.71)
$\quad\quad\quad\quad\quad\quad\;\; |$
$\quad\quad\quad\quad\quad\quad S.CH_2.CHNH_2.COOH$

4. γ-Aminobutyric acid has an isoionic point of 7.3. A solution of pH 3.24 contains 10% of the compound in its zwitterionic form. Calculate the pK_a values of γ-aminobutyric acid.

5. Ten cm^3 of a solution of histidine at pH 7.5 was titrated with 0.1 mol dm^{-3} hydrochloric acid. After addition of 36 cm^3 of acid, the pH was 6.0. Calculate the concentration of the histidine solution. (pK_a values of histidine are 1.8; 6.0; 9.3.)

6. 1.05 g of an α-amino acid with dissociation constants of 4.6×10^{-3} and 2.5×10^{-10}, when dissolved in 50 cm^3 of 0.2 mol dm^{-3} sodium hydroxide, gave a solution of pH 10. Calculate the molecular weight of the amino acid and suggest its formula.

7. The component amino acids in a mixture of histidine, valine and aspartic acid were separated by high voltage paper electrophoresis in pyridine-acetate buffer pH 5.2. Suggest the probable locations of the amino acid 'bands' on the final electrophoretogram. (pK_a values are: histidine: 1.8, 6.0 (Im.), 9.3; valine, 2.3, 9.7; aspartic acid, 2.0, 3.9 (β-COOH), 9.9.)

8. The protein, salmine, has the following amino acid composition:

Amino acid	Number of amino acid residues per molecule of salmine
Alanine	1
Glycine	3
Valine	2
Isoleucine	1
Proline	4
Serine	7
Arginine	40

Would salmine be anionic or cationic (overall) at the following pH's: (a) pH 2, (b) pH 7, (c) pH 9?

9. The protein, egg albumin, contains 51 aspartic plus glutamic acid, 15 arginine, 20 lysine and 7 histidine residues per molecule. Suggest a reasonable value for its isoelectric point.

Would egg albumin be expected to be electrostatically bound to β-lactoglobulin or ribonuclease at pH 7? (pI values: β-lactoglobulin, 5.1; ribonuclease, 9.5.)

10. How might you use ion-exchange celluloses to purify the following proteins:

(a) β-lactoglobulin, pI = 5.2;
(b) cytochrome c, pI = 10.1?

11. What is the maximum concentration of bicarbonate ions contained in (a) gastric juice, pH 1.1, and (b) pancreatic juice, pH 7.8, if the concentration of carbonic acid in each is 1 mmol dm^{-3}? (Assume pK_a of carbonic acid to be 6.1.)

12. At its normal pH of 7.4, the bulk of the blood plasma's content of inorganic phosphate (1 mmol dm^{-3}), will be in the form of $H_2PO_4^-$ and HPO_4^{2-} ions. Yet, for the calcification of bone, phosphate must be available

in the form of its tertiary anion PO_4^{3-}. What will be the concentration of PO_4^{3-} ions in blood plasma? (Assume apparent pK_a values of phosphoric acid to be 2, 6.8 and 12 at the ionic strength of blood plasma.)

13. When a wide range of narcotics was tested on a lug-worm, the minimum anaesthetic dose of some of these compounds was found to vary with the prevailing pH. A selection of the results obtained is given below.

pH	Minimum anaesthetic dose (mg/100 cm^3) iso-Amyl alcohol	Cocaine	Nembutal
7	100	10	3
8	100	5	6
9	100	2.5	13

Can you explain these findings (qualitatively) in terms of the effective ionic state of the narcotics? (Cocaine is a weak base, pK_b 5·6; nembutal is a weak acid, pK_a about 8.)

14. The following results were obtained when benzoic acid was tested at various pH's as an inhibitor of fungal growth in submerged culture. (Growth of the mould used was not affected by pH changes within the range pH 3 to pH 6, in the absence of added benzoic acid.)

pH	Minimum fungistatic concentration of benzoic acid in mol dm^{-3}
3·5	1.2×10^{-3}
4·0	1.63×10^{-3}
4·5	3.0×10^{-3}
5·0	7.3×10^{-3}
5·5	2.1×10^{-2}
6·0	6.4×10^{-2}

(a) Interpret these results quantitatively in terms of the dissociation of benzoic acid.

(b) If you wished to prevent the growth of moulds in a buffered solution of phosphates at pH 8, would you choose to add benzoic acid as a preservative? Assuming the pH to be maintained at pH 8, what would be the minimum fungistatic concentration of benzoic acid at this pH? (Assume acid dissociation constant of benzoic acid to be 6.32×10^{-5}.)

7

Background Thermodynamics

Although *thermodynamics* is generally defined as 'the study of the relationship of heat to mechanical and other forms of energy', this definition scarcely does justice to the scope and impact of this enquiry. Nor indeed does the definition of *thermochemistry* as 'that branch of thermodynamics that studies the heat changes associated with chemical reactions'. For although thermochemical measurements are measurements of heat changes in a system undergoing chemical change, they must be interpreted in terms of a redistribution of its energy content among the various forms in which energy is present in the system.

By identifying and assaying the energetic changes associated with a reaction, thermodynamics seeks to determine what it is that impels the reaction and what it is that determines its end. Classical (equilibrium) thermodynamics investigates the feasibility and extent of chemical reactions by measuring properties of matter in bulk. This type of thermodynamics is not concerned with the rates and mechanisms of these reactions, which must be studied by kinetic methods (Chapter 10). The interpretation of thermodynamic functions in terms of atomic and molecular properties is the goal of another branch of energetics, viz. *statistical thermodynamics*—a fascinating study, but not one that we can undertake in any depth in this elementary text.

Thermodynamic systems

Thermodynamics studies assemblages of matter and energy that are called 'systems'. Classical thermodynamics examines the alterations in the content of energy and its distribution that take place when a system passes from an initial, defined state into a terminal state at equilibrium.

The state of a system can be defined in terms of its pressure, temperature and composition, and when both the temperature and the pressure of a system are kept constant, energetic changes can be directly related to changes in material composition.

Two of the most important systems which are examined by classical thermodynamics are: (i) the isolated system, and (ii) the closed system. An *isolated* system is completely insulated from its surroundings and is both materially and energetically self-contained. On the other hand, a *closed* system, though it is materially self-contained, is freely able to exchange energy with its surroundings. A biologist's interest in classical thermodynamics is largely confined to its predictions concerning reactions that take place at a constant temperature and pressure. It is for this reason that we shall deal almost exclusively with isothermal, closed systems maintained at constant pressure, i.e. closed system (T and P constant).

Units of energy

To compare the quantities of energy present in different forms in the one system, it is necessary to select a single unit of energy in which all may then be expressed. The SI unit employed in thermodynamic studies is the joule (J); relatively large quantities of energy can be expressed in kilo-joules (1 kJ = 1000 J). Since the energy content of a system is an 'extensive' property† proportional to the 'size' of the system, terms having to do with the energy contents of systems and their components are related to a standard quantity of compound, namely 1 mol. All energy terms employed in this chapter, unless otherwise designated, will have the units J mol^{-1}.

Prior to the adoption of SI, the most usual unit of energy used in thermodynamics was the calorie (1 cal = 4.184 J), and in many biology textbooks this remains the unit in common usage. Values of energy terms reported in cal/mole are converted to values in J mol^{-1} by multiplying by 4.184; conversely, to convert J mol^{-1} to cal/mole, simply divide by 4.184.

CONSERVATION OF ENERGY

The ***1st Law of Thermodynamics*** states that 'the total energy of an isolated system is constant, though within that system it may change its form'; it means, in effect, that energy can neither be created nor be destroyed. Thus change in an isolated system can neither lead to an

† As opposed to an 'intensive' property, e.g. temperature, which is independent of the quantities of material involved.

increase nor a decrease in the internal (intrinsic) energy of the system, it can only redistribute what energy it contains, amongst different forms. Similarly, change in a closed system, though it can involve both redistribution of energy within that system and energy transfer between the closed system and its surroundings, is obviously still subject to the ruling of the 1st Law since a closed system plus its surroundings constitutes an isolated system. Thus whatever reaction takes place in a closed system, the summed intrinsic energies of that system and its surroundings must remain unchanged.

If a closed system (T and P constant) which initially possesses a total internal energy equal to U_1 undergoes some change whereby it attains a different intrinsic energy U_2, the 1st Law assures us that the change in intrinsic energy, $\Delta U = (U_2 - U_1)$, is accounted for by an equal but opposite change in the intrinsic energy content of its surroundings. The energetic interaction between the closed system and its surroundings is accomplished by (a) thermal transfer, and (b) performance of work. If the reaction in a closed system (T and P constant) produces a change in its volume, then by its expansion (ΔV m^{-3}) the system obligatorily performs work on the surroundings equal to $-w_{\text{obligatory}} = -P\Delta V$ J mol^{-1}. [The minus sign is introduced, since thermodynamics conventionally considers energy changes from the standpoint of the system, and work done by the system on its surroundings is energy expended by, and hence lost to, the system.] Various other types of work may optionally be performed by the reaction, so that the total energy expended by the system in the performance of all types of work (both obligatory and optional) may be summed in one term as $-w$ J mol^{-1}. Concurrent thermal transfer (heat exchange) between the system and its surroundings is designated as q J mol^{-1}. Thus in a closed system (T and P constant), any change in intrinsic energy (ΔU_{system} J mol^{-1}) will be the resultant of heat exchanged and work performed; i.e.

$$\Delta U_{\text{system}} = (q - w) \text{ J mol}^{-1}$$

Intrinsic energy

The intrinsic energy (U) of a system is an attribute of the system that depends on its present state and is independent of its previous history; i.e. it does not depend on how the system attained its present state. This means that intrinsic energy is a so-called *function of state*. Thus, if a closed system (T and P constant) originally in state X undergoes a spontaneous change which produces state Y, the change in the intrinsic energy of the system (ΔU) will be the same whatever mechanism was employed to effect the change. This follows from the fact that ΔU represents the difference between the intrinsic energies of the system in state X and in state Y, and however devious the reaction path, each route commences with the system in state X and ends with the system in state Y.

Since ΔU is 'path-independent' and equals $(q-w)$, it follows that $(q-w)$ must also be path-independent, even though the individual values of q and $-w$ will be unique for each route.

Can a reaction be detected and characterized by its associated heat change?

With rare exceptions, the occurrence of a chemical reaction will be associated either with the liberation, or with the absorption of heat. This means that, given sensitive enough methods for following the heat exchange between a closed system and its surroundings, reaction within the system can indeed be detected. This raises the further possibility that quantitative measurement of this heat exchange (by calorimetry) might be employed to determine the extent of the reaction. Yet we have just seen that the quantity of heat exchanged during the course of a reaction between a closed system (T and P constant) and its surroundings, will depend on how much work is concurrently performed by the system. This suggests that before investing the heat of reaction with unwarranted significance (or constancy) we should enquire into its origins. In fact, we must relate the quantity of heat exchanged between an isothermal, closed system and its surroundings, to the overall change in its content of energy, and to its distribution. To do this we must identify three further energetic functions of state of the system and its surroundings, namely, *enthalpy*, *entropy* and *free energy*.

ENTHALPY

We have seen that for a reaction in a closed system (T and P constant),

$$\Delta U = (q_p - w) \quad \text{where} \begin{cases} q_p = \text{heat acquired by the system from the surroundings (at constant pressure)} \\ -w = \text{work done by the system on its surroundings} \\ \quad = -(w_{\text{obligatory}} + w_{\text{optional}}) \end{cases}$$

Now, $-w_{\text{obligatory}} = -P\Delta V$, so that if the reaction occurred in such a way that no work was performed other than that obligatorily associated with the change in volume of the system (i.e. when $-w_{\text{optional}} = 0$),

then,
$$\Delta U = (q_p - P\Delta V)$$

or, $$q_p = (\Delta U + P\Delta V)\dagger$$

The thermodynamicist recognizes that $(\Delta U + P\Delta V)$ measures the change in another energetic function of state of the system which he terms its *enthalpy* (H);

i.e. $$(\Delta U + P\Delta V) = \Delta H$$

Thus $\Delta H = q_p$ and the change in enthalpy ΔH may be defined as 'the quantity of heat absorbed by a closed, isothermal system when at constant pressure it undergoes a change of state without performing any work save that associated with its change in volume'. If under these conditions heat is released into the surroundings, the system must decrease in enthalpy (ΔH is negative) and the reaction is said to be *exothermic*. Absorption of heat under these conditions is characteristic of an *endothermic* reaction and is evidence of an increase in the enthalpy of the system (ΔH is positive). For a chemical reaction, the sign and magnitude of ΔH can be shown to be largely attributable to the energy changes associated with the making and breaking of chemical bonds. However, we shall quickly be disillusioned if we suppose that exothermy is an indicator of spontaneous change. Reactions may occur spontaneously whose ΔH is negative, zero or positive, so that the sign of the ΔH term is no criterion of spontaneity (see p. 191).

Hess's Law

Since enthalpy is a function of state, the value of ΔH is invariant for any given change in a closed system (T and P constant). It depends only on the initial and final states of the system, and is completely independent of the mechanism of the reaction that accomplishes that change. This is the basis of ***Hess's Law of Constant Heat Summation***, which states that 'the heat change in a particular reaction is the same whether it is accomplished in one or in several stages'.

Hess's Law frequently makes it possible to calculate a heat of reaction that it is not feasible to measure directly. Suppose, for example, that the value of ΔH for the reaction $A \rightarrow D$ is -4.5 kJ mol^{-1} when this is accomplished at certain fixed temperature and pressure. If this same reaction is carried out in a stepwise fashion, then so long as the same fixed temperature and pressure are maintained throughout, the sum of the values of ΔH for the contributory reactions will equal the value of

† If the pressure *were* allowed to change while the volume of the system was kept constant (as in a bomb calorimeter), then when no optional work is performed the heat exchange $q_v = \Delta U$. (No $P\Delta V$ work is done, since $\Delta V = 0$.)

ΔH for the overall reaction. Thus if,

$$A \longrightarrow B \qquad \Delta H = \text{(i)}$$
$$B \longrightarrow C \qquad \Delta H = \text{(ii)}$$
$$C \longrightarrow D \qquad \Delta H = \text{(iii)}$$

since $\Delta H = -4.5$ kJ mol^{-1} for the overall reaction $A \rightarrow D$, then (i)+(ii)+(iii) $= -4.5$ kJ mol^{-1}.

EXAMPLE

The bacterium Acetobacter suboxydans *may obtain energy for growth by oxidizing ethanol, first to acetaldehyde and then to acetic acid.*

Calculate the values of ΔH (at 293 K and standard atmospheric pressure) for the reactions,

$$C_2H_5OH + \tfrac{1}{2}O_2 \longrightarrow CH_3CHO + H_2O$$

and

$$CH_3CHO + \tfrac{1}{2}O_2 \longrightarrow CH_3COOH$$

given the following heats of combustion at 293 K and standard atmospheric pressure, ethanol $= -1371$ kJ mol^{-1}, acetaldehyde $= -1168$ kJ mol^{-1}, acetic acid $= -876$ kJ mol^{-1}. (These values represent the enthalpy changes associated with the complete oxidation of these compounds at 293 K and standard atmospheric pressure.)

1. *Ethanol to acetaldehyde*

$$\text{(a)}\ C_2H_5OH + 3O_2 \longrightarrow 2CO_2 + 3H_2O \qquad \Delta H = -1371 \text{ kJ mol}^{-1}$$
$$\text{(b)}\ CH_3CHO + 2\tfrac{1}{2}O_2 \longrightarrow 2CO_2 + 2H_2O \qquad \Delta H = -1168 \text{ kJ mol}^{-1}$$

Subtract (b) from (a),

$$C_2H_5OH - CH_3CHO + \tfrac{1}{2}O_2 \longrightarrow H_2O$$
$$\Delta H = (-1371 - (-1168)) \text{ kJ mol}^{-1}$$

i.e. $$C_2H_5OH + \tfrac{1}{2}O_2 \longrightarrow CH_3CHO + H_2O \qquad \Delta H = -203 \text{ kJ mol}^{-1}$$

Thus, ΔH for the oxidation of ethanol to acetaldehyde under the stated conditions $= -203$ kJ mol^{-1}.

2. *Acetaldehyde to acetic acid*

$$\text{(b)}\ CH_3CHO + 2\tfrac{1}{2}O_2 \longrightarrow 2CO_2 + 2H_2O \qquad \Delta H = -1168 \text{ kJ mol}^{-1}$$
$$\text{(c)}\ CH_3COOH + 2O_2 \longrightarrow 2CO_2 + 2H_2O \qquad \Delta H = -876 \text{ kJ mol}^{-1}$$

Subtract (c) from (b)

$$CH_3CHO + \tfrac{1}{2}O_2 \longrightarrow CH_3COOH \quad \Delta H = (-1168-(-876))\ \text{kJ mol}^{-1} = -292\ \text{kJ mol}^{-1}$$

Thus, ΔH for the oxidation of acetaldehyde to acetic acid under the stated conditions is -292 kJ mol^{-1}.

ENTROPY

Entropy (given the symbol S) is, like enthalpy, a function of state, and any change in state of a system will be associated with a change of entropy $\Delta S = (S_{\text{final}} - S_{\text{initial}})$. However, unlike enthalpy, entropy is essentially a mathematical function with no simple physical analogue. Yet at the price of laying ourselves open to the charge of entertaining too simplistic a view of this abstraction, we can imagine the S value of an isolated system to be an index of its intrinsic stability. The greater its entropy the more stable is this system and the smaller is its capacity for spontaneous change.

The ***3rd Law of Thermodynamics*** states that 'at the absolute zero of temperature (0 K) perfect crystals of all compounds possess zero entropy'. Thereafter the entropy of each compound increases uniquely with temperature, its quantity being measured in J K^{-1} mol^{-1}. Statistical thermodynamics suggests that the entropy of a system is a measure of the randomness of energy distribution in the system and so associates entropy with the number of distinct energy levels available to the system. This is the origin of attempts to 'explain' entropy as a measure of the disorder of the system. For example, the more chaotic the system the greater is its entropy; the more orderly the system the smaller is its entropy. Or again, the greater is the informational content of the system the smaller is its entropy; nonsense (i.e. disorder) is associated with enhanced entropy. Though each of these analogies may reveal some aspect of the significance of entropy, none is wholly adequate, and none should be taken too literally.

One of our declared objectives in this chapter is to discover how classical thermodynamics enables us to predict the feasibility, or otherwise, of a proposed reaction in a given system. Where the 1st Law proved inadequate, the 2nd Law of Thermodynamics succeeds in providing us with the necessary criterion of feasibility, for the 2nd Law informs us that in a closed system the only reactions that can occur spontaneously are those that increase the total entropy of the system and its surroundings. The criterion for spontaneous change in a closed system is therefore revealed as

$$\Delta S_{\text{system}} + \Delta S_{\text{surroundings}} = \text{a positive value}$$

If $\Delta S_{system} + \Delta S_{surroundings} = 0$, the total (isolated) system remains throughout at equilibrium, so that any reaction which is associated with nil change in total entropy of the isolated system must occur under conditions of perfect thermodynamic reversibility. [Reversibility here means 'in such a manner that an infinitesimal reversal of the conditions would reverse the direction of energy flow'.] Perfect thermodynamic reversibility infers that the reaction has to take place infinitely slowly so that its components may constantly be maintained in perfect temperature and pressure equilibrium with their surroundings. It is therefore an ideal not attainable in practice though it can be approached in certain situations; e.g. the reaction in an electrochemical cell at its potentiometric null point (p. 331). Even so, the concept of thermodynamic reversibility is none the less important, for it can be shown that if a spontaneous change in a closed system (T and P constant) could take place under conditions of thermodynamic reversibility then the change in entropy of the system (ΔS_{system}) would equal q_{rev}/T where q_{rev} is the heat acquired by the system by thermal transfer taking place reversibly at T K. Since under these conditions $\Delta S_{system} + \Delta S_{surroundings} = 0$, the corresponding $\Delta S_{surroundings}$ must equal $-q_{rev}/T$. As we shall see later (p. 194), this relationship enables us to obtain approximate values of ΔS for reactions for which a value of q_{rev} can be measured under nearly perfect conditions of thermodynamic reversibility.

It follows from this description of the stringent conditions of thermodynamic reversibility that all natural processes that take place at a significant rate are thermodynamically irreversible; it also follows that only if such a process enhances the net entropy of the closed system and its surroundings will it occur spontaneously.

FREE ENERGY

The 2nd Law's criterion of spontaneity (feasibility of independent occurrence) of reaction in a closed system (T and P constant) as stated above, is a somewhat inconvenient one in that it compels us to take account of entropy changes in both the system itself and its surroundings. It would be more satisfactory if we could derive the same information from changes in the closed system alone, and so be relieved of the necessity to evaluate changes in its environment. Fortunately we are able to do just this, for it can be shown that the 1st Law requires that,

$$\Delta S_{surroundings} = \frac{\text{heat absorbed by surroundings}}{T}$$

whence, $$\Delta S_{surroundings} = \frac{-\Delta H_{system}}{T}$$

The criterion of spontaneous change under thermodynamically irreversible (non-equilibrium) conditions in a closed system (T and P constant), i.e.

$$\Delta S_{\text{system}} + S_{\text{surroundings}} > 0,$$

becomes,

$$\Delta S_{\text{system}} - \frac{\Delta H_{\text{system}}}{T} > 0$$

whence,

$$\Delta H_{\text{system}} - T\Delta S_{\text{system}} < 0$$

(Hereafter we can omit the suffix 'system' since it is understood that we are solely considering changes in the properties of the closed system.)

Since ΔH and ΔS are changes in functions of state, the term ($\Delta H - T\Delta S$) is a measure of the change in another function of state named the ***Gibbs Free Energy*** (so named, and given the symbol G, after John Willard Gibbs, who drew attention to its significance in systems maintained at constant pressure).

Thus,

$$\Delta H - T\Delta S = \Delta G$$

and the criterion of spontaneous reaction in a closed system (T and P constant) can be expressed quite succinctly as follows:

(i) system not already at equilibrium (i.e. reactions proceeding at measurable rates under conditions of thermodynamic irreversibility)

$$\Delta G < 0$$

(ii) at equilibrium (conditions of thermodynamic reversibility)

$$\Delta G = 0$$

This means that in any closed system (T and P constant) which is not already at equilibrium, only exergonic reactions (ΔG negative) can occur spontaneously. If under given conditions a proposed reaction $A+B \rightarrow C+D$ is endergonic (say, $\Delta G = +5$ kJ mol^{-1}), then the reverse reaction $C+D \rightarrow A+B$ will be exergonic ($\Delta G = -5$ kJ mol^{-1}) and only this reverse reaction could proceed spontaneously.†

† The Gibbs free energy of a closed system (T and P constant) has been frequently represented as the measure of the ability of the system to perform useful (i.e. optional) work. An exergonic reaction is theoretically able to perform useful work ($-w_{\text{optional}}$) to a maximum of $-\Delta G$ J mol^{-1}, but in fact any spontaneous process which proceeds at a measurable rate is less than completely efficient in its performance of useful work; i.e. under thermodynamically irreversible conditions $-w_{\text{optional}} < -\Delta G$ (see p. 188).

EXAMPLE

The hydrolysis of adenosine triphosphate which liberates its terminal phosphate group is a reaction of considerable biochemical significance, and many attempts have been made to measure the values of ΔH, ΔS and ΔG for this reaction at 'physiological' temperatures and pH values. In one such determination at 309 K (i.e. 36°C) and pH 7 in the presence of Mg^{2+} ions, it was calculated that when ΔH was -20.08 kJ mol^{-1} ΔS was $+35.21$ J K^{-1} mol^{-1}. Calculate the corresponding value of ΔG of the reaction.

For an isothermal reaction at constant pressure in a closed system,

$$\Delta G = \Delta H - T\Delta S$$

For the given reaction,

$$\begin{aligned} \Delta G &= \text{? J mol}^{-1} \\ \Delta H &= -20\,080 \text{ J mol}^{-1} \\ T &= 309 \text{ K} \\ \Delta S &= +35.21 \text{ J K}^{-1}\text{ mol}^{-1} \end{aligned}$$

Substituting these values in the above equation,

$$\Delta G = -20\,080 - (309 \times 35.21) \text{ J mol}^{-1}$$

Whence,

$$\begin{aligned} \Delta G &= -20\,080 - 10\,880 \text{ J mol}^{-1} \\ &= \underline{-30.96 \text{ kJ mol}^{-1}} \end{aligned}$$

When analysing chemical changes in multi-component systems (T and P constant), it is frequently useful to sub-divide the total free energy change of the system into the contributions made by changes in its individual components. If an individual component (i) changes in amount by dn_i mol, while the other components (as well as T and P) remain unchanged, then the resultant contribution made to the total free energy change of the system (dG) is given by $\mu_i dn_i$, where μ_i is termed the 'chemical potential' of that component. Considering the dG of a system as the resultant of the μdn contributions of its components can considerably simplify appraisal of the energetics of open systems, and it is in statistical thermodynamics that the concept of chemical potential has proved most rewarding. Even so, for any system to be at equilibrium the chemical potential of each component must be identical in all parts of the system. As an illustration, consider the system (described in Chapter 4, p. 71) in which an osmotic pressure is evidenced. Addition of n_2 mol of solute to a volume V m^3 of pure solvent decreases the chemical potential of that solvent by n_2RT J. If this solution is separated by a semi-permeable membrane from pure solvent (at the same T and P) an unstable system is created which can achieve equilibrium only when the chemical potential of the solvent is equalized in all parts of the system. Spontaneous transfer of solvent from that part of the system wherein it possesses the higher chemical potential (pure solvent) to that in which it has lesser chemical potential (solution) is the expression of this drive to a new equilibrium. But application of additional pressure (ΔP Pa) to the solution will raise the chemical potential of its solvent by an amount $V\Delta P$ J, and if ΔP is such that $V\Delta P = n_2RT$ then the lowering of the chemical potential of the solvent by solute addition is exactly countered by that elevation in its chemical potential resulting from the application of pressure. It is this value of ΔP that is the osmotic pressure of the solution (π), which explains the statement (p. 74) that 'the osmotic pressure of a solution is that pressure which must be exerted on the solution to raise the chemical potential of its solvent so that it equals that of the pure solvent at the same temperature'.

SPONTANEOUS REACTIONS

We are made aware of spontaneous processes by their apparent inevitability. Almost intuitively we accept that heat should be transferred from a hot to a cold body or that bodies should fall from a height. It is a feature of these events that they are apparently unidirectional and may (optionally) be harnessed to perform useful work. We now have some understanding of what it is that causes such reactions in closed systems (T and P constant). The motivation behind such reactions is revealed as the tendency to increase the total entropy of the closed system plus its surroundings until this achieves its maximum value. Whether or not a proposed reaction can contribute to this end determines whether it can occur spontaneously. Limiting ourselves to an appraisal of the closed system alone, we find that the two factors which together determine whether a proposed reaction will be spontaneous are (a) the enthalpy change ΔH of the system, and (b) the entropy change ΔS of the system. Since $\Delta H - T\Delta S = \Delta G$, when considering the sign and magnitude of ΔG (the change in Gibbs free energy of the system) we are examining the resultant of the ΔH and $-T\Delta S$ contributions, and are able to declare that only reactions with a negative ΔG (i.e. exergonic reactions) can occur spontaneously.† One word of warning: recognition of the exergonic nature of a reaction whilst enabling one to predict that it *can* occur spontaneously does not allow one to infer that it *will* proceed rapidly or

† Isothermal reactions are:
Exothermic if ΔH is negative; *Endothermic* if ΔH is positive;
Exergonic if ΔG is negative; *Endergonic* if ΔG is positive.

even at a measurable rate. As we shall see (Chapter 10), how fast even the most exergonic reaction proceeds is determined by other factors, e.g. the activation energy of the rate-limiting step in its mechanism.

Since every spontaneous change in an isolated system must enhance its entropy, we can confidently predict that through the agency of spontaneous processes the energy content of every isolated system not already in equilibrium is being redistributed (however slowly) in such a way that its entropy is increasing. *Clausius's Law* states this by declaring that 'the energy of the universe is constant but its entropy is increasing to a maximum'. Accordingly, the prospect for the universe is of inevitable progress to 'entropic doom'—the terminal state of equilibrium where entropy is at its maximum and the capacity for spontaneous change is exhausted.

To summarize: we have identified in classical thermodynamic terms both what it is that drives a spontaneous reaction in a closed system (T and P constant) and what determines its end.

What makes it go?—the tendency to increase the net entropy of the closed system plus its surroundings. [Expressed in terms of the closed system alone, as its tendency to decrease in free energy.]

When does it stop?—when equilibrium is achieved, in which state the net entropy of the system plus its surroundings has its maximum value. [When the free energy of the closed system has fallen to its minimum value.]

THERMODYNAMIC STANDARD STATES AND STANDARD FUNCTIONS

In order that we may compare the energetic changes associated with different reactions, we must first standardize the conditions of reaction. By convention, the standard reaction conditions are defined as follows: temperature, *298.15 K*; pressure, *1 atmosphere* (*i.e. 101 325 Pa*); composition, *all components in their defined standard states.*

The standard state of a substance is decreed to be the pure substance as its exists at 298 K and standard atmospheric pressure, e.g. pure gaseous ammonia, liquid water and solid glucose. For reactants in solution, the standard state of a solvent is generally defined as unit activity (ideal mole fraction of 1); for solutes, too, the standard state is unit activity so that if molarities are used, the standard state of the solute is its (hypothetical) ideal, 1 mol dm^{-3} solution (p. 65).

A special suffix identifies the thermodynamic properties of reactions under standard conditions. Thus $\Delta G^{\ominus}$ designates the standard change in

Gibbs free energy that is associated with a reaction in which 1 mole of reactant gives rise to product(s) when all are maintained throughout in their standard states at 298 K and standard atmospheric pressure. This and other symbols used in thermodynamic calculations are listed in Table 7.1, though the meaning of many of these is only explained later in this chapter.

Table 7.1

ΔG =	Change in Gibbs free energy when the reaction takes place at constant temperature and pressure under arbitrary, but reported, conditions.
$\Delta G^{\ominus}$ =	ΔG under standard conditions of 298.15 K, standard atmospheric pressure (101 325 Pa), components in standard states throughout.
$\Delta G^{\ominus\prime}$ =	ΔG for a reaction in solution under standard conditions with the components in their standard states except for the H^+ ion activity (taken to be pH 7, unless otherwise stated).
ΔH, $\Delta H^{\ominus}$, $\Delta H^{\ominus\prime}$ / ΔS, $\Delta S^{\ominus}$, $\Delta S^{\ominus\prime}$	suffixes denote reaction conditions as detailed for ΔG, $\Delta G^{\ominus}$ and $\Delta G^{\ominus\prime}$
K_{eq} =	Thermodynamic (true) equilibrium constant—calculated from the activities of the components as they coexist at equilibrium.
K_c =	Equilibrium constant—calculated from the concentrations of the components at equilibrium.
R =	Gas constant (8.314 J K^{-1} mol^{-1})
T =	Absolute temperature (K)

Calculation and measurement of values of ΔH and ΔS

Since we shall chiefly be concerned with values of ΔG, methods of determining values of ΔH and ΔS will only be briefly summarized. Any elementary textbook of thermochemistry will provide further details.

1. *Enthalpy change* (ΔH)

The standard molar heat of formation ($\Delta H_f^{\ominus}$) of a compound is 'the heat change associated with the formation of 1 mole of the compound in its standard state from its elements which are also in their standard states'. The enthalpy of any element in its standard state is arbitrarily defined as zero, and this means that the $\Delta H_f^{\ominus}$ of a compound is equal to the enthalpy of 1 mole of that compound in its standard state. Tables are available that list the values of $\Delta H_f^{\ominus}$ of most common organic compounds, and from these values the $\Delta H_f^{\ominus}$ of a reaction may be obtained by difference, i.e.

$$\Delta H^{\ominus} = (\text{sum of values of } \Delta H_f^{\ominus} \text{ of products}) - (\text{sum of values of } \Delta H_f^{\ominus} \text{ of reactants})$$

Experimental methods for determining values of ΔH (including values of $\Delta H_f^\ominus$) obtain their values:

(a) from 'direct' calorimetry at constant pressure ($\Delta H = q_p$);
(b) from 'indirect' calorimetry at constant volume in a bomb calorimeter; this gives a value for q_v from which ΔH can be calculated;
(c) from the heats of combustion of the reaction components;
(d) by application of Hess's Law (method (c) is one example of this);
(e) from the value of the equilibrium constant, together with the way in which this changes with temperature (p. 221).

2. *Entropy change* (ΔS)

Again, tables are available which list the values of the standard molar entropies ($S^\ominus$) of common substances. For reactions whose components appear in such tables, the $\Delta S^\ominus$ is obtainable by difference, i.e.

$$\Delta S^\ominus = (\text{sum of } S^\ominus \text{ values of products}) - (\text{sum of } S^\ominus \text{ values of reactants})$$

Non-standard values of ΔS can be obtained for isothermal, thermodynamically reversible reactions as being equal to q_{rev}/T (see p. 188). Alternatively, the value of ΔS for any isothermal reaction can be calculated from its values of ΔH and ΔG by substituting these in the equation $\Delta G = \Delta H - T\Delta S$.

$\Delta G^\ominus$; FREE ENERGY CHANGE UNDER STANDARD CONDITIONS

The change in Gibbs free energy that is associated with any reaction will depend on the temperature and pressure and on the activities of the reaction components that prevail throughout.† By fixing these at 298 K and standard atmospheric pressure, with all components in their standard states (at unit activities), reactions can be considered as they would occur under identical conditions in standard systems; the then standard free energy change of each reaction ($\Delta G^\ominus$) is strictly defined. A useful corollary is that these values of $\Delta G^\ominus$ are additive, for example,

if	$A+B \rightleftharpoons C+D$	$\Delta G^\ominus = -7000\ \text{J mol}^{-1}$
and	$C+D \rightleftharpoons X+Y$	$\Delta G^\ominus = +5000\ \text{J mol}^{-1}$
then,	$A+B \rightleftharpoons X+Y$	$\Delta G^\ominus = -2000\ \text{J mol}^{-1}$

† When considering values of $\Delta G^\ominus$ or ΔG, remember that only for reactions conducted isothermally *at constant pressure* does the 'change in Gibbs free energy' have a simple interpretation.

This treatment may be compared with Hess's Law which deals with values of ΔH in a like manner (p. 185).

If a value is reported for the standard Gibbs free energy change of a reaction at a temperature other than 298 K, the actual temperature must be indicated. Thus, for a reaction at 310 K, the modified value of $\Delta G^{\ominus}$ should be reported as '$\Delta G^{\ominus}$ at 310 K equals . . . J mol^{-1}'. According to this notation, the primary standard value of $\Delta G^{\ominus}$, which is usually written without further qualification, would be written as '$\Delta G^{\ominus}$ at 298 K'. Alternatively, the prevailing temperature can be recorded as a subscript in K, e.g. '$\Delta G^{\ominus}$ at 288 K' equals '$\Delta G^{\ominus}_{288}$'. The important thing to remember about such values is that they inform us about the behaviour of reactions at a single temperature only. Thus if values of $\Delta G^{\ominus}$ are to be summed, they must all have been determined at the one temperature, e.g. a value of $\Delta G^{\ominus}$ at 298 K for one reaction cannot be added to a value of $\Delta G^{\ominus}$ at 303 K for another reaction and still give a meaningful net result.

The meaning of yet another modified $\Delta G^{\ominus}$ term will be discussed later in this chapter (p. 199); this is $\Delta G^{\ominus\prime}$ which refers to reactions involving H^+ ions which are carried out under otherwise standard conditions but at a pH other than zero. Values of $\Delta G^{\ominus\prime}$ determined for reactions at temperatures other than 298 K are 'doubly modified' standard values, for example '$\Delta G^{\ominus\prime}$ at 310 K and pH 7.5'.

EXAMPLE

$\Delta G^{\ominus}$ of glucose and ethanol in aqueous solution equals −917.0 and −181.6 kJ mol^{-1} respectively, and $\Delta G_f^{\ominus}$ of carbon dioxide as a gas is −394.5 kJ mol^{-1}. Deduce $\Delta G^{\ominus}$ for the net reaction of alcoholic fermentation.

$$\text{Glucose} \longrightarrow 2\ \text{Ethanol} + 2CO_2$$

in aqueous solution (at 298 K) with the evolution of gaseous CO_2.

Since values of $\Delta G^{\ominus}$ are additive,

$\Delta G^{\ominus}$ = (sum of values of $\Delta G_f^{\ominus}$ of products)—(sum of values of ΔG_f of reactants)

In the present example,

$\Delta G^{\ominus}$ = ($2 \times \Delta G_f^{\ominus}$ ethanol in aqueous solution + $2 \times \Delta G_f^{\ominus}$ CO_2 as gas) *minus* ($\Delta G_f^{\ominus}$ glucose in aqueous solution).

$$\begin{aligned} \therefore\ \Delta G^{\ominus} &= ((2\times -181.6)+(2\times -394.5))-(-917.0)\ \text{kJ mol}^{-1} \\ &= ((-363.2)+(-789.0))+917.0\ \text{kJ mol}^{-1} \\ &= -1152.2+917.0\ \text{kJ mol}^{-1} \\ &= -235.2\ \text{kJ mol}^{-1} \end{aligned}$$

$\therefore$ $\Delta G^{\ominus}$ for the net reaction = −235.2 kJ mol^{-1}.

Some methods for determining the value of ΔG° of a reaction

1 From the standard free energies of formation of products and reactants, ΔG°=(sum of values of ΔG_f° of products)−(sum of values of ΔG_f° of reactants).
2 From the known values of ΔG° for reactions whose net outcome is the reaction whose value of ΔG° is required.
3 From the known values of ΔH° and ΔS° for the reaction, by substituting these in the equation, $\Delta G^{\circ}=\Delta H^{\circ}-T\Delta S^{\circ}$.
4 From the value of the equilibrium constant for the reaction (see below).
5 For oxidation-reduction reactions, from the difference between the standard electrode potentials of the component redox couples (see below and p. 336).

The relationship between the value of ΔG° and the value of the equilibrium constant K_{eq} of a chemically reversible reaction

A chemically reversible reaction in an isothermal, closed system at constant pressure, proceeds until the system achieves a state of equilibrium which is characterized thermodynamically by its minimal free energy content (p. 192). The chemical composition of this equilibrium state is defined by the equilibrium constant for the reaction (Chapter 8, p. 212). Thus the isothermal reaction $A+B \rightleftharpoons C+D$ proceeds to an equilibrium at which the activities of the components of the reaction are uniquely related according to the equation,

$$K_{eq} = \frac{(C)(D)}{(A)(B)} \quad \text{where} \begin{cases} K_{eq} = \text{true (thermodynamic) equilibrium constant for the reaction at } T \text{ K} \\ (A) \text{ etc., are the activities of the reaction components at equilibrium at } T \text{ K} \end{cases}$$

If, instead of measuring the activities of the components at equilibrium, one measures their concentrations and uses *these* values to calculate the equilibrium constant, one obtains an approximate equilibrium constant (K_c) which equals the thermodynamic constant K_{eq} only in ideal solution (infinitely dilute; see pp. 54 and 218).

It can be shown that the equilibrium constant of an isothermal reaction is related to its standard Gibbs free energy change at that temperature (and at standard atmospheric pressure) in the way defined by the equation,

$$\Delta G^{\circ} = -RT \ln K_{eq} \quad \text{where} \begin{cases} R = \text{gas constant} = 8.314 \text{ J K}^{-1} \text{ mol}^{-1} \\ T = \text{temperature in K} \\ K_{eq} = \text{true equilibrium constant at } T \text{ K} \\ \Delta G^{\circ} = \Delta G^{\circ} \text{ at } T \text{ K} \end{cases}$$

Since, for reactions performed in very dilute solutions, K_{eq} and K_c are approximately equal, for these reactions,

$$\Delta G^\ominus \text{ at } T \text{ K approximately equals } -RT \ln K_c$$

The usefulness of this most important relationship is illustrated in the following example.

EXAMPLE

During glycolysis, fructose 1,6-diphosphate is broken down to yield glyceraldehyde 3-phosphate plus dihydroxyacetone phosphate. Yet during glucogenesis, fructose 1,6-diphosphate is synthesized from these triose phosphates. A single enzyme, aldolase, catalyzes both these processes which are due to the chemically reversible reaction,

$$\textit{Fructose 1,6-diphosphate} \rightleftharpoons \textit{Glyceraldehyde 3-phosphate} + \textit{Dihydroxyacetone phosphate}$$

If the thermodynamic equilibrium constant for this reaction (from left to right as written) equals 8.91×10^{-5} mol dm^{-3}, calculate the value of $\Delta G^\ominus$ for the cleavage of fructose diphosphate ($R = 8.314$ J K^{-1} mol^{-1}).

Since no temperature is explicitly mentioned, we must assume that the values of K_{eq} and $\Delta G^\ominus$ refer to the reaction performed isothermally at 298 K. Enzyme catalysis does not alter the value of K_{eq} for a reaction, and we may apply the following equation,

$$\Delta G^\ominus = -RT \ln K_{eq} \qquad \text{where} \begin{cases} R = 8.314 \text{ J K}^{-1} \text{ mol}^{-1} \\ T = 298 \text{ K} \\ K_{eq} = 8.91 \times 10^{-5} \text{ mol dm}^{-3} \end{cases}$$

Substituting these values in the equation,

$$\Delta G^\ominus = -(8.314 \times 298 \times 2.303 \log K_{eq}) = -5706 \log K_{eq}$$

But $$\log K_{eq} = \log (8.91 \times 10^{-5}) = \bar{5}.95 = -4.05$$

$$\therefore \Delta G^\ominus = -(5706 \times -4.05) \text{ J mol}^{-1} = 23\,120 \text{ J mol}^{-1}$$

Thus, $\Delta G^\ominus$ for the reaction is 23.12 kJ mol^{-1}.

The relationship between $\Delta G^\ominus$ and $\Delta E^\ominus$ for an oxidation-reduction reaction

When oxidizing and reducing redox couples of differing standard electrode potentials ($E^\ominus$) interact (Chapter 12, p. 336), $\Delta G^\ominus$ of this interaction is related to $\Delta E^\ominus$ by the following equation,

$$\Delta G^\ominus = -nF\Delta E^\ominus$$

where n = number of electrons transferred per molecule from reductant to oxidant

$\boldsymbol{F}$ = faraday constant = 96 487 C mol^{-1} = 96 487 J V^{-1} mol^{-1}

$\Delta E^{\ominus}$ = e.m.f. = difference between the standard electrode potentials of the contributory redox couples (in V)

The practical standard electrode potential, $E_{\ominus}$(p. 335), is negligibly different from $E^{\ominus}$ for a reduction-oxidation couple in relatively dilute solutions, so that it is usual in such cases to employ the more approximate equation,

$$\Delta G^{\ominus} = -n\boldsymbol{F}\Delta E_{\ominus} \text{ (see p. 336)}$$

EXAMPLE

Much of our present knowledge of electron transport processes in mitochondria derives from the study of the complex of enzymes and cofactors known as succinoxidase. The members of this complex cooperate in linking the oxidation of succinate with the reduction of oxygen to water,

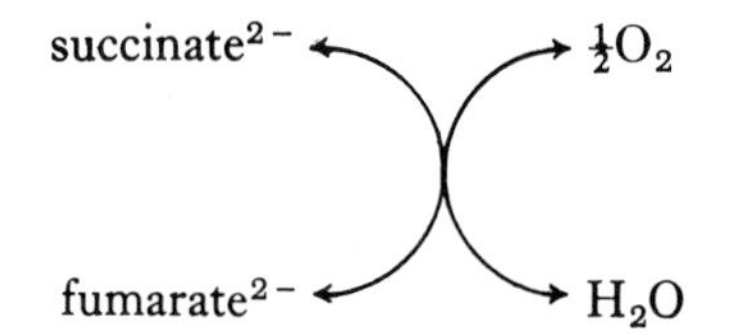

If $\Delta G^{\ominus\prime}$ at pH 7 for this reaction is 151.2 kJ mol^{-1}, how much greater is the value of the standard electrode potential ($E_{\ominus}'$) of the $O_2|H_2O$ redox couple than the corresponding value of $E_{\ominus}'$ for the fumarate|succinate redox couple (at 298 K and pH 7)?

We may apply the following equation,

$$\Delta G^{\ominus\prime} = -n\boldsymbol{F}\Delta E_{\ominus}' \quad \text{where} \begin{cases} \Delta G^{\ominus\prime} = -151.2 \text{ kJ mol}^{-1} \\ n = 2 \\ \boldsymbol{F} = 96.487 \text{ kJ V}^{-1}\text{ mol}^{-1} \\ \Delta E_{\ominus}' = ?\text{ V} \end{cases}$$

Substituting these values in the equation,

$$-(151.2\times 10^3) = -2\times(96.487\times 10^3)\times \Delta E_{\ominus}',$$

whence

$$\Delta E_{\ominus}' = \frac{151.2}{192.97} = 0.78 \text{ V}$$

Thus at 298 K and pH 7, the $E^\circ{}'$ of the oxidizing couple ($O_2|H_2O$) is more positive than the $E^\circ{}'$ of the reducing couple (fumarate|succinate) by 0.78 V(see Fig. 12.6, p. 358).

The meaning of $\Delta G^{\circ\prime}$ (applied to a reaction which involves H^+ ions)

If H^+ ions participate in a reaction, the pH of the system must be maintained at 0 for the standard conditions to prevail to which ΔG° refers. This highly acidic state is particularly unrealistic for biological systems, so for these we generally employ $\Delta G^{\circ\prime}$ values, where $\Delta G^{\circ\prime}$ is the Gibbs free energy change under standard conditions save that the pH is not maintained at 0 but at some other, reported, value.

The value of $\Delta G^{\circ\prime}$ for a reaction at a given pH differs from its value of ΔG° (at pH 0) by an amount determined by the value of the non-standard pH. Consider the reaction,

$$A + B \rightleftharpoons C + D + H^+$$

then, $$\Delta G^{\circ\prime} = \Delta G^\circ + RT \ln \frac{(C)(D)(H^+)}{(A)(B)}$$ see p. 204

where (A), (B), (C) and (D) are the maintained, standard activities of these components. By definition, all these components are at unit activity, but $(H^+) \neq 1$

$$\therefore \quad \Delta G^{\circ\prime} = \Delta G^\circ + RT \ln (H^+)$$

$$= \Delta G^\circ + RT 2.303 \log (H^+)$$

But $pH = -\log (H^+)$

$$\therefore \quad \Delta G^{\circ\prime} = \Delta G^\circ - 2.303\, RT\, pH$$

EXAMPLE

Nicotinamide adenine dinucleotide (NAD) is one of the most important 'electron transport agents' in living cells. It can exist in oxidized (NAD^+) and reduced (NADH) forms and the latter could, theoretically at least, be oxidized to give the former according to the equation,

$$NADH + H^+ \rightleftharpoons NAD^+ + H_2$$

If ΔG° for this reaction is -21.84 kJ mol^{-1}, calculate the value of $\Delta G^{\circ\prime}$ at pH 7. (Gas constant $R = 8.314$ J K^{-1} mol^{-1}.)

For the oxidation of NADH by H^+ at pH 7,

$$\Delta G^{\circ\prime} = \Delta G^{\circ} + RT \ln \frac{(NAD^+).p_{H_2}}{(NADH)(H^+)}$$

Since (NAD^+), p_{H_2}, and (NADH) are by definition standard, they all equal 1.

$$\therefore \quad \Delta G^{\circ\prime} = \Delta G^{\circ} + 2.303RT \log \frac{1}{[H^+]}$$

But $pH = \log \frac{1}{[H^+]}$

$$\therefore \quad \Delta G^{\circ\prime} = \Delta G^{\circ} + 2.303RT\,pH \qquad \text{where} \begin{cases} \Delta G^{\circ} = -21.84 \text{ kJ mol}^{-1} \\ R = 8.314 \text{ J K}^{-1} \text{ mol}^{-1} \\ T = 298 \text{ K} \\ pH = 7 \end{cases}$$

$$\begin{aligned} \therefore \quad \Delta G^{\circ\prime} &= -21\,840 + (2.303 \times 8.314 \times 298 \times 7) \text{ J mol}^{-1} \\ &= -21\,840 + 39\,940 \text{ J mol}^{-1} \\ &= +18\,100 \text{ J mol}^{-1} \end{aligned}$$

Thus $\Delta G^{\circ\prime}$ equals $+18.1$ kJ mol^{-1} for the oxidation of NADH by H^+ at pH 7, even though for the same reaction at pH 0, ΔG° equals -21.84 kJ mol^{-1}.

The temperature-dependence of ΔG°

Biologists are often interested in values of ΔG° for reactions at temperatures other than 298 K (particularly at their favourite incubation temperatures of 30°C and 37°C; i.e. 303 K and 310 K). Fortunately, the way in which ΔG° changes with temperature is predictable for a given reaction, since it depends on the ΔH° of the reaction. The relationship between ΔH° and the rate of change of ΔG° with temperature is defined by the following form of the Gibbs–Helmholtz equation:

$$\frac{d(\Delta G^{\circ}/T)}{dT} = -\frac{\Delta H^{\circ}}{T^2}$$

When this equation is integrated (assuming that ΔH° is independent of temperature) we obtain the following relationship,

$$\frac{\Delta G^{\circ}}{T} = \frac{\Delta H^{\circ}}{T} + \text{constant}$$

This is the equation of a straight line, $y = ax + b$ (p. 11) where $y = \Delta G^{\ominus}/T$, $x = 1/T$, $a = \Delta H^{\ominus}$ = a constant, and b = constant. Thus if values of $\Delta G^{\ominus}/T$, for a reaction at constant pressure, are plotted against corresponding values of $1/T$, a straight line is obtained whose slope is $\Delta H^{\ominus}$.

The value of $\Delta G^{\ominus}(\Delta G_2^{\ominus})$ at temperature T_2 K, can therefore be derived from the known value of $\Delta G^{\ominus}$ $(\Delta G_1^{\ominus})$ at T_1 K if $\Delta H^{\ominus}$ is also known, for these values can be substituted in the following equation (obtained from the integrated form of the Gibbs–Helmholtz equation),

$$\frac{\Delta G_2^{\ominus}}{T_2} - \frac{\Delta G_1^{\ominus}}{T_1} = \Delta H^{\ominus}\left(\frac{1}{T_2} - \frac{1}{T_1}\right)$$

This equation can be rearranged to give the following more convenient form,

$$\frac{\Delta G_2^{\ominus}}{T_2} = \frac{\Delta G_1^{\ominus}}{T_1} - \Delta H^{\ominus}\left(\frac{T_2 - T_1}{T_1 T_2}\right)$$

Since $\Delta G^{\ominus} = -RT \ln K_{eq}$, the value of the equilibrium constant K_{eq} is similarly affected by the temperature in a manner related to the value of $\Delta H^{\ominus}$ of the reaction. The relevant equation (known as the van't Hoff equation, p. 221) is $\frac{d(\ln K_{eq})}{dT} = \frac{\Delta H^{\ominus}}{RT^2}$, and is merely another expression of the Gibbs–Helmholtz equation used above. The van't Hoff equation and its implications will be discussed in Chapter 8 (p. 220–3), but it will be obvious from what we have deduced above concerning the change in the value of $\Delta G^{\ominus}$ with temperature, that from the values of K_{eq} at just two temperatures it is possible to determine an approximate value of $\Delta H^{\ominus}$ for a chemically reversible reaction; $\Delta G^{\ominus}$ is simultaneously calculable since it equals $-RT \ln K_{eq}$, and $\Delta S^{\ominus}$ can be obtained from the values of $\Delta H^{\ominus}$ and $\Delta G^{\ominus}$ by substitution in the equation $\Delta G^{\ominus} = \Delta H^{\ominus} - T\Delta S^{\ominus}$.

EXAMPLE

It is frequently stated that the hydrolysis of ATP to ADP plus inorganic phosphate under 'nearly physiological conditions' at 36°C is associated with a $\Delta G^{\ominus\prime}$ of -30.96 kJ mol^{-1} and a $\Delta H^{\ominus\prime}$ of -20.08 kJ mol^{-1}. This temperature may indeed be 'nearly physiological' for a warm-blooded mammal, but what would be the value of $\Delta G^{\ominus\prime}$ for this reaction in the muscle of a North Sea codfish at 5°C?

We may apply the integrated form of the Gibbs–Helmholtz equation,

$$\frac{\Delta G_2^{\ominus\prime}}{T_2} = \frac{\Delta G_1^{\ominus\prime}}{T_1} - \frac{\Delta H^{\ominus\prime}(T_2 - T_1)}{T_1 T_2}$$

where $\Delta G_2^{\ominus\prime} = ?\ \text{J mol}^{-1}$

$$T_2 = 5°\text{C} = 278\ \text{K}$$
$$\Delta G_1^{\ominus\prime} = -30\,960\ \text{J mol}^{-1}$$
$$T_1 = 36°\text{C} = 309\ \text{K}$$
$$\Delta H^{\ominus\prime} = -20\,080\ \text{J mol}^{-1}$$

$$\frac{\Delta G_2^{\ominus\prime}}{278} = \frac{-30\,960}{309} - \frac{-20\,080(278-309)}{309 \times 278}\ \text{J mol}^{-1}\ \text{K}^{-1}$$

$$\frac{\Delta G_2^{\ominus\prime}}{278} = \frac{-30\,960}{309} - \frac{-20\,080(-31)}{309 \times 278}\ \text{J mol}^{-1}\ \text{K}^{-1}$$

$$\frac{\Delta G_2^{\ominus\prime}}{278} = \frac{-30\,960}{309} - \frac{622\,480}{309 \times 278}\ \text{J mol}^{-1}\ \text{K}^{-1}$$

Multiplying both sides of this equation by (309×278) we obtain,

$$309\ \Delta G_2^{\ominus\prime} = -(30\,960 \times 278) - 622\,480\ \text{J mol}^{-1}$$

$$\therefore\ \Delta G_2^{\ominus\prime} = \frac{-8\,608\,000 - 622\,480}{309}\ \text{J mol}^{-1}$$

$$= \frac{-9\,230\,480}{309} = -29\,860\ \text{J mol}^{-1}$$

Thus if $\Delta G^{\ominus\prime}$ at 36°C of ATP hydrolysis is -30.96 kJ mol^{-1}, at the lower temperature of the fish muscle (5°C), $\Delta G^{\ominus\prime}$ of ATP hydrolysis is -29.86 kJ mol^{-1}.

ΔG; THE FREE ENERGY CHANGE UNDER NON-STANDARD CONDITIONS

From the value of $\Delta G^{\ominus}$ of a proposed reaction in a closed system (T and P constant) we can obtain its equilibrium constant, but if any one of its components is present at a non-standard activity (e.g. solute at any concentration other than ideal 1 mol dm^{-3}), then we cannot, from the sign and magnitude of $\Delta G^{\ominus}$ alone, conclude whether the reaction is able to proceed spontaneously.

Consider for example the proposed reaction $A \rightleftharpoons B$ in aqueous solution. If $\Delta G^{\ominus} = -1.2$ kJ mol^{-1}, this means that at 278 K and standard atmospheric pressure the equilibrium mixture of A and B would consist of these substances in solution in the ratio of activities $(B)/(A) = 2.06$. But if A was provided at 10^{-7} mol dm^{-3} and B at 10^{-5} mol dm^{-3} could A be spontaneously converted into B? The fact that $\Delta G^{\ominus}$ for this reaction has a negative sign is not directly relevant—for only if A and B had been provided at their standard conditions of unit activity could we instantly declare that since $\Delta G^{\ominus}$ is negative the spontaneous conversion of A to B under these standard conditions is feasible. In fact, in the non-standard conditions described ($A = 10^{-7}$ mol dm^{-3} and $B = 10^{-5}$ mol dm^{-3}), spontaneous conversion of $A \rightarrow B$ is *not* feasible. This is obvious from the fact that the ratio of initial concentrations of $[B]/[A]$, which is 100, is greater than the ratio at equilibrium (2.06) so that the spontaneous process would obviously be the reverse conversion of $B \rightarrow A$. The free energy change (ΔG) of the reaction under the given non-standard conditions confirms this conclusion, for when calculated for the 'forward' reaction $A \rightarrow B$ it turns out to have the value $+10.2$ kJ mol^{-1}. Thus only the actual (practical) value ΔG which takes into account the non-standard composition of the system will indicate whether a proposed reaction is feasible 'in the given circumstances'. However, if we know the activities (or in dilute solution, the concentrations) of the reaction components, we can calculate this required value of ΔG from the value of $\Delta G^{\ominus}$ of the reaction at the same temperature and pressure.†

The free energy G of a compound is related to its activity (a), the relationship being,

$$G = RT \ln a + \text{constant}$$

Thus when a compound changes in state from a_1 to a_2 its free energy will change by an amount ΔG, where,

$$\Delta G = (G_2 - G_1) = RT \ln a_2 - RT \ln a_1 = RT \ln \frac{a_2}{a_1}$$

Since the standard state (for a solute) is defined as unit activity (i.e. $a = 1$), change in state of any solute from its standard state to its non-standard actual activity (say in its physiological concentration c_{physiol}) causes its free energy to change by an amount ΔG, where

$$\Delta G = RT \ln \left(\frac{c_{\text{physiol}}}{1}\right) = RT \ln (c_{\text{physiol}})$$

The free energies of the reaction components in a non-standard state system will therefore differ by a calculable quantity from their free energy contributions to a system wherein they are present in their standard states. It is understandable therefore that ΔG should differ from $\Delta G^{\ominus}$ for a proposed reaction by a quantity calculable from the activities (concentrations) of the reaction components in the non-standard state system.

† Values of ΔG for an oxidation–reduction reaction are relatively easily *measured*, since the null-point e.m.f. generated by the reaction under the non-standard conditions (at a given temperature and pressure) will equal ΔE, where $-nF\Delta E = \Delta G$.

The relationship of ΔG to $\Delta G^{\ominus}$

Consider the reaction $A+B \rightleftharpoons C+D$

If (A) and (B) are the activities of the reactants, and (C) and (D) are the activities of the products present in solution at 298 K and standard atmospheric pressure, ΔG represents the change in Gibbs free energy that is associated with the reaction of 1 mole of A with B to produce C plus D while the activities of these compounds are maintained at (A), (B), (C) and (D). It can be shown that the value of ΔG differs from the value of $\Delta G^{\ominus}$ for this reaction (at the same temperature) by an amount equal to $RT \ln \frac{(C)(D)}{(A)(B)}$ so that,

$$\Delta G = \Delta G^{\ominus} + RT \ln \frac{(C)(D)}{(A)(B)}$$

If the reaction takes place in dilute solution, concentrations may be used in place of activities in this equation. Then for a reaction at a given temperature and standard atmospheric pressure,

$$\Delta G = \Delta G^{\ominus} + RT \ln \frac{[C][D]}{[A][B]}$$

If H^+ ions are consumed or produced by the reaction and the pH is buffered at some value other than 0, then by using the appropriate modified, standard free energy value ($\Delta G^{\ominus\prime}$) for this pH, account is taken of the involvement of H^+ without having to introduce a separate $[H^+]$ term into the equation, i.e.

$$\Delta G = \Delta G^{\ominus\prime} + RT \ln \frac{[C][D]}{[A][B]}$$

This most important equation emphasizes the concentration-dependence of the value of ΔG, and should act as a warning against the indiscriminate use of values of $\Delta G^{\ominus}$ when very different values of ΔG might be appropriate.

EXAMPLE

In a previous example (p. 197), we calculated that $\Delta G^{\ominus}$ is $+23.12$ kJ mol^{-1} for the cleavage of fructose 1,6-diphosphate to yield two molecules of triose phosphate. Yet if fructose diphosphate (0.01 mol dm^{-3}) is added to a solution containing each of these triose phosphates at 10^{-5} mol dm^{-3}, the tendency is for the fructose diphosphate to yield these triose phosphates in a spontaneous reaction.

Explain this apparent paradox, in thermodynamic terms. (Gas constant $R = 8.314$ J K^{-1} mol^{-1}.)

Fructose diphosphate (FDP) $\rightleftharpoons$

Glyceraldehyde phosphate (G 3-P) + Dihydroxyacetone phosphate (DHAP)

When this reaction proceeds under the specified, non-standard conditions,

$$\Delta G = \Delta G^{\ominus} + RT \ln \frac{[\text{G 3-P}][\text{DHAP}]}{[\text{FDP}]}$$

(Since concentration terms are used, the calculated value of ΔG will be to some extent approximate.)

Substituting the given values of $\Delta G^{\ominus}$ and reactant and product concentrations,

$$\Delta G = 23\,120 + RT 2.303 \log \frac{10^{-5} \times 10^{-5}}{10^{-2}} \text{ J mol}^{-1}$$

At 298 K, $2.303RT = (2.303 \times 8.314 \times 298) = 5706 \text{ J mol}^{-1}$

$$\begin{aligned} \therefore \quad \Delta G &= 23\,120 + (5706 \log 10^{-8}) \text{ J mol}^{-1} \\ &= 23\,120 + (5706 \times (-8)) \text{ J mol}^{-1} \\ &= 23\,120 - 45\,648 \text{ J mol}^{-1} \\ &= -22\,528 \text{ J mol}^{-1} \\ &\simeq -22.53 \text{ kJ mol}^{-1} \end{aligned}$$

Thus although $\Delta G^{\ominus}$ for the reaction has a positive value, ΔG is negative for the reaction under the given conditions; consequently, under these conditions, the formation of the triose phosphates from fructose diphosphate *would* spontaneously proceed to equilibrium.

Since the value of $\Delta G^{\ominus}$ is related in a known manner to the values of the equilibrium constant of a reaction (K_{eq}), and to the standard e.m.f. of an oxidation-reduction process ($\Delta E_{\ominus}$) it follows that when either of these is known, its value can be employed to calculate values of ΔG. Thus,

since $\quad \Delta G^{\ominus} = -RT \ln K_{eq}, \qquad \Delta G = -RT \ln K_{eq} + RT \ln \dfrac{(C)(D)}{(A)(B)}$

whence

$$\Delta G = -RT \ln \left[K_{eq} \times \frac{(A)(B)}{(C)(D)} \right]$$

Since $\quad \Delta G^{\ominus} = -nF\Delta E_{\ominus}, \qquad \Delta G = -nF\Delta E_{\ominus} + RT \ln \dfrac{(C)(D)}{(A)(B)}$

Incidentally, the value of ΔG changes with temperature in the same way that $\Delta G^{\ominus}$ changes with temperature, i.e. according to the Gibbs–Helmholtz equation $\frac{d(\Delta G/T)}{dt} = -\frac{\Delta H}{T^2}$, where ΔH is measured under the same reaction conditions as ΔG.

HOW INFORMATIVE IS THE VALUE OF ΔG?

Save in the ideal situation of thermodynamic reversibility, ΔG does not measure the useful work actually performed by the reaction. Its sign does, however, indicate whether in a closed system at constant temperature and pressure the reaction will be spontaneous (ΔG is negative, reaction is exergonic and spontaneous; ΔG is positive, reaction is endergonic and cannot be spontaneous). Its magnitude indicates how removed is the initial state of the system from ultimate equilibrium (where G is minimal and ΔG is zero; see Chapter 8). In short, it answers the questions, 'Will the reaction proceed unaided by any net energy contribution from its surroundings?', and 'How far will it go?'. It gives no answer to the equally relevant queries, 'How long will the reaction take?', or 'What route will the reaction follow?'. Indeed, these are questions that classical thermodynamics is inherently unqualified to answer.

ΔG might therefore be likened to a flight number in an airline brochure —it neither describes the route or speed of travel, nor indeed is it by itself any assurance of actual take-off.

THERMODYNAMICS OF REACTIONS IN AQUEOUS SOLUTIONS

It is particularly important for a biologist to be aware of complications that may arise in the calculation of values of ΔG of reactions performed in aqueous solution, since it is with such reactions that he is generally concerned.

(a) *What is the standard activity of water in an aqueous solution?*

Imagine a solution of the water-soluble ester CH_3X, whose hydrolysis can be catalysed by the addition of a little strong acid,

$$CH_3X + H_2O \rightleftharpoons HX + CH_3OH$$

With water as solvent and reactant, are we to consider it as (i) pure liquid and therefore in its standard state of unit activity (p. 192), or (ii) an aqueous solution of pure water whose concentration would be approximately

55.5 mol kg^{-1}? Which we choose is not entirely an exercise in semantics for, depending on our choice, the value of ΔG that we ascribe to the reaction will be increased or decreased considerably (by about 10.5 kJ mol^{-1} at 311 K). The only restriction on our choice is that the same value (either 1 or 55.5) must be used when calculating a value of ΔG, as was initially used in the definition of $\Delta G^{\ominus}$ for the reaction. Generally this is unity, and this too is the activity of water assumed in calculations of the equilibrium constants of hydrolytic reactions. Biochemists conventionally assign an activity of 1 to water in the dilute aqueous solutions with which they are concerned.

(b) *The possible ionization of reactants and products*

This complication may also be illustrated by referring to the hydrolysis of the ester CH_3X,

$$CH_3X + H_2O \rightleftharpoons HX + CH_3OH \qquad \Delta G^{\ominus} = (1)$$

The hydrolytic products are methyl alcohol (non-ionizable) and the possibly weak acid HX. At appropriate pH's this acid will ionize in a reaction which is associated with its own $\Delta G^{\ominus}$,

$$HX \rightleftharpoons H^+ + X^- \qquad \Delta G^{\ominus} = (2)$$

The actual value of ΔG of hydrolysis of CH_3X will therefore be pH-dependent and will obviously be a composite term which may include a significant contribution made by the ionization of HX. Obviously, therefore, the greater is the number of ionizable participants in a reaction in aqueous solution, the more complicated will be its energetics.

(c) *$\Delta G_f^{\ominus}$ of a compound in aqueous solution is not identical with $\Delta G_f^{\ominus}$ of the compound in its standard state*

The standard free energy of formation of a compound from its elements ($\Delta G_f^{\ominus}$) is identical for the compound as it exists in its pure standard state and in saturated aqueous solution. This derives from the fact that in its saturated solution the undissolved compound (in its pure, standard state) is in equilibrium with the dissolved form of the compound (solute). Since it is characteristic of the state of equilibrium that $\Delta G = 0$ for any reversible change effected while maintaining equilibrium, the isothermal conversion of compound in its standard state to compound as solute in its saturated solution involves no change in free energy. Thus $\Delta G_f^{\ominus}$ for the compound in its defined standard state must equal the value of $\Delta G_f^{\ominus}$ of the compound in saturated solution at the same temperature and pressure.

It follows that the difference between $\Delta G_f^{\ominus}$ of a compound in its pure, standard state and $\Delta G_f^{\ominus}$ of that compound in unit activity in aqueous solution, is the free energy change associated with the dilution of a saturated solution to yield the solution of unit activity. This is called the '$\Delta G^{\ominus}$ of solution' of that compound.

$$\Delta G^{\ominus} \text{ of solution} = -RT \ln \text{(saturated solution)}$$

where (saturated solution) represents the activity of the solute in its saturated solution. Thus,

$$\Delta G_f^{\ominus} \text{ in solution} = \Delta G_f^{\ominus} \text{ in standard state} + \Delta G^{\ominus} \text{ of solution}$$

Similarly, if the compound is ionized in solution, then to obtain the value of $\Delta G_f^{\ominus}$ of its ionized form in aqueous solution, the $\Delta G^{\ominus}$ of ionization must be added to the value of $\Delta G_f^{\ominus}$ of the unionized compound in aqueous solution. Thus,

$$\Delta G_f^{\ominus} \text{ of ionized form in aqueous solution} = \Delta G_f^{\ominus} \text{ of unionized form in aqueous solution} + \Delta G^{\ominus} \text{ of ionization}$$

EXAMPLE

A saturated aqueous solution of L*-glutamic acid is 0.0595 molal at 298 K and the acid is 3.8% dissociated. Given that* $\Delta G_f^{\ominus}$ *of* L*-glutamic acid equals −728.3 kJ mol*$^{-1}$*,* $\Delta G^{\ominus}$ *of ionization of* L*-glutamic acid equals 24.64 kJ mol*$^{-1}$*, and that the molal activity coefficient of a 0.057 molal solution of undissociated* L*-glutamic acid equals 0·55, calculate the value of* $\Delta G^{\ominus}$ *in aqueous solution of (i)* L*-glutamic acid, (ii)* L*-glutamate ion. (Note: Activity = Concentration × Activity coefficient.)*

(i) $\Delta G_f^{\ominus}$ *of* L*-glutamic acid in aqueous solution*

$\Delta G^{\ominus}$ of solution of undissociated L-glutamic acid

$= -RT \times \ln$(Activity of L-glutamic acid in saturated solution at 298 K)

But L-glutamic acid was 3·8% dissociated = 96·2% undissociated in its saturated solution. As its total concentration was 0·0595 molal, the concentration of undissociated L-glutamic acid in the saturated solution was 0·0595 × 96.2/100 = 0.057 molal.

Since Activity = Concentration × Activity coefficient, and since the molal activity coefficient of a 0.057 molal solution of L-glutamic acid = 0·55, then,

activity of undissociated L-glutamic acid in its saturated solution at 298 K = 0.057 × 0.55 = 0.0314.

Substituting these values in the opening equation, we obtain, $\Delta G^{\ominus}$ of solution of undissociated L-glutamic acid $= -RT \ln \times 0{\cdot}0314$

$$\begin{aligned}
&= -(8.314 \times 298 \times 2.303 \times \log 0.0314) \text{ J mol}^{-1} \\
&= -(5706 \times (\bar{2}.4969)) \text{ J mol}^{-1} \\
&= -(5706 \times (-1.5031)) \text{ J mol}^{-1} \\
&= -(-8578) = +8578 \text{ J mol}^{-1} \\
&= +8.578 \text{ kJ mol}^{-1}
\end{aligned}$$

But ΔG_f° in solution of undissociated L-glutamic acid

$$= \Delta G_f^{\circ} \text{ in standard state} + \Delta G^{\circ} \text{ of solution}$$
$$= -728.3 + 8.58 \text{ kJ mol}^{-1}$$
$$\simeq -719.7 \text{ kJ mol}^{-1}$$

Thus, ΔG_f° of undissociated L-glutamic acid in aqueous solution

$$= \underline{-719.7 \text{ kJ mol}^{-1}}$$

(ii) *ΔG_f° of L-glutamate ion in aqueous solution*

This will equal ΔG_f° of undissociated L-glutamic acid in aqueous solution *plus* ΔG° of ionization

ΔG_f° of undissociated L-glutamic acid in aqueous solution

$$= -719.7 \text{ kJ mol}^{-1}$$

ΔG° of ionization (given) $= 24.64 \text{ kJ mol}^{-1}$

$\therefore$ ΔG_f° of L-glutamate ion in aqueous solution

$$= (-719.7 + 24.64) \text{ kJ mol}^{-1}$$
$$= \underline{-695.1 \text{ kJ mol}^{-1}}$$

Conclusion

We stated at the outset that classical thermodynamics seeks to determine what it is that impels a reaction, and what it is that determines its end. During the course of this chapter we have seen that, isothermally and at constant pressure, the heat change associated with a reaction (the change in enthalpy, ΔH) reflects the redistribution of energy in the forms of Gibbs free energy and entropy ($\Delta H = \Delta G + T\Delta S$). Furthermore, we have discovered that in a closed system (T and P constant), what impels a spontaneous reaction is the tendency of the intrinsically unstable system to lose free energy and so increase the total entropy of the system and its surroundings. The end of the reaction (chemical equilibrium) has also been identified—as the state in which that system possesses no further capacity for spontaneous change ($\Delta G = 0$) and the total entropy of the system plus its surroundings has its maximum value.

In Chapter 8 we shall examine some of the thermodynamic implications of the fact that exergonic reactions proceed spontaneously to a state of stable, but dynamic, equilibrium.

PROBLEMS

(*Assume throughout that the value of the gas constant $R = 8.314$ J K^{-1} mol^{-1}.*)

1. If ΔH° of combustion of citric acid (solid) is -1986 kJ mol^{-1}, calculate the heat liberated when 10 g of solid citric acid is totally combusted at

298 K. (a) at constant pressure, (b) at constant volume. (Citric acid = $C_6H_8O_7$; molecular weight = 192.)

2. If the heats of combustion ($\Delta H^\ominus$ of combustion) of the monosaccharide glucose (solid), and the disaccharide maltose (solid) are -2816 and -5648 kJ mol^{-1} respectively at 293 K and standard atmospheric pressure, calculate the standard enthalpy change accompanying the conversion of 18 g of glucose to maltose at this temperature and pressure according to the equation,

$$2\ \text{Glucose (s)} \longrightarrow \text{Maltose (s)} + H_2O\ \text{(l)}$$

3. Bacteria of the genus *Nitrobacter* play an important role in the 'Nitrogen Cycle' in nature, by oxidizing soil nitrite to nitrate. From this simple oxidation they obtain all the energy required for growth,

$$NO_2^-\ \text{(aq.)} + \tfrac{1}{2}O_2\ \text{(g)} \longrightarrow NO_3^-\ \text{(aq.)}$$

Calculate the value of $\Delta G^\ominus$ of this reaction, given that $\Delta G_f^\ominus$ in aqueous solution of NO_2^- is -34.5 kJ mol^{-1}, and of NO_3^- is -110.5 kJ mol^{-1}.

4. The amino acid L-alanine, in the presence of a suitable amino acid oxidase, is converted to pyruvate according to the following equation,

$$\text{L-alanine} + O_2 + H_2O \rightleftharpoons \text{pyruvate} + NH_4^+ + H_2O_2$$

Calculate $\Delta G^{\ominus\prime}$ at pH 7 of this reaction from the known values of $\Delta G^{\ominus\prime}$ at pH 7 of the following reactions,

$$H_2O_2 \rightleftharpoons O_2 + H_2 \qquad \Delta G^{\ominus\prime} = +136.8\ \text{kJ mol}^{-1}$$

$$\text{L-alanine} + H_2O \rightleftharpoons \text{pyruvate} + NH_4^+ + H_2 \qquad \Delta G^{\ominus\prime} = +54.4\ \text{kJ mol}^{-1}$$

5. (a) If $\Delta H^\ominus$ for the dissociation of acetic acid in aqueous solution is -385 J mol^{-1}, and $\Delta S^\ominus$ is -92.5 J K^{-1} mol^{-1}, calculate

(i) the thermodynamic dissociation constant of acetic acid at 298 K, and
(ii) $\Delta G^{\ominus\prime}$ for its dissociation at pH 7.

(b) Monochloroacetic acid is a stronger acid than acetic acid. Which, then, of the following is likely to be the $\Delta G^\ominus$ of dissociation of monochloroacetic acid: (i) 48.1, (ii) 27.1, or (iii) 16.3 kJ mol^{-1}?

(c) Trichloroacetic acid is a much stronger acid than acetic acid, yet $\Delta H^\ominus$ of its dissociation virtually equals $\Delta H^\ominus$ of dissociation of acetic acid. Which of the following is likely to be the value of $\Delta S^\ominus$ for the dissociation of trichloroacetic acid: (i) -8.4, (ii) -168.2 or (iii) -90 J K^{-1} mol^{-1}?

6. The enzyme triose phosphate isomerase, catalyses the interconversion of glyceraldehyde 3-phosphate (G 3-P) and dihydroxyacetone phosphate (DHAP),

$$\text{G 3-P} \rightleftharpoons \text{DHAP}$$

If the thermodynamic equilibrium constant for the reaction, proceeding in the direction of DHAP formation, is 22 at 298 K in aqueous solution at atmospheric pressure, calculate the value of $\Delta G^\ominus$ for this reaction.

7. In the presence of the enzyme lactate dehydrogenase, pyruvate is readily reduced by NADH at pH 7 in aqueous solution,

$$\text{pyruvate} + \text{NADH} + \text{H}^+ \rightleftharpoons \text{lactate} + \text{NAD}^+$$

Determine the value of $\Delta G^{\ominus\prime}$ at pH 7 of this reaction, given that $E_{\ominus}{}'$ at pH 7 of pyruvate|lactate is -0.19 V, and $E_{\ominus}{}'$ at pH 7 of NAD^+|NADH is -0.32 V. (Faraday constant is 96 487 J V^{-1} mol^{-1}.)

8. The biochemically important hydrolysis of acetyl coenzyme A is an exergonic reaction in the living cell. If it takes place according to the equation,

$$\text{acetyl.CoA} + H_2O \rightleftharpoons \text{acetate}^- + H^+ + \text{CoA} \qquad \Delta G^{\ominus} = -15.48 \text{ kJ mol}^{-1}$$

what will be the value of ΔG for this reaction at 298 K and pH 7, when acetate$^-$, CoA and acetyl CoA are all present at 0.01 mol dm^{-3} concentration? (Assume activity coefficients of unity for the purposes of this calculation.)

9. Malic enzyme, purified from mammalian liver or from bacteria, will catalyse the oxidative decarboxylation of malate,

$$\text{malate}^{2-} + \text{NADP}^+ \rightleftharpoons \text{pyruvate}^- + \text{NADPH} + CO_2\ (g)$$

If $\Delta G^{\ominus\prime}$ for this reaction at pH 7 is -1.5 kJ mol^{-1}, calculate its ΔG at pH 7 with the components at 0.01 mol dm^{-3} concentration except for CO_2 at one twentieth of standard atmospheric pressure. (Assume activity coefficients of unity.)

10. The addition of water to fumarate to yield malate,

$$\text{fumarate}^{2-} + H_2O \rightleftharpoons \text{malate}^{2-}$$

is essential to the operation of the tricarboxylic acid cycle in the tissues both of man and of the frog. When this reaction was performed at 25°C at an 'approximately physiological pH', values of -3.68 kJ mol^{-1} and $+14.89$ kJ mol^{-1} were determined for $\Delta G^{\ominus\prime}$ and $\Delta H^{\ominus}$ respectively. Calculate the values of $\Delta G^{\ominus\prime}$ for this reaction at the same pH but, (a) in man at 37°C, and (b) in a frog at 7°C (presumably on a brisk spring day).

11. A saturated solution of the amino acid, L-aspartic acid, at 298 K contained 0.0355 molal undissociated aspartic acid. If:

(a) the molal activity coefficient of undissociated L-aspartic acid in its 0.0355 molal aqueous solution is 0.45,
(b) $\Delta G_f^{\ominus}$ of L-aspartic acid (solid) is -721.4 kJ mol^{-1},
(c) $\Delta G_f^{\ominus}$ of L-aspartate ion in this aqueous solution is -699.2 kJ mol^{-1};

calculate the value of $\Delta G^{\ominus}$ of ionization of L-aspartic acid.

8

Chemical Equilibrium and the Coupling of Reactions

Chemically reversible reactions proceed to a stable state of chemical equilibrium which has characteristic chemical, kinetic and thermodynamic properties.

THE NATURE OF CHEMICAL EQUILIBRIUM

1. The chemical composition of the equilibrium mixture

Suppose that two compounds, A and B, are interconverted in the reaction $A \rightleftharpoons B$. Whatever quantities of A and B are present at the outset, at equilibrium, at a given temperature, they will be present in concentrations whose ratio is characteristic of the reaction. The composition of this equilibrium mixture will not change in an isothermal system.

By convention, the ratio, $\dfrac{\text{Concentration of Product at equilibrium}}{\text{Concentration of Reactant at equilibrium}}$, is termed the equilibrium constant (K_c) of the reaction. Thus,

(i) the equilibrium constant K_c for the reaction $A \rightleftharpoons B$ as it forms B from A equals $[B]/[A]$, where $[B]$ and $[A]$ are the concentrations at equilibrium of B and A. Similarly, the reaction as it forms A from B will have as its equilibrium constant $[A]/[B]$, which is therefore equal to $1/K_c$;

(ii) for $A+B \rightleftharpoons C+D$, the equilibrium constant of the reaction proceeding from left to right as written equals $\dfrac{[C][D]}{[A][B]}$;

(iii) for more complex reactions, e.g.

$$aA+bB+cC \rightleftharpoons xX+yY+zZ$$

K_c of the reaction proceeding from left to right is

$$\frac{[X]^x[Y]^y[Z]^z}{[A]^a[B]^b[C]^c}.$$

The equilibrium constant K_c calculated in this way, is generally only approximately equal to the *true* equilibrium constant of the reaction K_{eq}, and could be very different from it. The value of this true equilibrium constant (also called the activity-based equilibrium constant, K_a) is calculable from the *activities* of the components at chemical equilibrium instead of from their equilibrium concentrations (see later). Though its value changes with temperature, the equilibrium constant defines the position of chemical equilibrium for a chemically reversible reaction at a given temperature, and in this sense the equilibrium state is definable in terms of the composition of the equilibrium mixture.

2. Chemical equilibrium as a dynamic state

Kinetic studies (Chapter 10) demonstrate that the velocity of an isothermal reaction is proportional to the active concentrations of its reactants (this is the Law of Mass Action). Therefore, the velocity of reaction $A+B \rightarrow C+D$ equals $k_1[A][B]$, while the velocity of the reverse reaction $C+D \rightarrow A+B$ is $k_{-1}[C][D]$, where k_1 and k_{-1} are the rate constants of these reactions at the given temperature (p. 263). The unchanging composition of the equilibrium mixture can thus be explained by the fact that, at equilibrium, the active concentrations of the components are such that the rate at which C and D combine to form A and B is exactly equal to the rate at which A and B conversely interact to re-form C and D.

Thus,

$$A+B \underset{k_{-1}}{\overset{k_1}{\rightleftharpoons}} C+D,$$

and at equilibrium,

$$k_1[A][B] = k_{-1}[C][D], \quad \text{or} \quad \frac{k_1}{k_{-1}} = \frac{[C][D]}{[A][B]} = K_c.$$

The equilibrium constant is therefore equal to the ratio of the rate constants of the opposed and nullifying reactions that maintain the equilibrium state.

3. The thermodynamics of chemical equilibrium

In Chapter 7 we identified the driving force behind a spontaneous isothermal reaction, in a closed system at constant pressure, with its tendency to lose Gibbs free energy ($\Delta G = -$ve). This in turn was seen to be the resultant of the dual tendency to change in enthalpy and in entropy so that $-\Delta G = -\Delta H + T\Delta S$. At constant temperature and pressure, the Gibbs free energy of this system tends to decrease to the minimum value which it has at the terminal chemical equilibrium. Since no further (spontaneous) decrease in free energy is possible in the closed system at equilibrium, any subsequent change which takes place whilst equilibrium is maintained can involve no change in Gibbs free energy, i.e. $\Delta G = 0$ and $\Delta H = T\Delta S$.

Let us reconsider the isothermal reaction $A \rightleftharpoons B$. Supposing that the reaction in the direction of synthesis of B has an equilibrium constant of 4, then chemical equilibrium is attained when the concentrations of A and B attain a ratio of 1:4. If it were possible to measure the Gibbs free energy of the system and then to plot this against the chemical composition of the system, a curve would be obtained of the type shown in Fig. 8.1. The Gibbs

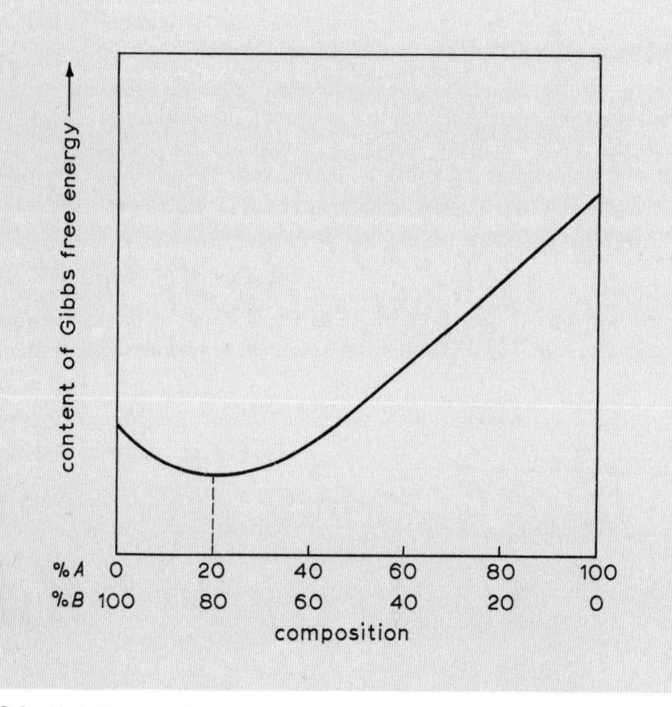

Fig. 8.1 Variation of Gibbs free energy content with composition of reaction mixture.

free energy would be greatest when the mixture contained 100% A, and would be least when only 20% A was present. If the reaction mixture initially consisted of 100% B, the Gibbs free energy (though not now as great as at 100% A) would decrease until it reached the same minimum, when again the mixture would contain 20% A. At chemical equilibrium, when the system contains 20% A and 80% B, it possesses minimal Gibbs free energy. The shape of this 'free energy profile' (Fig. 8.1) further illustrates that no spontaneous change is possible in the system at chemical equilibrium, since any change must yield a mixture of higher free energy content and must therefore be associated with a positive ΔG.

From these considerations, chemical equilibrium at constant temperature and pressure emerges as a state of defined chemical composition in which the Gibbs free energy of the system is at a minimum. Though stable, it is a dynamic state of perfectly opposed, thermodynamically reversible reactions, the ratio of whose rate constants is the equilibrium constant.

Enzyme-catalysed reactions

Since we shall here be chiefly concerned with biochemical reactions catalysed by specific enzymes, it must be mentioned at the outset that catalysis in no way alters the 'overall' thermodynamic characteristics of a reaction. Though an enzyme hastens the attainment of the predestined chemical equilibrium, it does not change it, for catalysis does not alter the value of the equilibrium constant of an isothermal reaction. Nor does an enzyme in any way affect the values of ΔG, ΔS and ΔH of the reaction that it catalyses.

THE RELATIONSHIP BETWEEN THE EQUILIBRIUM CONSTANT AND THE STANDARD FREE ENERGY CHANGE

In Chapter 7 it was stated that the standard free energy change (ΔG°) of a chemical reaction is a function of its thermodynamic equilibrium constant, so that $\Delta G^{\circ} = -RT \ln K_{eq}$, when ΔG° and K_{eq} are measured at T K.

Thus, at 298 K, $\Delta G^{\circ} = -5706 \log K_{eq}$ J mol^{-1}

We may now, and more properly, reinterpret this relationship as meaning that the equilibrium position is defined by the standard free energy change of the reaction. When ΔG° is negative, K_{eq} must be greater than 1 and the equilibrium 'lies to the right-hand side of the reaction as written'. Conversely, a positive value for ΔG° does *not* mean that 'this reaction will not proceed from left to right', but rather that K_{eq} is less than 1 and the equilibrium 'lies to the left-hand side of the equation as written'.

EXAMPLE

The enzyme L-*glutamate-pyruvate aminotransferase catalyses a transamination reaction between glutamate and pyruvate to yield α-ketoglutarate and* L-*alanine,*

$$\text{L-}glutamate + pyruvate \rightleftharpoons \alpha\text{-}ketoglutarate + \text{L-}alanine$$

(*a*) *If the equilibrium constant K_{eq} for the synthesis of* L-*alanine is 1.107 at 298 K, calculate the value of $\Delta G^{\ominus}$.*

(*b*) *Does the result obtained in* (*a*) *mean that the synthesis of alanine from pyruvate plus glutamate is always spontaneous?*

(*c*) *If 10^{-4} mol dm^{-3} each of* L-*glutamate and pyruvate were mixed with 10^{-2} mol dm^{-3} each of α-ketoglutarate and* L-*alanine at 298 K in the presence of the aminotransferase,* (*i*) *what is the value of ΔG for* L-*alanine formation?* (*ii*) *will alanine be formed spontaneously under these conditions?*

(a) For alanine synthesis,

$$\Delta G^{\ominus} = -RT \ln K_{eq} \text{ where } \begin{cases} R & = 8.314 \text{ J K}^{-1} \text{ mol}^{-1} \\ T & = 298 \text{ K} \\ K_{eq} & = 1.107 \end{cases}$$

$$\therefore \Delta G^{\ominus} = -RT \times 2.303 \log K_{eq}$$

$$\therefore \quad \Delta G^{\ominus} = -(8.314 \times 298 \times 2.303 \times \log 1.107) \text{ J mol}^{-1}$$
$$= \underline{-252 \text{ J mol}^{-1}}$$

(b) No. The value of ΔG depends upon the actual conditions of reaction, including the temperature and the concentrations of all reactants and products. This is amplified in (c) below.

(c) (i) $$\Delta G = \Delta G^{\ominus} + RT \ln \frac{[\alpha\text{-ketoglutarate}][\text{alanine}]}{[\text{pyruvate}][\text{glutamate}]}$$ (see p. 204)

(An approximate value of ΔG is obtained using this equation, since it assumes that the ratio of the activity coefficients of the reaction components (multiplied in the same way as the corresponding concentrations) equals unity.)

Substituting the given values for the concentration terms,

$$\Delta G = -252 + \left(8.314 \times 298 \times 2.303 \times \log \frac{10^{-2} \times 10^{-2}}{10^{-4} \times 10^{-4}}\right) \text{ J mol}^{-1}$$
$$= -252 + (5706 \times \log 10^{4}) \text{ J mol}^{-1}$$
$$= -252 + (5706 \times 4) \text{ J mol}^{-1}$$
$$= -252 + 22\,824 \text{ J mol}^{-1}$$
$$= \underline{+22\,572 \text{ J mol}^{-1}}$$

(ii) No. Alanine synthesis under these conditions is endergonic, so that the spontaneous reaction will be the reverse, net conversion of alanine to pyruvate.

Apparently unidirectional reactions

By discussing chemically reversible reactions, we have inferred that there also exist reactions that are chemically irreversible (i.e. unidirectional), of a type that we would represent as $A+B \rightarrow C+D$. Yet these are only extreme cases of chemically reversible reactions in which the equilibrium 'lies so far over to the right-hand side' that the reaction appears to proceed to completion in this direction. In this view, all chemical reactions are chemically reversible, though some are more reversible than others.

The apparently undirectional reaction will be characterized by the very large negative value of its $\Delta G^{\ominus}$ (and hence by the very large value of its K_{eq}).

EXAMPLE

The oxidation of α-ketoglutarate proceeds according to the equation,

$$\alpha\text{-}ketoglutarate + \tfrac{1}{2}O_2 \rightarrow succinate + CO_2$$

and is deemed an essentially unidirectional reaction with a $\Delta G^{\ominus}$ of -286.6 kJ mol^{-1}. What is the equilibrium constant at 298 K of the reverse reaction which yields α-ketoglutarate and oxygen by the carboxylation of succinate?

Since $\Delta G^{\ominus} = -286.6$ kJ mol^{-1} for the oxidation of α-ketoglutarate, $\Delta G^{\ominus} = +286.6$ kJ mol^{-1} for the reverse reaction, namely for the carboxylation of succinate according to the equation,

$$CO_2 + \text{succinate} \rightarrow \alpha\text{-ketoglutarate} + \tfrac{1}{2}O_2$$

$$-RT \ln K_{eq} = \Delta G^{\ominus}$$

$$\therefore \quad -2.303RT \log K_{eq} = \Delta G^{\ominus}, \quad \text{or} \quad \log K_{eq} = \frac{\Delta G^{\ominus}}{2.303RT}$$

whence at 298 K

$$\log K_{eq} = -\frac{286\,000}{5\,706}$$

$$= -50.2$$

$$= \overline{51}.8$$

$$\therefore \quad K_{eq} = \text{antilog } \overline{51}.80 = \underline{6.31 \times 10^{-51}}$$

Activities and activity coefficients

The value of the equilibrium constant of a reaction should be independent of the magnitude of the concentrations of the components if it is only related to the *ratio* of these at equilibrium. Thus, if K_{eq} equals 5 for the reaction $A \rightleftharpoons B$, then it should always equal 5 (at the same temperature), irrespective of whether A and B are present, at equilibrium, in mol dm^{-3} or in μmol dm^{-3} concentrations. Indeed, this must be so, since $-RT \ln K_{eq} = \Delta G^{\circ}$ and ΔG° is a constant for the reaction, whose value refers to the behaviour of the reaction under defined, standard conditions. Yet, when the 'concentration equilibrium constant' K_c is calculated from the concentrations of the components at equilibrium, its value *is* found to depend on their magnitudes. In fact, for aqueous solutions of electrolytes, it is only when a reaction takes place in extremely dilute solution (of small ionic strength, p. 83), that the measured value of K_c equals the value of K_{eq} computed from the value of ΔG° of the reaction. We conclude therefore that just as real gases may depart from ideal behaviour, so too may solutions behave non-ideally, and that, in general, the more concentrated is the solution the less likely is it to behave ideally (see Chapter 4).

To assay the concentration of a solute in a non-ideal solution is merely to enquire 'How much solute is present in a unit volume?' What is often more relevant is to ask 'From the manner in which it behaves in this solution, how much solute appears to be present in a unit volume?' As explained in Chapter 4, the latter value is the *activity* of the solute in the solution, so that the less ideally it behaves, the greater will be the discrepancy between the concentration of the solute and its activity. You will also recall from Chapter 4 (p. 65) that an 'activity coefficient' may be used to express quantitatively this departure from ideality on the part of a solute, i.e.

$$\text{activity coefficient} = y = \frac{\text{activity}}{\text{concentration (in mol dm}^{-3})}$$

$$\text{molal activity coefficient} = \gamma = \frac{\text{activity}}{\text{molal concentration}}$$

(For dilute aqueous solutions the difference between f and γ is negligible; see p. 18 if in doubt concerning the distinction between mol dm^{-3} and molal concentration.) Remember also that the solvent may behave non-ideally in a solution (p. 64), and that the activity of a solute is not always less than its concentration in a solution (p. 91); for example, the molal activity coefficient of HCl is 0.76 in its 0.5 molal aqueous solution at 298 K, but is 1.76 for HCl in its 4 molal aqueous solution at the same temperature.

To allow for this non-ideality, two courses may be followed when studying reactions in aqueous solution.

1. CIRCUMVENTION: by choosing to study reactions in solutions so dilute that they can be assumed to behave ideally. Theoretically, such solutions would have to be infinitely dilute (or of zero ionic strength for electrolytes) but, fortunately, many reactions can be studied in 'real' solutions which are nevertheless sufficiently dilute for it to be possible to

assume that the ratio of the activity coefficients is unity in the equation which yields the value of the equilibrium constant from the composition of the system at equilibrium (p. 212).

2. EXPERIMENTAL DETERMINATION OF THE ACTIVITY COEFFICIENTS: the activity coefficients of solutes in a solution can be deduced from the discrepancy between their defined, ideal behaviour and their experimentally determined behaviour. In practice, measurement of the colligative properties of a solution can yield this information (Chapter 4).

As has been pointed out, the true equilibrium constant K_{eq} of a reaction is related to the *activities* of the reaction components at equilibrium. Since these are 'thermodynamic activities' and since only K_{eq} is calculated from ΔG°, the true, activity-based K_{eq} is also called the ***thermodynamic equilibrium constant***. Its value can therefore differ significantly from the value of the approximate equilibrium constant K_c which is calculated from the concentrations of the components at equilibrium.

Thus, for the reaction $A+B \rightleftharpoons C+D$, the thermodynamic equilibrium constant K_{eq} equals $\frac{(C)(D)}{(A)(B)}$ where (A), (B), etc., are the activities at equilibrium of A, B, etc., but the approximate equilibrium constant K_c equals $\frac{[C][D]}{[A][B]}$ where $[A]$, $[B]$, etc., are the equilibrium concentrations of A, B, etc.

EXAMPLE

Ammonium glutamate is formed by the hydrolysis of glutamine,

$$glutamine^{+-} + H_2O \rightleftharpoons NH_4^{+} + glutamate^{+2-}$$

In an experiment in which chemical equilibrium was established at 298 K, the mixture contained 0.92 millimolal glutamine and 0.98 molal ammonium glutamate.

Given that the molal activity coefficients of the components at equilibrium were as follows: glutamine = 0.94, ammonium glutamate = 0.54 and water = 0.97, calculate the values of: (a) the approximate equilibrium constant K_c, (b) the thermodynamic equilibrium constant K_{eq}.

The hydrolytic reaction is,

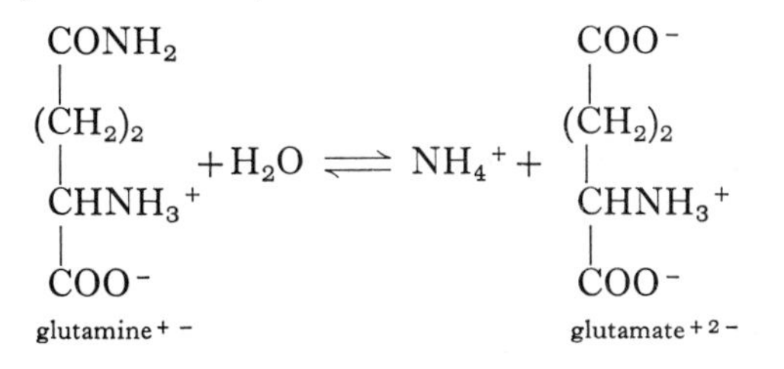

$$\therefore \qquad K_c = \frac{[NH_4^+][\text{glutamate}^{+2-}]}{[\text{glutamine}^{+-}][H_2O]}$$

But $\quad [NH_4^+] = [\text{glutamate}^{+2-}] = [\text{ammonium glutamate}]$

$$\therefore \qquad K_c = \frac{[\text{ammonium glutamate}]^2}{[\text{glutamine}^{+-}][H_2O]}$$

At equilibrium,

$$[\text{ammonium glutamate}] = 0.98 \text{ molal}$$
$$[\text{glutamine}^{+-}] = 0.92 \times 10^{-3} \text{ molal}$$
$$[H_2O] = 1 \text{ (by convention, see p. 206)}$$

Substituting these values in the above equation,

$$K_c = \frac{(0.98)^2}{(0.92 \times 10^{-3}) \times 1} = \frac{0.96}{0.92 \times 10^{-3}} \quad \therefore K_c = \underline{1044}$$

While $\quad K_c = \dfrac{[\text{ammonium glutamate}]^2}{[\text{glutamine}^{+-}][H_2O]},$

$$K_{eq} = \frac{(\gamma_{\text{amm. glut.}} \times [\text{ammonium glutamate}])^2}{(\gamma_{\text{glutamine}} \times [\text{glutamine}^{+-}]) \times (\gamma_{H_2O} \times [H_2O])}$$

So that, $\quad K_{eq} = \dfrac{\gamma^2_{\text{amm. glut.}}}{\gamma_{\text{glutamine}} \times \gamma_{H_2O}} \times K_c$

Since, $\quad \gamma_{\text{amm. glut.}} = 0.54 \qquad \gamma_{H_2O} = 0.97$

$\quad \gamma_{\text{glutamine}} = 0.94 \qquad K_c = 1044$

$$K_{eq} = \frac{(0.54)^2}{0.94 \times 0.97} \times 1044$$

$$= \frac{0.292}{0.91} \times 1044$$

$$\therefore K_{eq} = \underline{334}$$

Note the discrepancy between the value of $K_c = 1044$ and the value of the true equilibrium constant $K_{eq} = 334$.

Note too that in the above, accurate, calculation of the value of K_{eq}, the activity coefficient of H_2O $(=0.97)$ was taken into account. This was made necessary by the high concentration of ammonium glutamate present at equilibrium (0.98 mol kg^{-1}). In more dilute solutions it is usual to assume that the water, as solvent, is present at unit activity.

HOW TEMPERATURE AFFECTS THE VALUE OF THE EQUILIBRIUM CONSTANT

K_{eq} is temperature-dependent, and the amount by which its value is altered by a given change in temperature is related to the standard change in enthalpy associated with the reaction. The way in which the equilibrium responds to changes in temperature is in fact predicted by Le Chatelier's Principle which declares that 'A system in equilibrium reacts to any change in its conditions in a manner that would tend to abolish this change'. This means that the value of the equilibrium constant of an exothermic reaction is *decreased* by an increase in its temperature (under standard conditions less heat would be released in proceeding to the new equilibrium). Conversely, raising the temperature of an endothermic reaction increases the value of its equilibrium constant.

Van't Hoff deduced a quantitative relationship between the values of the equilibrium constant and the absolute temperature of a reaction at constant pressure, which showed that the rate at which the value of K_{eq} changes with temperature is related to the standard enthalpy change ($\Delta H^{\ominus}$) of the reaction as follows:

$$\frac{d(\ln K_{eq})}{dT} = \frac{\Delta H^{\ominus}}{RT^2}$$

When this equation is integrated (assuming that $\Delta H^{\ominus}$ is independent of temperature) we obtain the following relationship†,

$$\ln K_{eq} = \text{constant} - \frac{\Delta H^{\ominus}}{RT}$$

or

$$\log K_{eq} = \text{constant} - \frac{\Delta H^{\ominus}}{2.303RT}$$

Whence, since $R = 8.314\ \text{J K}^{-1}\ \text{mol}^{-1}$, $2.303R = 19.14\ \text{J K}^{-1}\ \text{mol}^{-1}$.

$$\therefore \qquad \log K_{eq} = \text{constant} - \frac{\Delta H^{\ominus}}{19.14} \cdot \frac{1}{T}$$

The value of $\Delta H^{\ominus}$ for a reaction can therefore be obtained graphically from the values of its true equilibrium constant at different temperatures, for if $\log K_{eq}$ is plotted against $1/T$, a straight line is obtained whose slope equals $-\Delta H^{\ominus}/19.14$ (Fig. 8.2). Alternatively, if values of K_{eq} are known at only two temperatures, the value of $\Delta H^{\ominus}$ can be calculated by substitu-

† An alternative derivation of this equation is as follows:

$$\Delta G^{\ominus} = -RT \ln K_{eq}$$

and

$$\Delta G^{\ominus} = \Delta H^{\ominus} - T\Delta S^{\ominus}$$

∴

$$-RT \ln K_{eq} = \Delta H^{\ominus} - T\Delta S^{\ominus}$$

and

$$\ln K_{eq} = \frac{-\Delta H^{\ominus}}{RT} + \frac{\Delta S^{\ominus}}{R}$$

whence

$$\ln K_{eq} = \text{constant} - \frac{\Delta H^{\ominus}}{RT}$$

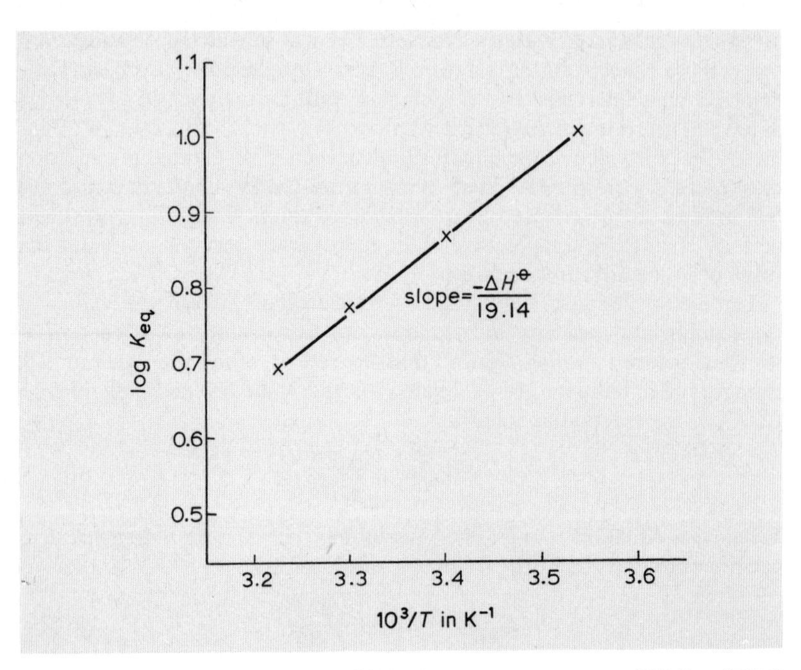

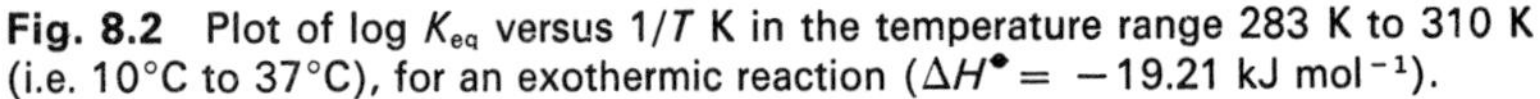
Fig. 8.2 Plot of log K_{eq} versus $1/T$ K in the temperature range 283 K to 310 K (i.e. 10°C to 37°C), for an exothermic reaction ($\Delta H^{\ominus} = -19.21$ kJ mol^{-1}).

tion in the following equation (derived from the integrated form of the van't Hoff equation),

$$\log \left[\frac{K''_{eq}}{K'_{eq}}\right] = \frac{-\Delta H^{\ominus}}{19.14}\left(\frac{1}{T_2} - \frac{1}{T_1}\right)$$

or,

$$\log K''_{eq} - \log K'_{eq} = \frac{+\Delta H^{\ominus}}{19.14}\left(\frac{T_2 - T_1}{T_1 T_2}\right)$$

whence,

$$\Delta H^{\ominus} = \frac{19.14\, T_1 T_2 (\log K''_{eq} - \log K'_{eq})}{(T_2 - T_1)}$$

This means that if we know the values of the equilibrium constant of a reaction at just two temperatures, we can calculate values of $\Delta H^{\ominus}$, $\Delta G^{\ominus}$ and $\Delta S^{\ominus}$ for this reaction at any given temperature T K within this range,

$\Delta H^{\ominus}$ ... from the integrated van't Hoff equation, as above,
$\Delta G^{\ominus}$ ... from the equation $\Delta G^{\ominus} = -2.303 RT \log K_{eq} = -19.14 T \log K_{eq}$,
$\Delta S^{\ominus}$ ... from the values of $\Delta H^{\ominus}$ and $\Delta G^{\ominus}$, since,

$$\Delta S^{\ominus} = \frac{(\Delta H^{\ominus} - \Delta G^{\ominus})}{T}$$

EXAMPLE

An enzyme was specifically activated by the nucleotide AMP when this was bound by the enzyme protein in a chemically reversible reaction,

$$\text{Enzyme} + \text{AMP} \rightleftharpoons \text{Enzyme-AMP}$$

The thermodynamic equilibrium constant of this association reaction was determined at two temperatures with the following results:

Temperature/°C	K_{eq} *of association/dm³ mol⁻¹*
22	1.83×10^3
38	5.78×10^3

Calculate the value of $\Delta H^{\ominus}$ *at 30°C for this AMP-binding reaction.*

Assuming that the value of $\Delta H^{\ominus}$ remains constant over the small range of temperature examined, then,

$$\Delta H^{\ominus} = \frac{19.14\, T_1 T_2(\log K''_{eq} - \log K'_{eq})}{(T_2 - T_1)}$$

where $K'_{eq} = 1.83 \times 10^3 \text{ dm}^3 \text{ mol}^{-1}$, when $T_1 = 22°\text{C} = 295 \text{ K}$
$K''_{eq} = 5.78 \times 10^3 \text{ dm}^3 \text{ mol}^{-1}$, when $T_2 = 38°\text{C} = 311 \text{ K}$

$$\therefore \quad \Delta H^{\ominus} = \frac{19.14 \times 295 \times 311(\log 5.78 \times 10^3 - \log 1.83 \times 10^3)}{(311 - 295)} \text{ J mol}^{-1}$$

$$= \frac{17.56 \times 10^5(3.7619 - 3.2625)}{16} \text{ J mol}^{-1}$$

$$= \underline{54\,810 \text{ J mol}^{-1}}$$

Thus the binding of AMP by the enzyme has a $\Delta H^{\ominus}$ of 54.81 kJ mol⁻¹ at 30°C.

Since the 'best straight line' fitted to several points will minimize the effect of an experimental error in any one point, the graphical method of determining $\Delta H^{\ominus}$ from the values of K_{eq} at three or more temperatures, is always to be preferred to the above method of calculation based on the values of K_{eq} determined at only two temperatures.

REACTIONS THAT INVOLVE PROTONS, PROCEEDING IN BUFFERED MEDIA

Since the many biochemical reactions that liberate or utilize H^+ ions normally take place in buffered media, it is necessary to enquire how this affects their thermodynamic characteristics.

Consider the following proton-liberating reaction,

$$A \rightleftharpoons B + H^+$$

If this reaction takes place in a buffer of high pH (low concentration of H^+ ions), the efficient removal of H^+ ions tends to 'drive the reaction from left to right'. On the other hand, if the reaction consumes protons,

$$A + H^+ \rightleftharpoons B$$

then the higher the pH of the buffer, the less will be the tendency for the reaction to proceed from left to right.

This may be summarized by stating that, as the pH of the buffered medium is increased:

(a) a proton-yielding reaction becomes more exergonic ($\Delta G \rightarrow -$ve),
(b) a proton-consuming reaction becomes more endergonic ($\Delta G \rightarrow +$ve).

If the reaction is studied calorimetrically, then it is important to consider the fate of the protons neutralized by buffer action. As discussed in Chapter 5, a pH buffer is an equilibrium mixture of a weak acid and its conjugate base. So long as this chemical equilibrium is maintained (that is to say, the capacity of the buffer is not exceeded), the uptake of H^+ ions by the buffer cannot be associated with any free energy change (since while equilibrium is maintained $\Delta G = 0$). Yet the process is attended by an enthalpy change whose magnitude depends on the standard heat of neutralization of the buffer. This varies greatly from one buffer to another and can be quite large. Thus if the true value of ΔH for a proton-yielding reaction is to be obtained, ΔH of neutralization of H^+ by the buffer must be subtracted from the total ΔH determined calorimetrically in buffered solution.

EXAMPLE

Calorimetric measurement of the heat evolved when ATP was hydrolysed in 'tris' buffer pH 8 gave a value of ΔH of -68.45 kJ mol^{-1},

$$ATP^{4-} + H_2O \rightleftharpoons ADP^{3-} + HPO_4^{2-} + H^+$$

When the same hydrolysis was performed in phosphate buffer pH 8, the measured value of ΔH was -28.51 kJ mol^{-1}.

If the heat of proton neutralization by the phosphate buffer was -7.53 kJ mol^{-1}, calculate the heat of proton neutralization by the 'tris' buffer.

In phosphate buffer:

'Total' measured ΔH of ATP hydrolysis $= -28.51$ kJ mol^{-1}

But, ΔH of H^+ neutralization by the buffer $= -7.53$ kJ mol^{-1}

$\therefore$ true ΔH of hydrolysis of ATP $= -28.51-(-7.53)$ kJ mol^{-1}
$= -20.98$ kJ mol^{-1}

In 'tris' buffer:

'Total' measured ΔH of ATP hydrolysis $= -68.45$ kJ mol^{-1}

But, true ΔH of hydrolysis of ATP $= -20.98$ kJ mol^{-1}

$\therefore$ required ΔH of H^+ neutralization by 'tris' buffer

$$= -68.45-(-20.98) \text{ kJ mol}^{-1}$$
$$= \underline{-47.47 \text{ kJ mol}^{-1}}$$

Thus ΔH of H^+ neutralization by 'tris' buffer at pH 8 is almost 40 kJ mol^{-1} more negative than the ΔH of neutralization of H^+ in phosphate buffer at the same pH.

True and apparent equilibrium constants of reactions involving protons

Consider a reaction which yields 1 g equiv. of H^+ ions, e.g.

$$A+B \rightleftharpoons C+D+H^+$$

At a given temperature and pH, this reaction proceeds to a state of equilibrium which is defined by its true, thermodynamic equilibrium constant K_{eq} at this temperature, i.e.

$$K_{eq} = \frac{(C)(D)(H^+)}{(A)(B)}$$

where the bracketed terms symbolize the activities of the components of the reaction (including H^+ ions) at equilibrium. Whatever is the value of (H^+) in this equation (i.e. whatever the pH of the buffered medium in which the reaction takes place), the value of K_{eq} remains unchanged, so that K_{eq} is *pH-independent.*

If, however, we omit the (H^+) term from the equilibrium equation, we obtain a new ratio of equilibrium activities equal to K'_{eq}, where,

$$K'_{eq} = \frac{(C)(D)}{(A)(B)}$$

This value of K'_{eq} only defines the equilibrium position that is achieved at a single pH (corresponding to the (H^+) term omitted from the true equilibrium expression). Thus the value of *the apparent equilibrium constant* K'_{eq} *varies with the pH*. It follows that whenever a value of an apparent equilibrium constant is reported, not only the temperature, but also the pH at which it was determined, must be stated.

The values of K_{eq} and K'_{eq} of a reaction are quite simply related to the values of ΔG° and $\Delta G^{\circ\prime}$ at the same temperature (and pH). Thus:

$\Delta G^{\ominus}$: the true standard Gibbs free energy change associated with the reaction at a given temperature. Its value is related to the value of the true equilibrium constant of the reaction at the same temperature,

$$\Delta G^{\ominus} = -RT \ln K_{eq}$$

Values of $\Delta G^{\ominus}$ and K_{eq} are independent of pH.

$\Delta G^{\ominus\prime}$: the modified standard free energy change associated with the reaction at a given temperature and pH (usually pH 7 in biological work). Its value is related to the value of the apparent equilibrium constant of the reaction at the same temperature and pH,

$$\Delta G^{\ominus\prime} = -RT \ln K'_{eq}$$

Values of $\Delta G^{\ominus\prime}$ and K'_{eq} are pH-dependent.

We have seen above that, for the reaction $A+B \rightleftharpoons C+D+\mathrm{H}^+$ at a given temperature and pH,

$$K'_{eq} = \frac{K_{eq}}{(\mathrm{H}^+)}$$

Taking natural logarithms and multiplying throughout by $-RT$, we obtain the equation,

$$-RT \ln K'_{eq} = -RT \ln K_{eq} + RT \ln (\mathrm{H}^+)$$

whence, $$\Delta G^{\ominus\prime} = \Delta G^{\ominus} + 2.303RT \log (\mathrm{H}^+)$$

But, $\log (\mathrm{H}^+) = -\mathrm{pH}$,

$$\therefore \qquad \Delta G^{\ominus\prime} = \Delta G^{\ominus} - 2.303RT\,\mathrm{pH}.$$

This is precisely the relationship between $\Delta G^{\ominus\prime}$ and $\Delta G^{\ominus}$ that we deduced previously for this reaction on the basis of the defined meanings of $\Delta G^{\ominus\prime}$ and $\Delta G^{\ominus}$ (p. 199).

In Chapter 7 (p. 207) we mentioned the possibility that the free energy change associated with a reaction could be pH-dependent because the pH would determine the extent of ionization of one or more of its components. As a specific example, the hydrolysis of an ester CH_3X was considered, since it would result in the formation of the weak acid HX, the extent of whose dissociation would be pH-determined. If we suppose that the pK_a of HX at 298 K is 4, then in aqueous media buffered at pH's lower than about pH 2, HX will be undissociated, and hydrolysis of CH_3X will yield no H^+ ions,

$$CH_3X + H_2O \rightleftharpoons CH_3OH + HX$$

It follows that, at these low pH's, all values of $\Delta G^{\ominus\prime}$ will equal $\Delta G^{\ominus}$.

In media whose pH is greater than pH 6, HX will be fully dissociated,

$$CH_3X+H_2O \rightleftharpoons CH_3OH+X^-+H^+$$

Values of $\Delta G^{\ominus\prime}$ determined in this pH range will differ markedly from $\Delta G^{\ominus}$; for every unit increase in pH above pH 6, the value of $\Delta G^{\ominus\prime}$ becomes more negative by $2.303RT$ J mol^{-1} (equals 5706 J mol^{-1} at 298 K).

In the range of pH values (pH 2 to pH 6) where HX is dissociated to an extent determined by the pH and its acid dissociation constant, the change in the value of $\Delta G^{\ominus\prime}$ that results from a given change in pH, must be calculated using a more complicated equation which takes into account the concurrent change in the extent of dissociation of HX. Equations have been developed which relate $\Delta G^{\ominus\prime}$ to $\Delta G^{\ominus}$ and to the pH and the acid dissociation constants of the components of a reaction (see reference[12]). Here we need merely point out that only when a reaction involves a simple stoichiometric uptake or release of protons, can its values of $\Delta G^{\ominus\prime}$ and K'_{eq} at different pH's be simply related to the corresponding values of $\Delta G^{\ominus}$ and K_{eq}.

COUPLING OF REACTIONS

The 'coupling' of chemical reactions that share a component

The chemical reaction,

$$A+B \rightleftharpoons C+D \qquad (1)$$

achieves equilibrium, at a given temperature and pressure, when the activities of its components attain the particular fixed ratio that is its true equilibrium constant at this temperature; i.e.

$$K_1 = \frac{(C)(D)}{(A)(B)}, \quad \text{and} \quad \Delta G_1^{\ominus} = -RT \ln K_1$$

Similarly for the different reaction,

$$C \rightleftharpoons X+Y \qquad (2)$$

$$K_2 = \frac{(X)(Y)}{(C)}, \quad \text{and} \quad \Delta G_2^{\ominus} = -RT \ln K_2$$

Should these reactions take place simultaneously in the same isothermal system, the final equilibrium is a 'joint equilibrium' which must satisfy the dictates of both equilibrium constants, since C is a reactant common to both reactions.

$$\text{Since } K_1 = \frac{(C)(D)}{(A)(B)}, \quad \text{then } (C) = \frac{K_1(A)(B)}{(D)}$$

and, since $K_2 = \frac{(X)(Y)}{(C)}$, then $(C) = \frac{(X)(Y)}{K_2}$

$$\therefore \quad \frac{K_1(A)(B)}{(D)} = \frac{(X)(Y)}{K_2}$$

whence, $K_1 \times K_2 = \text{a constant} = \frac{(D)(X)(Y)}{(A)(B)}$,

where the bracketed terms symbolize the activities of the components at equilibrium. This new constant, equal to the product of the equilibrium constants of the component reactions, is the equilibrium constant (K_3) of the net reaction,

$$A+B \rightleftharpoons D+X+Y \tag{3}$$

which is the overall result of the 'collaboration' of reactions (1) and (2). Note that although C is present in the equilibrium mixture at an activity determined by the values of K_1 and K_2, it is not represented in the stoichiometric equation of the net reaction (3).

Thus, merely because the reversible reactions (1) and (2) possess one reactant in common (i.e. C), the equilibrium attained by their joint action (defined by K_3) is quite different from either equilibrium that would have been achieved by the same reactions proceeding independently.

The outcome of the concerted action of such 'coupled' reactions is predictable from their individual thermodynamic properties. In the example given above,

$$K_3 = K_1 \times K_2$$

By taking natural logarithms and multiplying throughout by $-RT$, we obtain the following equation,

$$-RT \ln K_3 = -RT \ln K_1 - RT \ln K_2$$

and, since $-RT \ln K_{eq} = \Delta G^{\ominus}$,

$$\Delta G_3^{\ominus} = \Delta G_1^{\ominus} + \Delta G_2^{\ominus}$$ (see p. 194)

Accomplishing reactions with positive values of $\Delta G^{\ominus\prime}$

Several reactions which, it can be calculated, possess a large positive value of $\Delta G^{\ominus\prime}$ at 310 K (i.e. 37°C) and pH 7, appear nevertheless to proceed quite readily at this temperature and pH in living cells. This is particularly true of many of the constituent reactions of biosynthetic pathways. At first sight, this performance of 'endergonic' reactions by the living cell might appear paradoxical, since we know that for a reaction to proceed spontaneously in an isothermal system it must be exergonic. Either or both of the following explanations could, however, account for the occurrence of these reactions:

1 The reaction as it takes place in the cell, does so under conditions so removed from the standard conditions (of unit activities of all components save H^+ at the given pH and temperature) that its actual ΔG is negative even though $\Delta G^{\ominus\prime}$ is positive.
2 The reaction that occurs in the cell is in fact different from the assumed reaction for which the positive value of $\Delta G^{\ominus\prime}$ was calculated. It could be that the assumed reaction forms only one of a coupled pair of reactions whose other member is so highly exergonic under standard conditions that even $\Delta G^{\ominus\prime}$ for the *net* reaction has a negative value. It is this net exergonic reaction, whose ΔG is negative, which proceeds spontaneously in the cell.

Let us suppose that under standard conditions the synthesis of compound D is an endergonic process proceeding according to the following reaction,

$$A+B \rightleftharpoons C+D \qquad \Delta G^{\ominus} = +17.2 \text{ kJ mol}^{-1}$$

Since $\Delta G^{\ominus}$ equals $-RT \ln K_{eq}$, the equilibrium constant K_{eq} for this reaction proceeding 'from left to right' at 298 K is 0.001, whence at equilibrium $(C)(D)/(A)(B) = 0.001$. To form substantial amounts of D from A plus B in a spontaneous reaction, either enormously high starting concentrations of A and B must be employed in the presence of very low concentrations of C and D, or some way of continuously removing the other product of the reaction (C) must be contrived, so that its concentration is kept at a very low level. In either of these situations one can visualize the forward reaction possessing a negative value of ΔG.

To achieve this, the synthesis of D from A plus B is frequently accomplished as part of a two-stage reaction which even under standard conditions is exergonic ($\Delta G^{\ominus}$ negative). This new reaction, which evidently must proceed to an entirely new equilibrium position, is the outcome of the coupling of reaction $A+B \rightleftharpoons C+D$ with another reaction of large, negative $\Delta G^{\ominus}$, which either produces one of the reactants (A or B), or utilizes the other product C. In this sense only, could the coupled exergonic reaction be considered to 'push' or 'pull' the endergonic reaction towards a more favourable equilibrium. It is important that you realize that there is no 'injection of free energy' from the exergonic into the endergonic reaction with which it is coupled. The effect of each upon the other is solely the consequence of their sharing a component and so proceeding to a common equilibrium in the manner described on p. 227. The extent to which the net reaction favours accumulation of the desired product will depend on the values of $\Delta G^{\ominus\prime}$ of the coupled reactions at the same temperature and pH, since,

$$\Delta G^{\ominus\prime}_{net} = \Delta G^{\ominus\prime}_1 + \Delta G^{\ominus\prime}_2$$

Consider the following pair of coupled reactions,

Endergonic:	$A+B \rightleftharpoons C+D$	$\Delta G^{\ominus} = +17.20$ kJ mol^{-1}; $K_{eq} = 1 \times 10^{-3}$
Exergonic:	$C \rightleftharpoons X$	$\Delta G^{\ominus} = -22.91$ kJ mol^{-1}
Net, exergonic reaction:	$A+B \rightleftharpoons D+X$	$\Delta G^{\ominus} = -5.71$ kJ mol^{-1}; $K_{eq} = 10$

Although the endergonic reaction retains its separate identity, coupling with the exergonic reaction results in the establishment of a new equilibrium ($K_{eq} = 10$), and means that now, even under standard conditions, it is feasible for D to be synthesized from A plus B.

Spontaneous reactions likely to remove the primary products of a reaction are quite common, e.g. ionization, decarboxylation, hydrolysis, even precipitation of very insoluble compounds. Such reactions would tend to enhance the feasibility (exergony) of the overall reaction and increase the yield at equilibrium of the desired product. This type of coupling via a shared product/reactant is frequently observed in biological systems, where the separate identity of the component coupled reactions is often evidenced by their being catalysed by different, separable enzymes.

EXAMPLE

L-Glutamate can be synthesized by transamination of its α-oxo-acid precursor (α-ketoglutarate) with L-alanine,

$$\textit{α-ketoglutarate} + \textit{alanine} \rightleftharpoons \textit{glutamate} + \textit{pyruvate}$$

At pH 7 and 298 K $\Delta G^{\ominus\prime}$ is 0.25 kJ mol^{-1} for the reaction proceeding in the direction of glutamate synthesis.

Will the spontaneous formation of glutamate at 298 K and pH 7 under otherwise standard conditions, (and hence its yield at equilibrium), be promoted by coupling this transamination reaction with:

(*a*) *an exergonic reaction in which pyruvate is oxidized to yield acetyl.CoA ($\Delta G^{\ominus\prime} = -258.6$ kJ mol^{-1})?*
(*b*) *an exergonic reaction which yields pyruvate from phospho*enol*pyruvate plus ADP ($\Delta G^{\ominus\prime} = -25.52$ kJ mol^{-1})?*

(a) Yes. Since pyruvate is the second product of the glutamate-forming transamination, any exergonic reaction that utilizes this pyruvate will thereby promote the spontaneous synthesis of glutamate.
(b) No. Coupling with a pyruvate-yielding exergonic reaction would diminish the yield of glutamate at equilibrium by driving the transamination reaction 'from right to left as written'.

The energetics of many a biochemical (particularly biosynthetic) process which is catalysed by a single enzyme is explicable in terms of the simultaneous occurrence of endergonic and more potently exergonic component reactions. Yet when the mechanism of such a process is investigated it becomes evident that these are fictional participants for they prove not to be component steps in the actual reaction, which, although it forms the same products, does so via a different route. Such processes remind us that classical thermodynamics by itself cannot predict the mechanism of any reaction.

A practical example is afforded by the biosynthesis of glutamine in living cells. This is apparently accomplished by the amidation of glutamic acid; a reaction which has a positive value of $\Delta G^{\ominus\prime}$ at 310 K and pH 7 in aqueous solution,

$$\text{Glutamate} + NH_4^+ \rightleftharpoons \text{Glutamine} \quad \Delta G^{\ominus\prime} = +15.69 \text{ kJ mol}^{-1}$$

Yet the actual biosynthetic reaction (catalysed by the enzyme glutamine synthetase) involves ATP as a reactant, and at 310 K and pH 7 in aqueous solution containing Mg^{2+} ions, has a negative $\Delta G^{\ominus\prime}$,

$$\text{Glutamate} + NH_4^+ + \text{ATP} \rightleftharpoons \text{Glutamine} + \text{ADP} + P_i \quad \Delta G^{\ominus\prime} = -15.36 \text{ kJ mol}^{-1}$$

To explain the spontaneity of this reaction we can assume that, in energetic terms, it is the equivalent of the amidation of glutamate plus the exergonic hydrolysis of ATP which yields ADP and P_i. *For the purposes of thermodynamic calculations*, the overall reaction may be regarded as follows (omitting H_2O and H^+ terms, and reporting values of $\Delta G\ '$ at 298 K and pH 7 in aqueous solution containing Mg^{2+} ions):

$$\text{Glutamate} + NH_4^+ \rightleftharpoons \text{Glutamine} \quad \Delta G^{\ominus\prime} = +15.69 \text{ kJ mol}^{-1}$$

$$\text{ATP} \rightleftharpoons \text{ADP} + P_i \quad \Delta G^{\ominus\prime} = -31.05 \text{ kJ mol}^{-1}$$

$$\text{Net: Glutamate} + NH_4^+ + \text{ATP} \rightleftharpoons \text{Glutamine} + \text{ADP} + P_i \quad \Delta G^{\ominus\prime} = -15.36 \text{ kJ mol}^{-1}$$

Yet it has been shown that in fact the reaction catalysed by glutamine synthetase does not proceed in the two separable stages of ATP hydrolysis and glutamate amidation (indeed, how could it, since these hypothetical steps do not share a common reactant/product?). Instead, its mechanism probably involves an intermediate glutamyl-phosphate-enzyme complex which breaks down by reaction with NH_4^+ to yield glutamine, inorganic phosphate, and enzyme.

Incidentally, this type of process, catalysed by a single enzyme, and divisible for purposes of thermodynamic analysis *only* into an exergonic and an endergonic component, is quite frequent in biological systems where the hypothetical exergonic component is generally the hydrolytic cleavage of ATP or of some alternative 'energy-rich' compound (see Chapter 9).

Summary

The relationship of $\Delta G^{\ominus}$ to K_{eq} provides us with a means of determining values of $\Delta G^{\ominus}$, $\Delta H^{\ominus}$ and $\Delta S^{\ominus}$ of reactions (p. 222). Conversely, it also allows us to predict the equilibrium which will be attained by a reaction of known thermodynamic properties, and the 'shift' in this equilibrium position that will result from changes in the conditions of the reaction. Of particular interest to biologists is the fact that suitably coupled isothermal reactions may co-operate to promote the spontaneous synthesis of a desired product. We shall have an opportunity to consider this interesting possibility more fully in Chapter 9, where we shall be particularly concerned with the relevance to biological systems of classical thermodynamic principles and methods.

PROBLEMS

(*Assume activity coefficients of unity throughout, and values of 8.314 J K^{-1} mol^{-1} for the gas constant, R, and 96 487 J V^{-1} mol^{-1} for the Faraday constant,* ***F***.)

1. An enzyme-catalysed reaction has a thermodynamic equilibrium constant (K_{eq}) of 1 at 298 K.

(a) What is the value of $\Delta G^{\ominus}$ of the reaction?
(b) What will be the value of K_{eq} at 298 K in the absence of the enzyme?
(c) If $\Delta H^{\ominus}$ has a positive value, will K_{eq} at 310 K be, (i) <1, (ii) 1, or (iii) >1?
(d) If H^+ ions are produced by the reaction, then at 298 K and pH 7,
  (i) will K'_{eq} be <1, 1, or >1?
  (ii) will $\Delta G^{\ominus\prime}$ have a positive or a negative value?
  (iii) will the reaction be spontaneous under standard conditions (but at pH 7)?

2. The reaction,

$$A+B \rightleftharpoons C+D \qquad \Delta H^{\ominus} = -\text{ve}$$

was allowed to attain chemical equilibrium at 298 K, and from the equilibrium concentrations of the components and their known activity coefficients at these concentrations, K_{eq} was found to equal 5.

(a) Would the formation of A and B from C plus D be an exergonic reaction under standard conditions?
(b) Will an increase in the temperature of reaction enhance the yield of C and D?

(c) Will the addition of component A to the equilibrium mixture increase the yield of C and D?

3. The enzyme, phosphoglucomutase, catalyses the interconversion of glucose 1-phosphate and glucose 6-phosphate,

$$\text{Glucose 1-phosphate} \rightleftharpoons \text{Glucose 6-phosphate}$$

If at chemical equilibrium at 298 K, 95% glucose 6-phosphate is present, calculate:

(a) K_{eq} and ΔG° of the reaction forming glucose 6-phosphate;
(b) ΔG of the reaction in the presence of 10^{-2} mol dm^{-3} glucose 1-phosphate and 10^{-4} mol dm^{-3} glucose 6-phosphate.

4. The following interconversions of phosphohexoses are catalysed by the enzymes phosphoglucomutase and phosphohexoisomerase, respectively,

$$\text{Glucose 1-phosphate} \rightleftharpoons \text{Glucose 6-phosphate} \quad \Delta G^{\circ\prime} = -7.20 \text{ kJ mol}^{-1}$$

$$\text{Glucose 6-phosphate} \rightleftharpoons \text{Fructose 6-phosphate} \quad \Delta G^{\circ\prime} = +2.09 \text{ kJ mol}^{-1}$$

If 0.1 mol dm^{-3} glucose 1-phosphate was added to a buffered solution of these enzymes at pH 7 and 298 K, calculate the concentrations of the three hexose phosphates that would be present in the final equilibrium mixture.

5. The enzyme, malate dehydrogenase, catalyses the oxidation of L-malate to oxaloacetate with the concomitant reduction of NAD^+, thus,

$$\text{malate}^{2-} + NAD^+ \rightleftharpoons \text{oxaloacetate}^{2-} + NADH + H^+$$

At pH 7 and 298 K, K'_{eq} of this reaction is 1.3×10^{-5} (in the direction of oxaloacetate synthesis). Calculate, (a) $\Delta G^{\circ\prime}$ of the reaction; (b) $E_{\circ}'$ of oxaloacetate^{2-}|malate^{2-} if $E_{\circ}'$ of $NAD^+|NADH + H^+ = -0.32$ V.

6. An enzyme of the tricarboxylic acid cycle catalyses the following reactions,

$$\text{citrate} \rightleftharpoons \textit{cis}\text{-aconitate} + H_2O \rightleftharpoons \text{isocitrate}$$

If the equilibrium mixture at 298 K and pH 7.4 contains 90.9% citrate, 2.9% *cis*-aconitate and 6.2% isocitrate, calculate values of $\Delta G^{\circ\prime}$ at pH 7.4 for,

(a) the formation of *cis*-aconitate from citrate;
(b) the formation of isocitrate from *cis*-aconitate;
(c) the formation of isocitrate from citrate.

7. The enzyme, phosphoglycerate mutase, catalyses the following reaction,

$$\text{2-phosphoglycerate} \rightleftharpoons \text{3-phosphoglycerate}$$

If at 298 K and pH 7, $K'_{eq} = 5.8$, and at 310 K and pH 7, $K'_{eq} = 5.45$, calculate the values of $\Delta G^{\circ\prime}$, $\Delta H^{\circ\prime}$ and $\Delta S^{\circ\prime}$ at 310 K and pH 7 for this reaction.

8. Fumarate reacts with water to yield malate in a reaction catalysed by the enzyme fumarase,

$$\text{fumarate}^{2-} + H_2O \rightleftharpoons \text{malate}^{2-}$$

If K'_{eq} at 310 K and pH 7 for this hydration reaction equals 3.3, and $\Delta H^{\ominus\prime} = -16.57$ kJ mol^{-1}, what will be the proportion of malate:fumarate in the equilibrium mixture at pH 7 and 290 K?

9. The following reactions have been demonstrated in mammalian liver at 310 K (i.e. 37°C) and pH 7.5,

$$\text{aspartate} + \text{citrulline} \rightleftharpoons \text{argininosuccinate} + H_2O \qquad \Delta G^{\ominus\prime} = +34.3 \text{ kJ mol}^{-1}$$

$$\text{argininosuccinate} \rightleftharpoons \text{arginine} + \text{fumarate} \qquad \Delta G^{\ominus\prime} = +11.7 \text{ kJ mol}^{-1}$$

$$\text{fumarate} + NH_4^+ \rightleftharpoons \text{aspartate} \qquad \Delta G^{\ominus\prime} = -15.5 \text{ kJ mol}^{-1}$$

Calculate $\Delta G^{\ominus\prime}$ at 310 K and pH 7.5 for the hydrolysis of arginine to citrulline plus NH_4^+,

$$\text{arginine} + H_2O \rightleftharpoons \text{citrulline} + NH_4^+$$

10. Fumarate may be both aminated to yield aspartate, and hydrated to yield malate. Thus at 310 K and pH 7.4,

$$\text{fumarate}^{2-} + NH_4^+ \rightleftharpoons \text{aspartate}^{+2-} \qquad \Delta G^{\ominus\prime} = -15.56 \text{ kJ mol}^{-1}$$

$$\text{fumarate}^{2-} + H_2O \rightleftharpoons \text{malate}^{2-} \qquad \Delta G^{\ominus\prime} = -\ 2.93 \text{ kJ mol}^{-1}$$

Calculate K'_{eq} at 310 K and pH 7.4 for the reaction which forms aspartate from malate,

$$\text{malate}^{2-} + NH_4^+ \rightleftharpoons \text{aspartate}^{+2-} + H_2O$$

11. An enzyme (ATP-sulphurylase), found in many micro-organisms and animal tissues, catalyses the following reaction,

$$ATP^{4-} + SO_4^{2-} + H^+ \rightleftharpoons APS^{2-} + HP_2O_7^{3-}$$

At 310 K and pH 8, $\Delta G^{\ominus\prime}$ for the synthesis of adenosine 5-phosphosulphate (APS) and inorganic pyrophosphate ($HP_2O_7^{3-}$) is +46 kJ mol^{-1}.

Calculate the concentration of APS that would be present at equilibrium at 310 K and pH 8 together with an equal concentration of pyrophosphate, and with ATP 10 μmol cm^{-3}, and SO_4^{2-} 20 μmol cm^{-3}.

12. The formation of APS by the reaction described in the preceding problem is promoted by coupling this reaction with the hydrolysis of pyrophosphate (catalysed by an inorganic pyrophosphatase),

$$HP_2O_7^{3-} + H_2O \rightleftharpoons HPO_4^{2-} + H_2PO_4^-$$

This is a quite common mechanism used in biological systems for promoting the spontaneous synthesis of a desired end product.

If K'_{eq} at 310 K and pH 8 for the hydrolysis of pyrophosphate is 3.3×10^4, calculate the values at this temperature and pH of:

(a) $\Delta G^{\ominus\prime}$ for the synthesis of APS from ATP and SO_4^{2-}, that is accomplished by the 'coupled' action of ATP-sulphurylase and inorganic pyrophosphatase;

(b) $\Delta G^{\circ\prime}$ for the biosynthesis of UPD-glucose from UTP and glucose 1-phosphate, that is catalysed by the concerted action of UDP-glucose pyrophosphorylase and pyrophosphatase, if the former enzyme catalyses the reaction,

$$\text{Glucose 1-phosphate} + \text{UTP} \rightleftharpoons \text{UDP-glucose} + \text{PP}_1$$

which has a K'_{eq} of approximately 1 at 310 K and pH 8.

9

The Application of Thermodynamics to Biochemistry

Until quite late in the nineteenth century, natural philosophers, envious of the synthetic versatility of living organisms, excused their comparative ineptitude in the laboratory by assuming that living organisms were possessed of a mysterious vital force. The energy derived from this vital force was available only to animate systems, and could accomplish synthetic tasks that the most gifted experimentalist could not emulate in his test tubes. After this palliative theory was discredited in 1828 by Wohler's synthesis of an organic compound (urea) from inorganic materials, organic synthetic chemistry grew progressively more accomplished until it was obvious that no compound existed whose synthesis need be the sole prerogative of living cells. The final refutation of the hypothesis of vital force, and the success of several of the new techniques of the organic chemist, were products of the newly emergent theories of classical thermodynamics. Yet even thermodynamicists remained unduly suspicious of living processes. They saw that the growth and reproduction of living cells involved the creation of order from chaos, and a decrease in entropy which seemingly defied their vision of a universe losing free energy and committed to ultimate stagnation at maximum entropy. Thus while living systems were acquitted of the charge of wizardry, new charges were laid of non-conformity ('Life swims against the stream of entropy') or non-ethical behaviour ('Life cheats in the game of entropy').

Today, the behaviour of living organisms is unlikely to arouse such comment. Over the years it has become obvious that the living cell is scarcely more efficient in its energetic transactions than many a man-made machine;

the synthesis and maintenance of cellular material are only made possible either by the degradation of much larger quantities of foodstuffs than the cell assimilates, or, in the case of photosynthetic organisms, by the receipt of solar energy. Living things do not so much 'swim against the stream of entropy' as avail themselves of the 'energetic wake' of greater, spontaneous processes. Fortunately, the thermodynamic problems posed by living organisms provoked more than emotive discussion; they were also a major stimulus for the thermodynamic evaluation of non-equilibrium, 'steady state' systems.

Thermodynamics of open systems

Chapters 7 and 8 dealt with isolated systems, and with closed systems, to which the classical laws of equilibrium thermodynamics are readily applicable (p. 182). Yet so-called 'open systems' also exist that can take in or expel both energy and matter, and whose confines are therefore merely geographical. A living cell constitutes an open system: for example, the bacterium that continuously acquires foodstuffs from its external environment and excretes waste products into its culture medium. Such an open system can achieve a 'steady state' in which the composition of the system is maintained constant by a balanced intake and outflow of matter and energy. A branch of thermodynamics has been evolved to deal with these systems, and while it is beyond the scope of these chapters, it is evident that any studied treatment of the thermodynamics of the intact living cell must proceed along these lines. But even the *irreversible thermodynamics* that deals with the energetics of open systems is severely limited in its treatment of living organisms since it is applicable to the throughput of matter through a sequence of linked chemical reactions only when these component reactions are very close to equilibrium, and it is known that this is certainly not the case for many key metabolic sequences. Indeed, to quote the aphorism of Gowland Hopkins that 'Life is a dynamic equilibrium in a polyphasic system' is not to summarize our conclusions, but rather to state the interpretative problem that is posed by the living system.† Fortunately, as in so many other fields of biochemical study, we can achieve some knowledge of the nature of the whole organism by disassembling it and examining its parts. To these non-animate fragments we can apply the precepts of classical thermodynamics whilst recognizing that the coordinated functioning of these parts in the whole organism requires a modified treatment.

† Even if, during a period of great biosynthetic activity, a living cell actually decreases in entropy, this would be only at the expense of an equal or greater increase in the entropy of its surroundings. In its dynamic steady state, the rate of net entropy production by an open system is at its minimum. Thus, by maintaining a steady state relationship with its environment, a living cell will produce entropy at a minimum rate and will operate most 'efficiently'.

Experimental determination of the thermodynamic characteristics of a biochemical reaction

Any of the methods already discussed in Chapter 7 which yield values of $\Delta G^{\ominus}$, $\Delta H^{\ominus}$ and $T\Delta S^{\ominus}$ for chemical reactions are, of course, equally applicable to biochemical reactions involving only metabolites of small molecular weight. That these reactions are catalysed by specific enzymes is irrelevant, since the enzyme does not alter the position of chemical equilibrium or change in any way the thermodynamic constants of the reaction. The difficulties that attend the thermodynamic study of a biochemical process generally stem from the fact that it might often only be possible to examine the reaction in extremely dilute solution at moderate temperatures, either because heat-labile compounds of very high molecular weight are involved, or because rare, and hence expensive, metabolites are consumed. In these circumstances the usual methods of calorimetry are frequently impracticable since these require the use of large quantities of reactants (grammes rather than milligrammes). Furthermore, it is not practicable to obtain $\Delta H^{\ominus}$ for the interaction between a protein and a metabolite of small molecular weight by determining the difference between the standard molar heats of formation of products and reactants; even if it were possible to obtain such values of $\Delta H_f^{\ominus}$, the desired $\Delta H^{\ominus}$ of interaction is likely to be much smaller than the error in determining the very large $\Delta H_f^{\ominus}$ of a protein of only modest molecular weight. Thus, either the thermodynamic characteristics of such biochemical reactions must be obtained by non-calorimetric methods, or by calorimetry of very great sensitivity.

(a) *Determination of K_{eq} or $\Delta E^{\ominus}$ at different temperatures*

If the relevant activity coefficients can be determined, measurement of the concentrations of the reaction components at chemical equilibrium gives the equilibrium constant (K_{eq}) for the reaction. Alternatively, the value of K_{eq} may be derived from kinetic measurements (Chapter 10). Analysis of the composition of the equilibrium mixture is usually limited to the more readily reversible reactions in which K_{eq} is neither very large nor very small, since otherwise either the reactant or product concentrations would be immeasurably small at equilibrium. Yet the use of radioactively labelled reactants, by increasing the sensitivity of assay of the components at equilibrium, has extended this method to reactions that would formerly have been deemed 'almost irreversible'.

From the values of the K_{eq} at different temperatures, the standard thermodynamic constants of the reaction can be obtained by three calculations:

(i) $\Delta G^{\ominus}$... $\Delta G^{\ominus} = -RT \ln K_{eq}$ (p. 196),

(ii) $\Delta H^{\ominus}$... if $\log K_{eq}$ is plotted against $1/T$, the slope of the resultant straight line equals $-\Delta H^{\ominus}/2.303R$ (p. 222),

(iii) $\Delta S^{\ominus}$ . . . obtained from the results of (i) and (ii) by applying the equation:

$$\Delta G^{\ominus} = \Delta H^{\ominus} - T\Delta S^{\ominus} \quad \text{(p. 189)}.$$

The values of $\Delta E^{\ominus}$ for an oxidation-reduction reaction at different temperatures are treated in a similar manner (p. 197).

An instructive paper by K. Burton and H. A. Krebs (*Biochem. J.* **54**, 94, 1953) well illustrates the application of such methods to reactions of biological interest.

(b) *Microcalorimetry*

A. V. Hill, in the 1920s, by using sensitive thermocouples to measure heat changes in muscle during contraction and relaxation, elegantly demonstrated the possibility of applying calorimetric methods to physiological situations. More recently, largely due to the pioneering work of Sturtevant and his colleagues,[34] and of Charlotte Kitzinger and T. H. Benzinger, both in the U.S.A., methods of microcalorimetry have been developed which measure the minute heat changes accompanying reactions between physiological concentrations of biochemical substances. The sensitivity of the method of 'heat burst microcalorimetry' used by Kitzinger and Benzinger is illustrated by the fact that it can record the heat change associated with the interaction of an enzyme with its substrate, or the heat generated when an antigen reacts with its specific antibody, when, since the protein antibody might be used at a concentration of only 10^{-5} mol dm^{-3}, the equipment has to be capable of detecting temperature changes of 10^{-6} K. It has further been proved possible to derive values of $\Delta G^{\ominus}$, $\Delta H^{\ominus}$ and $\Delta S^{\ominus}$ (and hence K_{eq}) for a reaction proceeding in very dilute solution, from just two heat measurements. As this equipment becomes more generally available, it is likely that the direct methods of microcalorimetry will more often be applied to biochemical problems. These techniques are fully explained in a review article by Kitzinger and Benzinger,[25] and in a more recent textbook of biochemical microcalorimetry.[18]

Assured that it is possible to measure the thermodynamic constants of biochemical processes, we can now enquire why and when it is desirable to do so.

BIOCHEMICAL RELEVANCE OF CLASSICAL THERMODYNAMICS

(a) Application to individual biochemical reactions

The thermodynamic properties of any chemical reaction are characteristics of that reaction which can be determined by calorimetric measurements. These measurements (i) do not interfere with the course of the reaction, and (ii) may be performed on any chemical reaction irrespective of the nature of its reaction components. This 'non-specificity' of microcalorimetry could be useful, for example, in determining the range of metabolites (substrates, inhibitors, activators) that react with a newly isolated enzyme; but besides indicating the occurrence of a reaction and aiding in its characterization, thermodynamic findings can disclose novel aspects of its chemistry. Thus, Laki and Kitzinger in 1956 measured ΔH for the action of the enzyme thrombin upon its substrate fibrinogen (a reaction involved in the clotting of blood). They obtained different values for ΔH depending upon the heat of neutralization of the buffer in which the reaction was conducted, and were able to attribute these differences to the liberation of 2 protons per molecule of fibrinogen utilized. From this observation they concluded that 2 peptide bonds were split during the reaction.

The determination of entropy changes in biochemical reactions is currently of even greater interest, since it has become evident that the three-dimensional shape of a macromolecule often determines its biological activity. Since changes in molecular shape are usually reflected in an attendant entropy change, this enhances the practical importance of the determination of entropy values. Thus certain compounds, 'allosteric effectors', specifically modify the catalytic activity of an enzyme when they combine with its molecule at a site different from that occupied by its normal substrate, and are thought to alter the configuration of the protein molecule to which they are attached (p. 306). Such distortion, if large, should be reflected in a detectable entropy change accompanying the interaction of effector and protein. An apt example is provided by the findings of Worcel, who in 1965 studied the activation of a respiratory enzyme ($NADH_2$ dehydrogenase of *Mycobacterium tuberculosis*) by the nucleotide adenosine 5′-phosphate (AMP), and found that it was (reversibly) bound to the surface of the enzyme molecule at a single site, i.e.

$$E + \text{AMP} \rightleftharpoons E\text{-AMP}$$

By measuring the equilibrium constant for this reaction at several temperatures, he was able to calculate $\Delta H^{\ominus}$ and $\Delta G^{\ominus}$ at 303 K for the binding of AMP as $+52.3$ and -20 kJ mol^{-1} respectively.

Since
$$\Delta G^{\ominus} = \Delta H^{\ominus} - T\Delta S^{\ominus}$$

then,
$$\Delta S^{\ominus} = \frac{\Delta H^{\ominus} - \Delta G^{\ominus}}{T}$$

whence, $$\Delta S^{\ominus} = \frac{+52\,300-(-20\,000)}{303}\ \text{J K}^{-1}\ \text{mol}^{-1}$$

$$= \frac{+72\,300}{303} = +238\ \text{J K}^{-1}\ \text{mol}^{-1}$$

Worcel concluded from these findings that 'the AMP activation of the $NADH_2$ dehydrogenase is an exergonic, endothermic reaction with a large increase in entropy'. In the same way as the heat denaturation of trypsin is accompanied by an increase in entropy of 891 J K^{-1} mol^{-1} and is interpreted as being due to an unfolding (randomization) of the molecule, it would appear that attachment of AMP to the molecule of this $NADH_2$ dehydrogenase caused a similarly pronounced conformational change in the molecule.

(b) Application to concerted biochemical reactions

Perhaps even more important than the information concerning individual biochemical reactions that can be derived from thermodynamic data is the insight which such information gives us into the cooperative functioning of these reactions during metabolism.

To a thermodynamicist, the criteria that distinguish a 'living' organism, viz. growth, reproduction, sensitivity, maintenance of a highly ordered structure in an alien environment, all represent the continual expenditure of free energy. This means, in turn, that the organism (which operates isothermally) must couple its apparently endergonic tasks to greater exergonic processes.

Imagine a living organism to contain a compound (Y) whose synthesis from a precursor (X) would be, under given conditions, an endergonic task, i.e.

$$X \longrightarrow Y \qquad \Delta G^{\ominus\prime} = +\text{ve}$$

This in turn means that, under the same conditions, the breakdown of Y to yield X is an exergonic reaction,

$$Y \longrightarrow X \qquad \Delta G^{\ominus\prime} = -\text{ve}$$

Now, every living cell relies for the maintenance of its viability on the occurrence of certain highly exergonic processes (viz. fermentation, respiration or the light reactions of photosynthesis). Were these processes to synthesize Y from X, and were the anabolic reactions of the cell then to proceed spontaneously as a consequence of the participation of Y as a reactant with the regeneration of X, the crucial linking (coupling) role of the $X \rightleftharpoons Y$ reaction becomes obvious.

Highly exergonic processes
(e.g. fermentation, respiration, photosynthesis)

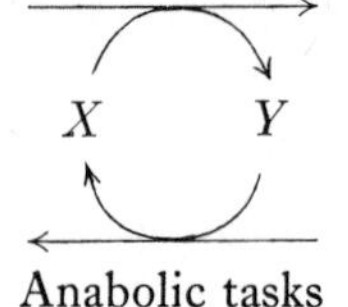

Anabolic tasks

Biochemists have identified the $X \rightleftharpoons Y$ system, and have found that it is essentially identical in all living creatures. The reaction involved is the synthesis of adenosine triphosphate (ATP) from adenosine diphosphate (ADP) and inorganic phosphate (P_i),

$$ADP + P_i \rightleftharpoons ATP \qquad \Delta G^{\ominus} = +ve$$

The 'driving' exergonic process is coupled with this reaction which then proceeds spontaneously in the direction of synthesis of ATP, while the otherwise endergonic task is also accomplished spontaneously, by being coupled with the hydrolysis of ATP to yield ADP and P_i (p. 246).

Further biochemical unity is disclosed by the finding that despite the apparent diversity of energy sources 'tapped' by living organisms (e.g. photosynthetic organisms that utilize sunlight, anaerobes that ferment foodstuffs, aerobes that respire) the generation of ATP from ADP and P_i is normally effected by coupling with an oxidation process. Thus the means by which an organism derives its energy can be represented by an equation such as:

$$AH_2 + ADP + P_i \rightleftharpoons A + ATP + (2H) \qquad \Delta G^{\ominus\prime} = -ve$$

Similarly, by representing their 'anabolic tasks' by the one equation,

$$\text{Substrates } (S) \rightleftharpoons \text{Products } (P) \qquad \Delta G^{\ominus\prime} = +ve,$$

the role of ATP might be designated as

$$S + ATP \rightleftharpoons P + ADP + P_i \qquad \Delta G^{\ominus\prime} = -ve$$

The function of ATP can therefore be expressed diagrammatically as in Fig. 9.1.

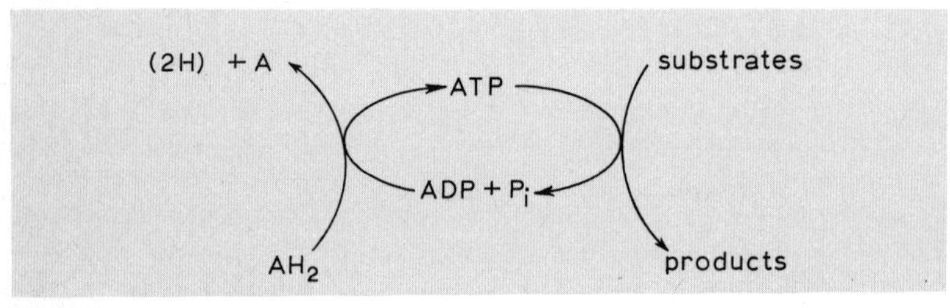

Fig. 9.1

Even to outline the known mechanisms of ATP generation and utilization in living cells would require several chapters, so that at this point I must refer you to other authors, e.g. Lehninger[10, 28], and Krebs and Kornberg.[27] You will find that although a great deal is now known concerning the pathways of electron transport whereby exergonic oxidation processes are accomplished, less has yet been discovered of the mechanism whereby these are coupled to the synthesis of ATP. Very much more information is available concerning the mechanisms whereby ATP is utilized in a host of reactions, though the situation may be complicated by the fact that ATP may first be employed to synthesize other 'activated', or 'high energy', compounds that will be the true substrates in the final 'free energy-requiring' process.

HIGH ENERGY COMPOUNDS

Early calculations of the value of $\Delta G^{\ominus\prime}$ at pH 7 for the hydrolysis of ATP to ADP and P_i suggested a figure of about -50 kJ mol^{-1}. This contrasted startlingly with the much lower values of $\Delta G^{\ominus\prime}$ for the hydrolysis of 'normal' phosphate esters, e.g. glycerol monophosphate or glucose 6-phosphate, which fell in a range of about -9 to -17 kJ mol^{-1}. Subsequently an 'activated form' of acetate was identified as acetyl phosphate whose $\Delta G^{\ominus\prime}$ of hydrolysis was again remarkably high (about -44 kJ mol^{-1}), and similarly high values of $\Delta G^{\ominus\prime}$ were found for the hydrolysis of certain other phosphate derivatives involved in 'energy-yielding' metabolic routes, e.g. 1,3-diphosphoglycerate, phospho*enol*pyruvate and creatine phosphate. These findings suggested that two distinct classes of phosphorylated metabolites could be distinguished. Those that were 'high energy compounds' possessed values of $\Delta G^{\ominus\prime}$ for hydrolysis of the order of -40 kJ mol^{-1}, whilst the 'low energy phosphates' had a 'normal' value of $\Delta G^{\ominus\prime}$ for hydrolysis of the order of -12 kJ mol^{-1}. It was further proposed that this classification could be extended to cover other 'activated' metabolites such as acyl thioesters of coenzyme A for which the value of $\Delta G^{\ominus\prime}$ for hydrolysis was again of the order of -40 kJ mol^{-1}.

The bond whose hydrolytic cleavage was attended by a 'high' negative value of $\Delta G^{\ominus\prime}$ became designated a 'high energy bond', a term popular for some time with biochemists, particularly as it could be represented symbolically as $\sim$. By the use of this shorthand, attention could be quickly drawn to the 'special' nature of, for example, acetyl phosphate, butyryl coenzyme A or ATP, by representing these compounds as acetyl$\sim$P, butyryl$\sim$CoA and ADP$\sim$P. But this introduction of the term 'high energy bond' was in many ways unfortunate. Chemists properly scorned its usage, firstly, because it could be confused with the term 'bond energy' which was already applied to the value of $\Delta H^{\ominus}$ for the breakage of a chemical bond, and secondly, because it falsely suggested that the chemical bond between the phosphate or CoA moiety and the remainder of the molecule was in some mysterious way 'stronger' or more highly 'energized' than the actual covalent bond that they knew to be involved. Indeed it is about as sensible to try to identify a 'high energy bond' in a 'high energy compound' as it is to attempt to attribute the tone of a violin to any one of its structural members. The tone of the instrument is a property of the instrument as a whole, evident only when it is played. So, too, the special property of the 'high energy compound' is an attribute of the molecule which is only in evidence when it reacts in a given manner. Thus the concept of the 'high energy bond' has been supplanted by a more meaningful assessment of 'high energy' metabolites (see Conn and Stumpf[1]). This has been facilitated by the more recent finding that even amongst the phosphorylated metabolites no exact division into 'low energy' and 'high energy' categories is possible, for the values of $\Delta G^{\ominus\prime}$ for their hydrolysis actually form a spectrum from about -10 to -50 kJ mol^{-1}. More exact estimates of the value of $\Delta G^{\ominus\prime}$ for hydrolysis of ATP have yielded a figure of -31 kJ mol^{-1} (at 309 K and pH 7 in the presence of Mg^{2+} ions) which is much less than the corresponding value of -54.4 kJ mol^{-1} for phospho*enol*pyruvate, though still considerably greater than the -13.8 kJ mol^{-1} of glucose 6-phosphate (Table 9.1). Furthermore, the enthalpy of hydrolysis of ATP at pH 8 to yield ADP and P_i is -20.1 kJ mol^{-1}, which is very close to an average value of $\Delta H^{\ominus}$ for the cleavage of any phosphoester bond.

Table 9.1 Values of the modified standard free energy of hydrolysis (at pH 7) of some biologically important substances

	$\Delta G^{\ominus\prime}$ / kJ mol^{-1}
Phospho*enol*pyruvate	−54.4
1,3-diphosphoglycerate	−49.4
Acetyl phosphate	−43.9
Creatine phosphate	−37.7
Acetyl coenzyme A	−32.2
ATP ($\xrightarrow{Mg^{2+}}$ AMP+PP_i)	−31.8 (at 310 K)
ATP ($\xrightarrow{Mg^{2+}}$ ADP+P_i)	−31.0 (at 310 K)
Phosphodiesters	−25.1
Glucose 1-phosphate	−20.9
Glucose 6-phosphate	−13.8
Glycerol 1-phosphate	−9.6

Fortunately, despite some initial confusion regarding the nature of 'high energy compounds', the basic theory of their bioenergetic role was not misconceived. It has now achieved maturity in the recognition that ATP plays its special role in metabolism, in part, because the value of $\Delta G^{\circ\prime}$ for its hydrolysis to ADP *is* intermediate between that of the very high energy compound and that of the low energy compound (see Lehninger,[10, 28]).

Klotz[26] has pointed out that the $\Delta G^{\circ\prime}$ of hydrolysis of ATP (or of any other so-called 'high energy compound') may quite usefully be considered as a 'group transfer potential' which is an index of the degree of readiness with which the molecule transfers a group (e.g. phosphate) to water (which serves the role of a standard group acceptor). Thus creatine phosphate ($\Delta G^{\circ\prime}$ of hydrolysis $= -37.7$ kJ mol^{-1}) has a greater phosphate group transfer potential than ATP ($\Delta G^{\circ\prime}$ of hydrolysis $= -31$ kJ mol^{-1}) which in turn has a considerably greater phosphate group transfer potential than glucose 6-phosphate ($\Delta G^{\circ\prime}$ of hydrolysis $= -13.8$ kJ mol^{-1}). Viewed in this way, immediate analogies are discernible with the indices which we more commonly use for (i) acidity—where the proton transfer potential is measured as pK_a which is $\propto \Delta G^{\circ}$ per mol H^+ transferred (since $pK_a = \Delta G^{\circ}/2.303RT$ for $AH + H_2O \rightleftharpoons A^- + H_3O^+$), and (ii) redox potential—where the electron transfer potential E° is $\propto -\Delta G^{\circ}$ for the reaction $A^- + H^+ \rightleftharpoons A + \frac{1}{2}H_2$ (since $E^{\circ} = -\Delta G^{\circ}/nF$).

When we assess the feasibility of certain oxidation reactions from the E_h values of the participating redox couples (Chapter 12), even though these electron transfer potentials are measured on the 'hydrogen scale' (in which the standard $H^+|\frac{1}{2}H_2$ electrode is the ultimate reference), we do not infer that hydrogen is an obligatory reactant/product in these reactions. Similarly, just because the phosphate group transfer potential of ATP is measured on a 'hydrolytic scale' with water as the standard reference acceptor, we must not infer that whenever ATP is employed to phosphorylate some metabolite the reaction must proceed via hydrolysis of ATP (see p. 246).†

Mechanisms of some reactions that synthesize and utilize ATP

One reaction used by many organisms as a source of ATP effects the synthesis of ATP by 'substrate phosphorylation' in which phospho*enol*pyruvate (PEP) acts as the phosphate donor. In order to assess the thermodynamic feasibility of this process we might justifiably consider it to be the resultant of the following components:

† We must also remember that ATP is not only employed as a phosphorylating agent. For example, in the synthesis of PRPP from ribose 5-phosphate it acts as a pyrophosphorylating agent, while in its reaction with methionine to yield S-adenosylmethionine it behaves as an alkylating agent.

Endergonic: $ADP + P_i \rightleftharpoons ATP + H_2O$ $\Delta G^{\ominus\prime} = +31.0$ kJ mol^{-1}

Exergonic: $PEP + H_2O \rightleftharpoons Pyruvate + P_i$ $\Delta G^{\ominus\prime} = -54.4$ kJ mol^{-1}

Net reaction: $PEP + ADP \rightleftharpoons Pyruvate + ATP$

$\Delta G^{\ominus\prime} = -23.4$ kJ mol^{-1}

This informs us that under modified standard conditions (pH 7) the synthesis of ATP from PEP and ADP would be a spontaneous process ($\Delta G^{\ominus\prime} = -23.4$ kJ mol^{-1}; $K'_{eq} = 1.26 \times 10^4$). Yet you must not imagine that in order to utilize PEP to synthesize ATP the living organism actually carries out the two separate reactions of PEP hydrolysis followed by ATP synthesis from ADP plus P_i, even though these reactions which 'share' P_i as a common reactant/product *could* be chemically coupled. In fact, no inorganic phosphate is liberated from PEP when the enzyme pyruvate kinase catalyses the overall process by quite a different mechanism, viz: transference of a phosphate group from PEP to ADP.

Similarly, if ATP were used in modified (pH 7) standard conditions to effect the spontaneous synthesis of glucose 6-phosphate from glucose, the reaction could, for ease of thermodynamic analysis, be considered as the equivalent of phosphorylation of glucose coupled to the exergonic hydrolysis of ATP:

Endergonic: $Glucose + P_i \rightleftharpoons$ Glucose 6-phosphate

$\Delta G^{\ominus\prime} = +13.8$ kJ mol^{-1}

Exergonic: $ATP \rightleftharpoons ADP + P_i$

$\Delta G^{\ominus\prime} = -31.0$ kJ mol^{-1}

Net reaction: $Glucose + ATP \rightleftharpoons$ Glucose 6-phosphate + ADP

$\Delta G^{\ominus\prime} = -17.2$ kJ mol^{-1}

But again, no inorganic phosphate is actually produced or consumed when a phosphate group is transferred from ATP to glucose in the exergonic reaction ($\Delta G^{\ominus\prime} = -17.2$ kJ mol^{-1}) catalysed by hexokinase.

Yet another example is afforded by the participation of ATP in the synthesis of glutamine that is catalysed by the enzyme glutamine synthetase (p. 231).

Remember therefore, that although for the purposes of thermodynamic analysis a reaction may be treated as the resultant of two or more processes of known thermodynamic properties, it must not be assumed that this represents its actual reaction mechanism.

Both the potency and shortcomings of classical thermodynamics are evident in its application to the biological situation. Yet the manner in which living organisms obtain and expend energy can to some extent be interpreted in its terms. The picture that emerges, of the universal employment of ATP as the prime coupling agent which, by providing the necessary mechanistic link with highly exergonic processes renders feasible the accomplishment of otherwise implausible tasks, discloses a basic unity beneath the apparent diversity of form and physiology of living organisms. In return, the complexity of the thermodynamic situation in the intact living cell has directed attention to 'open systems' in general, and has fostered the development of steady state thermodynamics.

10

The Kinetics of Chemical Reactions

The feasibility and extent of a chemical reaction can be predicted from the thermodynamic properties of its reactants and products, but these tell us nothing of its speed and mechanism. This information must be obtained by kinetic measurements whose aim is to determine the manner and rate at which molecules interact under certain conditions, or would react were these conditions changed.

Consider a highly exergonic reaction $A+B \rightarrow P+Q$. This equation presents the stoichiometry of the reaction, stating that 1 mole of A reacts with 1 mole of B to yield 1 mole each of P and Q in a reaction that effectively proceeds to completion. But is the reaction fast or slow? Does it occur in one or in several stages? These questions are answered not by the stoichiometric equation but by experiment.

Measurement of the velocity of a reaction

The velocity of a chemical reaction can be measured either as the rate of formation of one or more of its products, or as the rate of utilization of one or more of its reactants. If the reaction is homogeneous and occurs entirely in the gaseous phase, the partial pressures of its components could be followed; if the reaction takes place in solution, concentrations could be measured. Thus the units of velocity of a reaction in solution will be units of concentration per unit time, e.g. nmol of reactant utilized cm^{-3} s^{-1}. Experimental conditions such as temperature, pressure, etc., will affect the velocity of the reaction, and before one can predict the new velocity of the reaction under changed conditions, the way in which it changes when each of these conditions is individually varied must first be determined.

INFLUENCE OF REACTANT CONCENTRATIONS ON THE VELOCITY OF A REACTION

It is reasonable to suppose that before two molecules can interact they must first meet. It is therefore not surprising to find that when the concentrations of its reactants are increased, the velocity of a reaction usually increases; the chance of different molecules meeting will be proportional to the number of each that is present in a given space (i.e. their concentrations).

It is found in practice that the velocity of a reaction can be expressed as a simple function of the concentrations of certain of its reactants. This expression is the ***rate equation*** for the reaction which has the general form,

$$\text{rate of reaction} = \text{constant} \times [\text{reactant(s)}]^n$$

The value of the exponent (n) in its rate equation is the ***order of the reaction.***

Thus, isothermal, homogeneous reactions at constant pressure between reactants A, B, C, etc., can be classified according to their kinetic behaviour by grouping together those which exhibit the same value of n in their experimental rate equations. For simple reactions, n is generally a small integral number, and they are 1st, or 2nd, or rarely, 3rd order reactions where:

1st order: the rate of reaction is proportional to the concentration of only one reactant

$$\text{velocity} = \text{constant} \times [\text{A}] = k[\text{A}]$$

2nd order: the rate of reaction is proportional to the product of the concentrations of two reactants or to the square of the concentration of a single reactant

$$\text{velocity} = k[\text{A}][\text{B}]$$

$$\textbf{or } \text{velocity} = k[\text{A}]^2$$

3rd order: the rate is proportional to the product of the concentrations of three reactants, **or** to the product of the square of the concentration of one reactant times the concentration of a second reactant, **or** to the cube of the concentration of a single reactant

$$\text{velocity} = k[\text{A}][\text{B}][\text{C}] \textbf{ or } k[\text{A}][\text{B}]^2 \textbf{ or } k[\text{A}]^3$$

The numerical order of a reaction is therefore equal to the sum of the powers of the concentration terms in its rate equation.†

† Although concentration terms are used in these rate equations and throughout the remainder of this chapter, their values are only precisely appropriate to reactions between ideal gases, or in very dilute (ideal) solutions; otherwise, activity terms would be more exact.

The rate equation which takes account of the concentrations of all rate-determining reactants is known as the *general* or *overall* rate equation; it is this overall rate equation that defines the overall order of the reaction in the manner described above. However, by maintaining fixed concentrations of all reactants save one (e.g. A), and measuring the initial velocity of the reaction over a range of concentrations of A, we can determine the order of the reaction with respect to this variable reactant (sometimes called the order of the reaction in A). Figure 10.1 shows how the order of the reaction, with respect to A, can be deduced from the appearance of the graph of the initial velocity of the reaction at different concentrations of A (but at fixed concentrations of all other participants in the reaction). The diagnostic shape of the graph derives from the form of the 'pseudo' or partial rate equation that relates v to [A] when A is the variable reactant, i.e.

zero order with respect to A,	$v = k$	(n equals 0)
1st order with respect to A,	$v = k[\mathrm{A}]$	(n equals 1)
2nd order with respect to A,	$v = k[\mathrm{A}]^2$	(n equals 2)

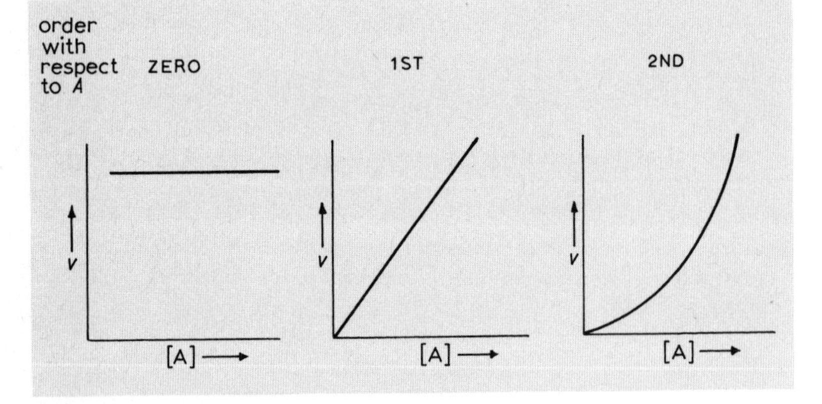

Fig. 10.1 Plot of initial velocity of reaction versus concentration of 'variable' reactant A, for reactions of zero, 1st and 2nd order with respect to A.

By repeating this experiment using each reactant in turn as the variable reactant, it is possible to determine the overall order of the reaction as the sum of the orders exhibited with respect to each reactant individually. For example, let us consider a reaction in which A and B are the only participants. If, when A is made the variable reactant and B is kept at a fixed concentration, the velocity of the reaction is directly proportional to $[\mathrm{A}]^2$, the reaction is 2nd order with respect to A. If, with B as the variable reactant and A at a fixed concentration, the velocity is directly proportional to [B], the reaction is 1st order with respect to B. Since it is 2nd order with respect to A and 1st order with respect to B, the reaction is 3rd order overall, and its overall rate equation is $v = k[\text{A}]^2[\text{B}]$.

The proportionality constant (k) in its rate equation is an important kinetic characteristic of an isothermal reaction. It is called the ***rate constant*** (or *velocity constant*, or *specific reaction rate*), and it measures the rate of the reaction when unit concentrations of reactants are present. The magnitude of its rate constant thus determines whether a reaction is intrinsically 'fast' or 'slow'.

How to determine the overall order and rate constant of a reaction

The overall order of a reaction cannot be determined from its stoichiometric equation. It must be obtained experimentally.

During the course of any reaction between reactants A, B, C, etc., their concentrations will progressively diminish. Except in the case of a zero order reaction, the reaction velocity will consequently also decrease with time in a way that could indicate the order of the reaction. However, instead of attempting a continuous assay of the rate of the reaction, it is often more practicable to measure the initial concentrations of the reactants, and then follow the reaction by measuring the concentrations of these reactants that remain at various times. By mathematical manipulation (integration) of the rate equation for a reaction, a kinetic equation can be obtained which defines the rate constant 'indirectly' in terms of the reactant concentrations present initially, and remaining after time t.

Derivation of the kinetic equation for a 1st order reaction

(Note: This derivation is given as an illustration of the way in which a kinetic equation can be derived from a rate equation. Since it employs calculus, this section can be passed over by those who would find it baffling. Such persons may just accept that it is possible to 'transform' rate equations involving v, into equivalent kinetic equations which convey the same information in terms of the changes in the concentrations of the rate-determining reactants during time t.)

Consider the simple reaction $A \rightarrow P$, which is 1st order overall. Let the initial concentration of A equal a mol dm^{-3}, and let the concentration of A that remains after time t be $(a-x)$ mol dm^{-3}. This means that during time t, x mol dm^{-3} of A have been converted into P.

At any instant, the rate at which A is being converted into P equals $v = \mathrm{d}x/\mathrm{d}t$, and as this is a first order reaction, $v = k[\text{A}]$.

Thus, at any time t, when $[\text{A}] = (a-x)$, $\mathrm{d}x/\mathrm{d}t = k(a-x)$.
Integrating this equation, we obtain,

$$-\ln(a-x) = kt + \text{constant}$$

The value of the constant in this equation is obtained by applying the equation to the situation at the start of the reaction when $t=0$ and $x=0$: then $\ln x$ and kt are both equal to zero, so that the constant equals $-\ln a$.

$$\therefore \qquad -\ln(a-x) = kt+(-\ln a)$$

whence,
$$\ln a - \ln(a-x) = kt$$

and
$$\ln \frac{a}{a-x} = kt$$

Substituting 2.303 log for ln, we obtain the kinetic equation for a 1st order process

$$2.303 \log \frac{a}{a-x} = kt \quad \text{or} \quad \log \frac{a}{a-x} = \frac{k}{2.303} \cdot t$$

This means that if we plot values of $\log a/(a-x)$ against t, a straight line is obtained whose slope equals $k/2.303$.

Kinetic equations for reactions of different orders (zero, 1st or 2nd)

In these equations, x represents the concentration of the rate-determining reactant that is utilized during time t of reaction. The initial concentration of reactant A is represented by a, so that after time t the remaining concentration of A is represented by $(a-x)$. Similarly, when a second reactant B is rate-determining, its initial concentration is represented by b. This means that in the case of the reaction $A+B \rightarrow$ Products, which is 1st order with respect both to A and to B, when concentration x of A has been utilized, the same concentration of B will have reacted. Therefore after time t, the concentrations of these reactants that remain will be $(a-x)$ and $(b-x)$ respectively.

The following rate equations will therefore apply to the velocity of the reaction at time t (N.B. velocity has the units concentration $\times$ time^{-1}):

zero order:
$$v = k$$

where k is the zero order rate constant of the reaction with the units of concentration time^{-1}

1st order
$$v = k(a-x)$$

where k is a 1st order rate constant with the units of time^{-1}

2nd order:
$$v = k(a-x)(b-x)$$

where k is a 2nd order rate constant with the units of concentration^{-1} time^{-1}.

Mathematical treatment (integration) of these rate equations transforms them into the following kinetic equations:

zero order:
$$x = kt$$

1st order:
$$\log \frac{a}{a-x} = \frac{kt}{2.303}$$

2nd order (i) when $a = b$
$$\frac{x}{a(a-x)} = kt$$
(ii) when $a > b$
$$\log \frac{b(a-x)}{a(b-x)} = \frac{k(a-b)t}{2.303}$$

Thus, by measuring the extent by which the concentrations of the reactants change in a certain time interval, both the order and the rate constant of a reaction can be determined. To do this, either the measured concentrations can be substituted in all of the theoretical kinetic equations in turn (when only that equation truly representing the order of the reaction will yield a constant value for k), or the same findings can be employed in the following graphical test of order.

Graphical determination of order and rate constant, of zero, 1st and 2nd order reactions

By choosing appropriate ordinates, all of the kinetic equations listed above can be represented as straight line plots against time (Fig. 10.2).

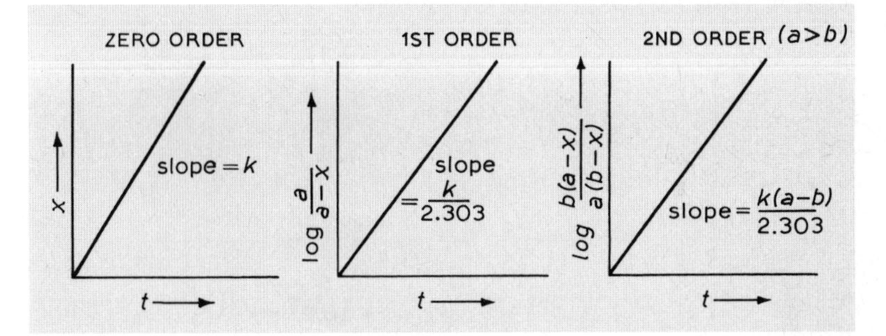

Fig. 10.2 Straight line plots of kinetic equations for zero, 1st and 2nd order reactions. From the slopes of these plots the values of the respective rate constants can be accurately assessed. (The symbols employed are described in the text.)

An illustration of the use of these graphs is provided by the following example.

EXAMPLE

A fast reaction in solution between compounds A and B was followed over a period of 60 s at 310 K by assaying the concentrations of A and B that remained at various times. The following results were obtained:

Time/s	*Concentration of A/mol dm*$^{-3}$	*Concentration of B/mol dm*$^{-3}$
0	0.2	0.1
10	0.166	0.066
20	0.146	0.046
30	0.134	0.034
60	0.114	0.014

Determine the order of the reaction and calculate the value of its rate constant.

1. Test for zero order

Calculate the concentrations of A and B that have been utilized at various times, i.e. calculate the values of x at various times. Since x is identical for A and B, we need only consider the change in concentration of one reactant, say A.

t/s	a/mol dm^{-3}	$(a-x)$/mol dm^{-3}	$\frac{[a-(a-x)]}{\text{mol dm}^{-3}} = \frac{x}{\text{mol dm}^{-3}}$
0	0.2	—	—
10		0.166	0.034
20		0.146	0.054
30		0.134	0.066
60		0.114	0.086

Plotting x against t, we obtain the following graph,

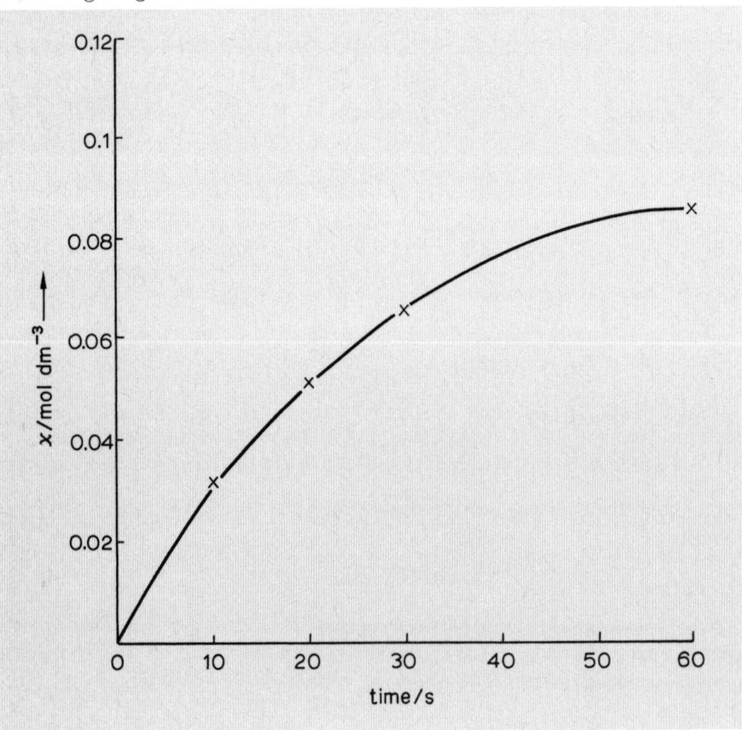

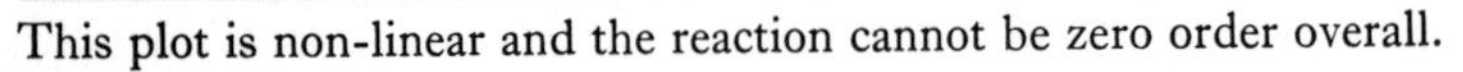
This plot is non-linear and the reaction cannot be zero order overall.

2. Test for 1st order

Calculate values of $\log \dfrac{a}{a-x}$ and $\log \dfrac{b}{b-x}$.

t/s	$\dfrac{a}{a-x}$	$\log \dfrac{a}{a-x}$	$\dfrac{b}{b-x}$	$\log \dfrac{b}{b-x}$
0	$\dfrac{0.2}{0.2} = 1.0$	0	$\dfrac{0.1}{0.1} = 1.0$	0
10	$\dfrac{0.2}{0.166} = 1.205$	0.081	$\dfrac{0.1}{0.066} = 1.53$	0.185
20	$\dfrac{0.2}{0.146} = 1.37$	0.137	$\dfrac{0.1}{0.046} = 2.17$	0.337
30	$\dfrac{0.2}{0.134} = 1.49$	0.174	$\dfrac{0.1}{0.034} = 2.94$	0.468
60	$\dfrac{0.2}{0.114} = 1.75$	0.244	$\dfrac{0.1}{0.014} = 7.14$	0.854

Plotting $\log \dfrac{a}{a-x}$ versus t, or $\log \dfrac{b}{b-x}$ versus t,

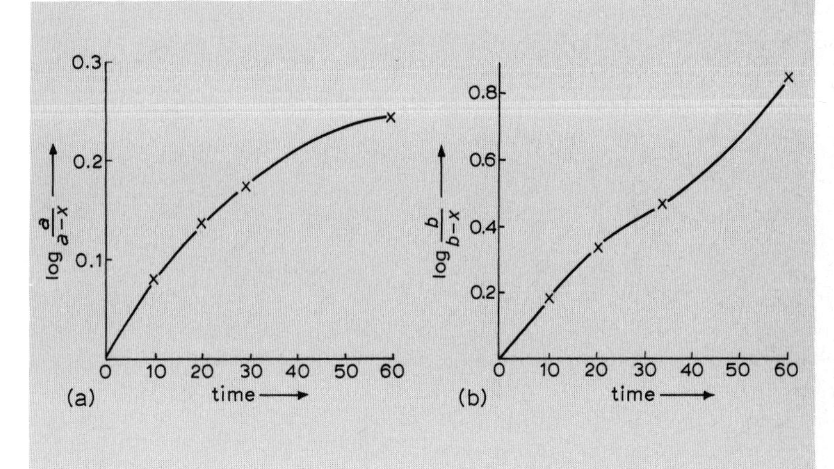

Neither plot is linear, so that the reaction does not demonstrate 1st order kinetics.

3. Test for 2nd order

Calculate values of $\log \frac{b(a-x)}{a(b-x)}$

t/s	$\frac{b(a-x)}{a(b-x)}$	$\log \frac{b(a-x)}{a(b-x)}$
0	$\frac{0.1 \times 0.2}{0.2 \times 0.1} = 1.0$	0
10	$\frac{0.1 \times 0.166}{0.2 \times 0.066} = 1.26$	0.100
20	$\frac{0.1 \times 0.146}{0.2 \times 0.046} = 1.59$	0.201
30	$\frac{0.1 \times 0.134}{0.2 \times 0.034} = 1.97$	0.295
60	$\frac{0.1 \times 0.114}{0.2 \times 0.014} = 4.07$	0.609

Plotting $\log \frac{b(a-x)}{a(b-x)}$ versus t,

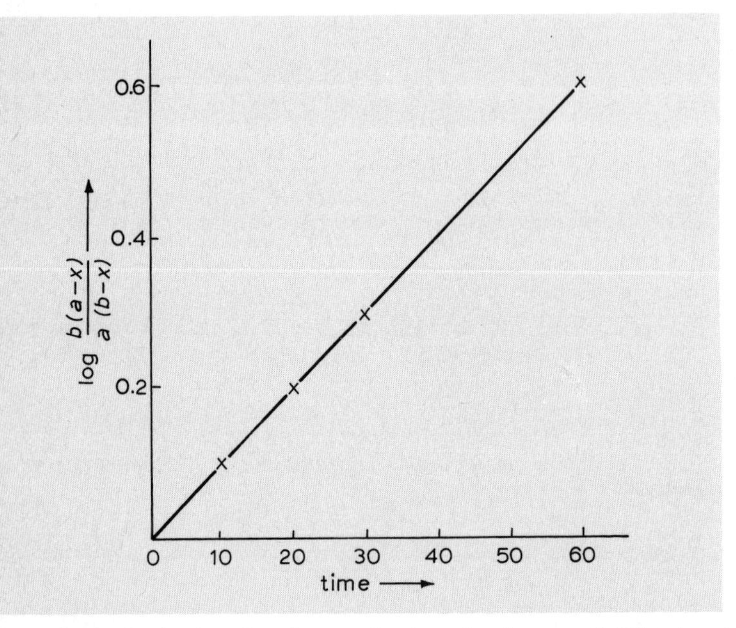

A linear plot is obtained so that the reaction is 2nd order overall. The slope of the line is 1×10^{-2},

$\therefore$ since the 2nd order rate constant $k = \dfrac{2.303}{(a-b)} \times \text{slope}$,

$$k = \frac{2.303 \times 10^{-2}}{(0.2-0.1)} = \underline{0.2303 \text{ dm}^3 \text{ mol}^{-1} \text{ s}^{-1}}$$

The half-life period of a reaction

The half-life period ($t_{\frac{1}{2}}$) of a reaction $A \rightarrow$ Products, is the time taken for the concentration of the reactant A to halve; it is therefore also called the 'half-life' of this reactant.

By substituting $t_{\frac{1}{2}}$ for t, and $\frac{1}{2}a$ for both x and $(a-x)$ in the kinetic equations for zero, 1st and 2nd order reactions, we find that the form of the relationship between the value of $t_{\frac{1}{2}}$ and the initial concentration of A is diagnostic of the order of the reaction.

Zero order: $$t_{\frac{1}{2}} = \frac{a}{2k} \quad \text{whence} \quad t_{\frac{1}{2}} \propto a$$

1st order: $$t_{\frac{1}{2}} = \frac{2.303}{k} \log \frac{a}{\frac{1}{2}a} = \frac{2.303 \log 2}{k} = \frac{0.693}{k}$$

whence $t_{\frac{1}{2}}$ is independent of a

2nd order: when $a = b$ or $v = k[\text{A}]^2$

$$t_{\frac{1}{2}} = \frac{\frac{1}{2}a}{ka(\frac{1}{2}a)} = \frac{1}{ka} \quad \text{whence} \quad t_{\frac{1}{2}} \propto \frac{1}{a}$$

The half-life period is most useful in defining the rate of a 1st order reaction, since its value for such a reaction is independent of the starting concentration of the reactant, and is a simple function of the 1st order rate constant ($t_{\frac{1}{2}} = 0.693/k$). The 'decay' of a radioactive isotope is a case in point, since this is a 1st order process whose rate is generally defined in terms of the half-life of the isotope.

EXAMPLE

^{32}P is much used as a radioactive tracer in biochemical and genetic studies. It decays by β-emission with a half-life of 14.2 days.

Cell hydrolysates containing 0.8 curies of ^{32}P were to be disposed of as waste once their total radioactivity had decreased to an acceptable level of 10 μCi (i.e. 1×10^{-5} curie). Calculate the period for which these hydrolysates should have to be stored before their radioactivity would have fallen to this level.†

Since radioactive decay is a 1st order process, $t_{\frac{1}{2}} = 0.693/k$. Thus, for ^{32}P, $k = 0.693/14.2 \text{ day}^{-1} = \underline{0.0488 \text{ day}^{-1}}$.

† Neither 'curie' nor 'day' is an SI unit; their use in this *Example* merely reflects current, common practice.

The kinetic equation for this 1st order reaction is,

$$\frac{kt}{2.303} = \log\frac{a}{a-x}, \quad \text{or} \quad t = \frac{2.303}{k}\log\frac{a}{a-x}$$

Since the logarithmic term is a *ratio* of two concentrations, these can be measured in any units so long as the same units are used for $(a-x)$ as are used for a. This means that these concentrations can be expressed in terms of any physical property that is directly proportional to concentration—in this case, radioactivity.

Hence,

$$\begin{aligned} a &= [\text{initial } ^{32}\text{P}] \equiv 0.8 \text{ Ci} \\ a-x &= [\text{final } ^{32}\text{P}] \equiv 10^{-5} \text{ Ci} \\ k &= 4.88\times 10^{-2} \text{ day}^{-1} \\ t &= \text{? days.} \end{aligned}$$

Substituting these values in the 1st order kinetic equation, we obtain,

$$t = \frac{2.303}{4.88\times 10^{-2}}\times\log\left(\frac{0.8}{10^{-5}}\right) = \frac{2.303}{4.88\times 10^{-2}}\times\log(8\times 10^{4})$$

$$= \frac{2.303\times 4.9031}{4.88\times 10^{-2}} \text{ days} = \underline{231.5 \text{ days}}$$

The cell hydrolysates could be disposed of after storing for 231.5 days.

Criteria that can be employed to determine (a) the order of a reaction (if this is zero, 1st or 2nd), and (b) the value of its rate constant, are summarized in Table 10.1.

Pseudo 1st order reactions

Consider the simple reaction $A+B \rightarrow$ Products, which is 1st order with respect to each of A and B, and is therefore 2nd order overall. If (a) the participation of B in this reaction is not realized, and (b) for some reason a constant concentration of B is always maintained in the reaction mixture, then the reaction will *appear* to be 1st order overall with an 'observed' rate constant k_{obs}, i.e. $v=k_{obs}[A]$. But this is a false or pseudo 1st order of reaction, since the reaction is in truth 2nd order, $v=k[A][B]$, and can easily be shown to be so (when the role of B is recognized) by determining how the rate of the reaction is affected by varying the concentration of B in the presence of a fixed concentration of A.

The determination of a pseudo 1st order of reaction is by no means uncommon and is particularly likely in the following situations.

(a) *One reactant present in great excess*

When, for example, an ester is hydrolysed in aqueous solution, water, as solvent, is present in excess at a virtually constant concentration, so that

Table 10.1

Order	Rate Equation at time t	Kinetic Equation	Graphical Test of overall order	Half-Life Period
Zero	$v = k$	$x = kt$ k units = mol dm^{-3} s^{-1}	x ↑ vs. t → k = slope	$t_{\frac{1}{2}} \propto a$
1st	$v = k(a-x)$	$\log \frac{a}{(a-x)} = \frac{kt}{2.303}$ k units = s^{-1}	$\log \frac{a}{a-x}$ ↑ vs. t → $k = 2.303 \times$ slope	$t_{\frac{1}{2}} = \frac{0.693}{k}$ ($t_{\frac{1}{2}}$ is independent of a)
2nd	$v = k(a-x)(b-x)$	$\log \frac{b(a-x)}{a(b-x)} = \frac{k(a-b)t}{2.303}$ k units = dm^3 mol^{-1} s^{-1}	$\log \frac{b(a-x)}{a(b-x)}$ vs. t → $k = \frac{2.303}{(a-b)} \times$ slope	When $a = b$ $t_{\frac{1}{2}} \propto \frac{1}{a}$

the reaction will very likely demonstrate 1st order kinetics with a velocity proportional to the concentration of the ester. Yet this is a pseudo 1st order of reaction ($v=k_{obs}$[Ester]), for the hydrolysis is bimolecular (p. 262), and in a non-aqueous solvent, when 'rate-limiting' concentrations, of water *can* be supplied, it will be 1st order with respect to both ester and water.

(b) *One reactant being continuously regenerated*

Consider the following two-step reaction,

$$A+B \xrightarrow{\text{slow}} C \xrightarrow{\text{fast}} B+\text{Products}$$

This is a cyclical process of a type well known to biochemists, i.e.

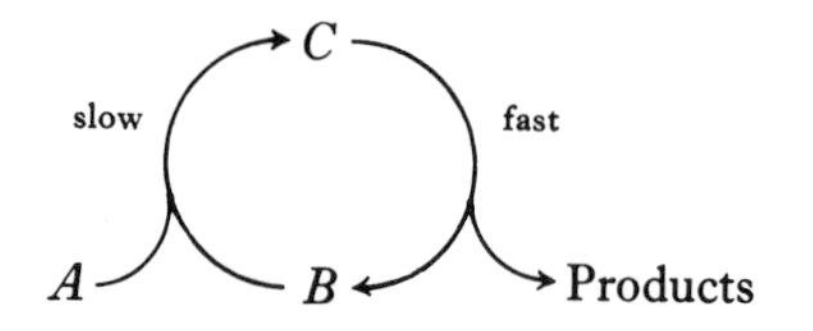

Since reactant B is regenerated by a fast reaction, its concentration will remain virtually constant, and the reaction will appear to be 1st order overall, ($v=k_{obs}$[A]). That this is a pseudo 1st order reaction is revealed by the fact that the value of the rate constant (k_{obs}) will depend on the concentration of B initially supplied. The true overall rate equation is $v=k$[A][B], so that, since v also equals k_{obs}[A], the pseudo 1st order rate constant k_{obs} must equal k[B].

The kinetics of multi-step reactions

It has been found that many reactions that can be represented by a simple stoichiometric equation actually proceed as a sequence of component reactions. Description of the mechanism of a complex reaction of this type therefore requires the identification and arrangement in true sequence of these individual reactions, each of which will be characterized by its own rate equation and rate constant. Thus a reaction, $A+B+C \rightarrow P+Q$, may be accomplished in two elementary steps, e.g.

$$A+B \xrightarrow{k_1} AB \xrightarrow[C]{k_2} P+Q$$

It should therefore be possible to predict the kinetic behaviour of the overall reaction as the resultant of the kinetic behaviour of its component steps (whose rate constants are k_1 and k_2). A rate equation for the whole reaction could thus be built up from the simpler rate equations of these contributory steps. The validity of the supposed mechanism could then be put to the test of determining whether this 'constructed' rate equation accurately describes the actual kinetic behaviour of the overall reaction.

Overall rate and kinetic equations that define the kinetic behaviour of multi-step reactions will inevitably be complex. It is therefore fortunate that in a series of consecutive steps there will often be one that proceeds much more slowly than all the others because its rate constant is much smaller than those relating to previous or subsequent steps. This one reaction thus determines the rate of conversion of the initial reactants into the final products, and is the 'rate-limiting step' of the whole sequence. Since the overall reaction must proceed at the rate of its slowest stage, the rate of formation of product will be the velocity of this rate-limiting step. We shall see later that the identification of the rate-limiting step can prove the key to the elucidation of the mechanism of many complex reactions. The situation is more complicated when there are two, or more, relatively slow steps in the one reaction sequence, but here we need merely recognize this as a possibility.

Why determine the order of a reaction?

We have seen that the experimentally observed relationship between the rate of a reaction and the current concentrations of its reactants can be expressed as an overall rate equation. Thus, the overall rate equation for a reaction $A \rightarrow \rightarrow \rightarrow$ Products, has the form $v = k[A]^n$, where k is the rate constant and n is the power to which the concentration of A must be raised to obtain a linear relationship between v and [A]. The value of n is the order of the reaction; but it is an empirical value which is just a shorthand way of describing the kinetic behaviour of the reaction under certain given conditions. If these conditions were changed, the value of n and hence the order of the reaction might very well be altered. It might be as well, therefore, to emphasize that, although we have particularly considered simple reactions whose kinetic behaviour is such that $n = 1$, or 2, or zero, n may in fact have a fractional value, e.g. 1.6. It is best to think of the order of a reaction as an experimentally measurable, kinetic feature of the reaction, whose value is explicable only when the mechanism of the reaction is known. In other words, the order of a reaction does not *define* its mechanism, though it can sometimes suggest a mechanism. On the other hand, knowledge of the order of a reaction is always helpful when it comes to deciding between alternative mechanisms, since those with which its experimentally determined value is incompatible cannot be correct.

Deductions of reaction mechanism based on simple kinetic findings generally assume:

(a) that in a stepwise reaction, the velocity of the overall reaction is the velocity of its rate-limiting step;
(b) that the velocity of any step in the reaction is a simple function of the concentrations of the reactants that participate in that step (and is therefore expressible as a simple rate equation).

According to these assumptions, the overall order of a reaction could tell us something of the nature of its rate-limiting step.

Consider, for example, the reaction whose stoichiometric equation is $A+B+C \rightarrow P+Q$, which is believed to proceed by one of the following mechanisms:

$$\text{(i)} \quad A+B+C \longrightarrow P+Q$$

$$\text{(ii)} \quad A+B \xrightarrow{\text{slow}} AB \xrightarrow[C]{\text{fast}} P+Q$$

$$\text{(iii)} \quad A \xrightarrow{\text{slow}} A' \xrightarrow[B+C]{\text{fast}} P+Q$$

If experiment showed that the reaction was 2nd order overall, then its mechanism should be (ii), for if it had mechanism (i) it should be 3rd order, and if it had mechanism (iii) it should be 1st order.

Unfortunately, as the following sections will show, the situation is rarely as straightforward as this and the order of a reaction is scarcely ever (and never by itself) a *decisive* index of its mechanism.

Since biologists are chiefly interested in enzyme-catalysed reactions, we shall postpone further discussion of the use of kinetic studies in determining the mechanisms of reactions until Chapter 11, when we will consider kinetic methods of investigating the mechanisms of enzymic processes.

The molecularity of a reaction

Any single-step reaction of known mechanism can be assigned a 'molecularity' which declares how many molecules are involved in its distinctive chemical act.† It follows that molecularity can be expressed only in whole units, e.g. unimolecular, bimolecular reactions. It also follows that one cannot talk of the molecularity of a multi-step reaction, only of the molecularity of one or other of its component steps, e.g. the rate-limiting step.

It is important to realize that the molecularity of a reaction need not be identical with its order. After all, we have seen that the order of a reaction is an empirical, possibly fractional, term, whose magnitude can change

† A more precise definition of molecularity is provided by the transition state theory (p. 266), according to which the molecularity of a reaction equals the number of molecules (or ions, or free radicals) that contribute to its transition state.

when the conditions of the reaction are altered, even though the mechanism of the reaction and the molecularity of its rate-limiting step undergo no change.

Remember therefore that while one can experimentally determine the order of a reaction and yet remain ignorant of its mechanism, only when its mechanism is known can the true molecularity of any of its component steps be declared.

Determination of the values of both rate constants of an elementary reversible reaction

Consider the reaction $A \underset{k_{-1}}{\overset{k_1}{\rightleftharpoons}} B$, where k_1 and k_{-1} represent the rate constants of the forward and back reactions respectively. The values of these rate constants can be obtained either independently or simultaneously.

(i) *Independent determination*

This involves separate measurement of the initial velocities of:

(a) the forward reaction when various concentrations of A (and no B) are supplied

Then initial velocity $v_1 = k_1[\text{A}] = k_1 a$

and (b) the back reaction ($B \rightarrow A$) when different concentrations of B (and no A) are supplied

Then initial velocity $v_{-1} = k_{-1}[\text{B}] = k_{-1}b$

By measuring the initial velocities only, we can assume that there is no significant decrease in the concentration of the supplied reactant, and that reaction proceeds in the absence of the reverse reaction (see p. 278).

(ii) *Simultaneous determination*

Let the concentration of A at zero time be a, and let its concentration at time t be $(a-x)$. If only A is present at the start of the reaction then, at time t, the concentration of B is x.

At any instant, the net rate of formation of B will be the difference between the rates of the forward and back reactions,

$$\therefore \text{ at time } t, \text{ net rate of formation of } B = k_1(a-x) - k_{-1}x$$
$$= k_1 a - x(k_1 + k_{-1}) \qquad (1)$$

At equilibrium, the rates of the forward and back reaction are equal (p. 213), so that the net rate of formation of B at equilibrium $= 0$.
Therefore, if the equilibrium concentration of B is x_e,

$$0 = k_1 a - x_e(k_1 + k_{-1})$$

$$\text{or,} \qquad k_1 a = x_e(k_1 + k_{-1})$$

Substituting $x_e(k_1+k_{-1})$ for k_1a in equation (*1*),

$$\text{net rate of formation of } B = x_e(k_1+k_{-1})-x(k_1+k_{-1})$$
$$= (x_e-x)(k_1+k_{-1}).$$

Mathematical treatment of this relationship yields the following equation

$$\frac{-\ln(x_e-x)}{t} = k_1+k_{-1} \qquad (2)$$

So that if $-\ln(x_e-x)$ is plotted against t, the slope of the line is the sum of the two rate constants (k_1+k_{-1}).
Moreover, at equilibrium,

$$\text{equilibrium constant} = \frac{k_1}{k_{-1}} = \frac{x_e}{a-x_e} \qquad (3)$$

which gives another measurable relationship between the two velocity constants.

Thus by measuring:
(i) the equilibrium constant of the reaction (equation *3*)
(ii) $-\ln(x_e-x)$ as a function of time (equation *2*)
the values of k_1 and k_{-1} can be obtained by solving the two simultaneous equations (*2*) and (*3*).

HOW THE TEMPERATURE AFFECTS THE VELOCITY OF A REACTION

Even when fixed concentrations of reactants are maintained, a reaction proceeds at different rates at different temperatures. This means that the magnitude of the rate constant must be temperature dependent. The relationship between the value of the rate constant and the absolute temperature is defined by the Arrhenius equation,

$$k = Ae^{-E/RT} \qquad \text{where} \begin{cases} k = \text{rate constant} \\ T = \text{temperature in K} \\ R = \text{gas constant} \end{cases}$$

A is a constant, and E is a second constant called the 'activation energy' of the reaction. The Arrhenius equation can also be written in a logarithmic form as

$$\ln k = \ln A - \frac{E}{RT}$$

which shows that the logarithm of the rate constant is a linear function of the reciprocal of the absolute temperature.

The meaning attributed to the constants A and E in the above equations is central to our view of the nature of chemical reaction. Were it possible to calculate them, the rate constant itself would be predictable; unfortunately, this has not yet proved possible. Precise kinetic studies have in fact shown that the values of A and E are not entirely independent of temperature, particularly for reactions in solution. Nevertheless, since their values for many reactions remain approximately constant over a reasonable range of temperatures, we shall continue to refer to them as 'Arrhenius constants'.

The Arrhenius equation was originally derived experimentally. As we shall see later (p. 268), if the rate of a reaction is measured at two or more temperatures, we can, by using the equation, predict the velocities of the reaction at other temperatures. Theoretical interpretations of the way in which the rate constant varies with temperature have been made from two viewpoints, namely that of the collision theory (developed by Arrhenius and van't Hoff), and that of the more modern transition state theory (developed by Eyring and co-workers).

1. The collision theory

This theory, strictly speaking, is applicable only to reactions between gases. In a gas, the molecules move with a range of velocities and therefore with a spectrum of kinetic energies; these kinetic energies are constantly being redistributed by the collisions that occur. Only a small fraction of the total number of collisions between reactant species will lead to a reaction event because, according to the collision theory, reaction occurs only if two molecules collide with more than a given amount of energy. This minimum energy that the molecules must acquire if they are to react is the activation energy E. The collision theory therefore sees the Arrhenius equation as,

$$k = Ze^{-E/RT}$$

where the term $e^{-E/RT}$ represents the fraction of the total number of collisions possessed of more than the critical energy that is needed to make their interaction productive. Z is the 'collision frequency'—the number of collisions per unit time and volume when there is only one molecule of each reactive species per unit volume.† Values of Z can be calculated using the simple kinetic theory of gases which treats their molecules as rigid spheres, but these calculated values are almost always very different from experimentally determined values of the Arrhenius constant A. One reason for this is that molecules that collide with the necessary activation energy might nevertheless not react if they are improperly orientated on collision.

A modified equation produced by the collision theory is therefore,

$$k = PZe^{-E/RT}$$

† By making use of the Avogadro number, the value of Z can be expressed as number of collisions $dm^{-3}\ s^{-1}$ in a gas mixture containing 1 mol dm^{-3} of each reacting gas.

where P is a constant called the 'steric factor', and allows for the occurrence of collisions with the necessary energy requirements which do not lead to reaction.

The equations derived by the collision theory contribute to our understanding of the events taking place during a chemical reaction. They are, however, unsatisfactory for a number of reasons, including:

(a) particularly for reactions in solution, the collision frequency Z does not have the value calculated for it by the simple kinetic theory;

(b) the equations do not allow the calculation of P and E from measurable properties of the interacting molecules, and therefore do not enable us to predict the rate constant of the reaction.

2. The transition state theory

The transition state theory interprets the manner in which a rate constant varies with temperature as indicating the existence of an initial energy barrier which reactants must surmount before reaction can take place. Thus if the 'energy profile' is plotted of the course of transformation of

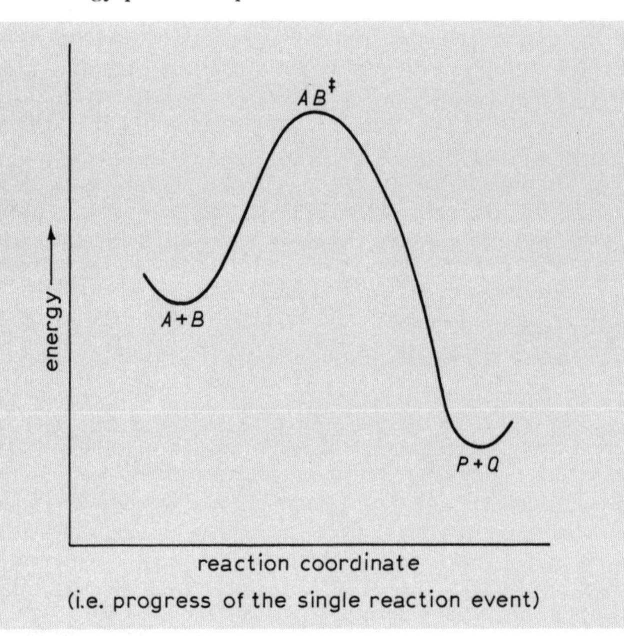

Fig. 10.3 Energy profile for a single-step reaction $A+B \rightarrow P+Q$. The reaction coordinate measures (in arbitrary units) progress along the reaction path from the reactants ($A+B$) to the products ($P+Q$). This two-dimensional energy profile of the reaction may be thought of as a cross section through the actual multi-dimensional energy contour map of the reaction.

reactants A and B into products P and Q in the single-step reaction $A+B \rightarrow P+Q$, it has the form shown in Fig. 10.3.

The conversion of A plus B into P plus Q is accomplished only by way of the intermediate formation of a 'transition state complex', $AB^\ddagger$, whose potential energy is greater than that of the reactants by a quantity that is the activation energy of this reaction. If reaction is to take place between A and B, the interacting molecules must acquire an activation energy E and form the transition state complex which can then decompose, either to yield the products of the reaction or to reproduce the reactants. The transition state theory treats the activated complex as a molecular entity with definable thermodynamic properties. It proposes that it is the concentration of this transition complex, and more particularly the rate of its breakdown, that determines the rate of the reaction. The overall reaction can thus be represented as follows,

$$A+B \rightleftharpoons AB^\ddagger \xrightarrow{\text{slow}} P+Q$$

in which the activated complex is effectively in equilibrium with the reactants.

The theory also shows that the rate constant of a reaction is governed primarily by the difference in standard Gibbs free energy between the transition state complex and the reactants A and B. This difference is the

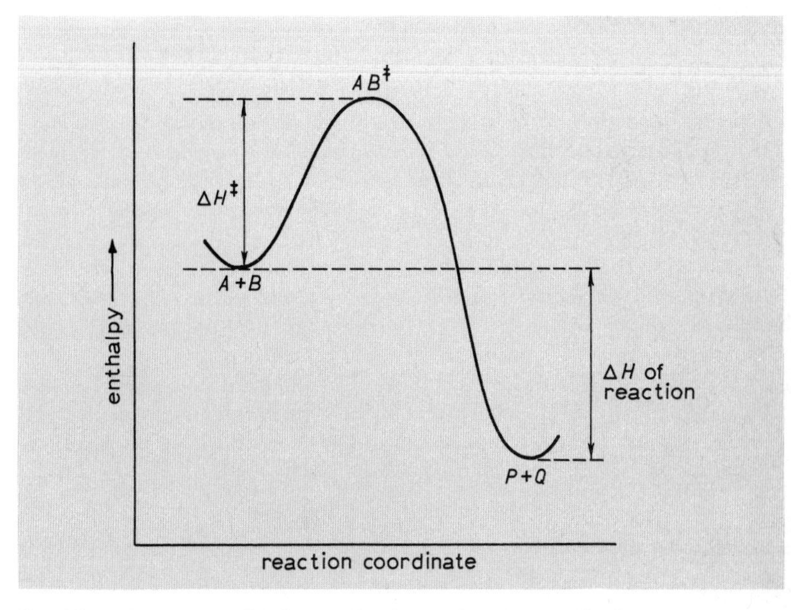

Fig. 10.4 Enthalpy profile for the simple reaction $A+B \rightarrow P+Q$.

'free energy of activation' $\Delta G^{\ddagger}$ which equals $-RT \ln K^{\ddagger}_{eq}$ where $K^{\ddagger}_{eq}$ is the equilibrium constant of that reaction which produces the transition state complex. Since changes in free energy are related to changes in enthalpy and entropy by the equation (p. 196)

$$\Delta G^{\ominus} = \Delta H^{\ominus} - T\Delta S^{\ominus}$$

the overall rate of reaction is governed by both standard entropy and enthalpy differences between the transition state complex and reactants. The theory proposes that the Arrhenius constant A involves the entropy of activation, whilst the value of E that is derived experimentally, is related to the enthalpy of activation by the expression,

$$E = \Delta H^{\ddagger} + RT$$

Thus an enthalpy profile for the reaction $A+B \rightarrow P+Q$, would have the appearance shown in Fig. 10.4.

Determination of the energy of activation of a reaction

Whatever is its theoretical basis, the energy of activation evidences itself as a very real energetic obstacle to reaction, whose magnitude to a large extent determines the rate of the reaction. The Arrhenius equation provides a means whereby its value can be determined for a reaction whose rate constant is k.

$$k = Ae^{-E/RT} \quad \text{becomes} \quad \ln k = \ln A - \frac{E}{RT} \quad \text{(see p. 264)}$$

$$\therefore \qquad \log k = \log A - \frac{E}{2.303R} \cdot \frac{1}{T}$$

This is the equation of the straight line obtained by plotting $\log k$ against $1/T$. The slope of this line equals $-E/2.303R$ (Fig. 10.5); as $2.303R$ equals 19.14 J K^{-1} mol^{-1}, the value of the activation energy can be determined from the 'Arrhenius plot' (Fig. 10.5) as being equal to $-(\text{slope} \times 19.14)$ J mol^{-1}.

A more approximate value for the energy of activation can be calculated from the values of the rate constant at just two temperatures. Suppose that k' and k'' are the values of the rate constant of the reaction at absolute temperatures T_1 and T_2 respectively; then by mathematical manipulation of the Arrhenius equation, the following relationship is derived (cf. the similar equations relating the values of the logarithms of the equilibrium constant at two different temperatures to the value of $\Delta H^{\ominus}$ of the reaction, p. 222):

$$\log \frac{k''}{k'} = -\frac{E}{2.303R}\left(\frac{1}{T_2} - \frac{1}{T_1}\right)$$

or

$$\log k'' - \log k' = \frac{E}{2.303R}\left(\frac{T_2 - T_1}{T_1 T_2}\right)$$

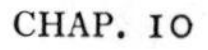

whence, $$E = \frac{19.14\, T_1 T_2 (\log k'' - \log k')}{(T_2 - T_1)}\ \text{J mol}^{-1}$$

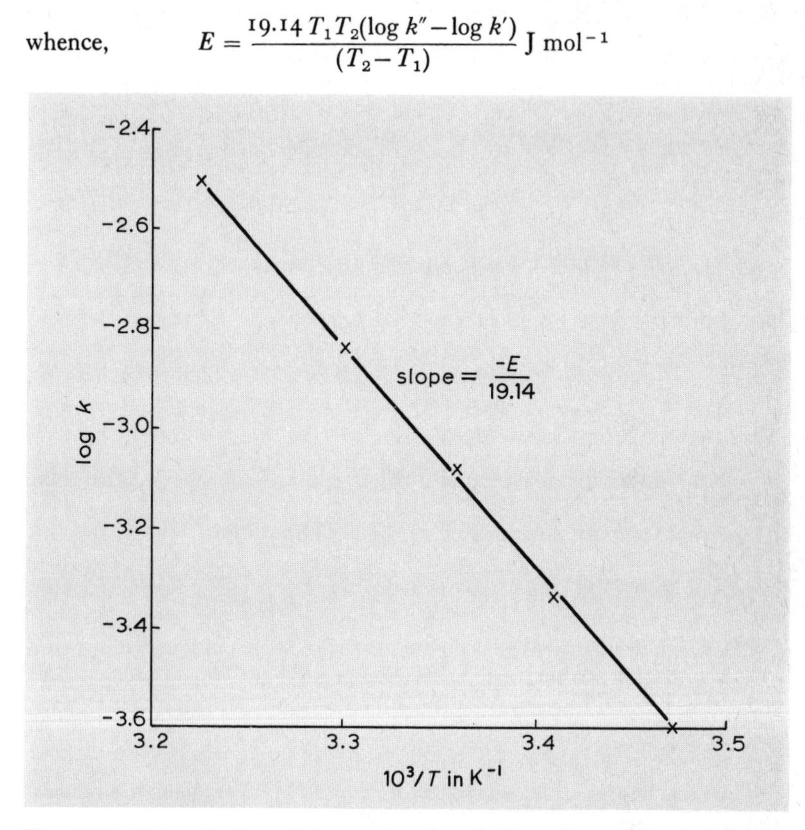

Fig. 10.5 Determination of the energy of activation of a reaction by plotting values of log k versus $1/T$ (the Arrhenius plot).

EXAMPLE

An unstable metabolite decomposed very rapidly in aqueous solution at pH 7. At all temperatures in the range 15 to 37°C, the decomposition proceeded with 1st order kinetics and the following values were obtained for its 1st order rate constant.

Temperature/°C	*Rate constant* k/s^{-1}
15	2.51×10^{-4}
20	4.57×10^{-4}
25	8.22×10^{-4}
30	1.445×10^{-3}
37	3.09×10^{-3}

Determine from these findings the energy of activation of the decomposition.

The energy of activation can be determined by 'fitting' these results into a plot of log k against $1/T$, when, if the reaction is straightforward, a single straight line will be obtained whose slope is related to E by the equation

$$E = -(\text{slope} \times 19.14)\ \text{J mol}^{-1}$$

Temperature/°C	*T*/K	$1/T$ in K^{-1}	k/s^{-1}	log k
15	288	3.472×10^{-3}	2.51×10^{-4}	$\bar{4}.3997 = -3.6003$
20	293	3.412×10^{-3}	4.57×10^{-4}	$\bar{4}.6599 = -3.3401$
25	298	3.356×10^{-3}	8.22×10^{-4}	$\bar{4}.9149 = -3.0851$
30	303	3.30×10^{-3}	1.445×10^{-3}	$\bar{3}.1599 = -2.8401$
37	310	3.225×10^{-3}	3.09×10^{-3}	$\bar{3}.4900 = -2.5100$

It is these results that are plotted in Fig. 10.5, and the slope of the graph is −4413.

Therefore, energy of activation of the decomposition

$$= -(-4413 \times 19.14)\ \text{J mol}^{-1}$$
$$= \underline{84.45\ \text{kJ mol}^{-1}}$$

It is worth noting that the ordinate in the Arrhenius plot uses a logarithmic scale. A length of this ordinate is thus (log k'' minus log k') = log (k''/k'), and so long as k'' is measured in the same units as is k', these units 'cancel each other out' in the calculation of the value of E from the slope of the plot.

The form of the Arrhenius equation ($k = Ae^{-E/RT}$) explains the remarkable sensitivity of the rate of a reaction to changes in temperature, for, since the relationship between k and T is an exponential one, a small change in the value of T causes a large change in the value of k and therefore a large change in the velocity of the reaction. This is most explicitly stated in the differential form of the Arrhenius equation viz.:

$$\frac{d(\ln k)}{dT} = \frac{E}{RT^2}$$

Similarly, the exponential relationship between k and E explains why an apparently small change in the energy of activation of a reaction will be reflected in a great alteration in its velocity. The relationship is such that a slight decrease in E causes a considerable increase in the value of k; for reactions at room temperature, a decrease of only 5.9 kJ mol^{-1} in the value of E is manifested as a tenfold increase in the magnitude of the rate constant.

When the velocity of a multi-step reaction is determined by the velocity of a single rate-determining step, then it is the energy of activation of this step that particularly determines the velocity of the overall reaction.

CATALYSIS

A ***catalyst*** accelerates a chemical reaction without changing its extent, and can be recovered chemically unchanged from amongst the end products of the reaction. Thus the changes in the thermodynamic functions of state (ΔG, ΔH and ΔS) that are characteristic of the reaction, are not affected by catalysis. This means that the same equilibrium is achieved by a chemically reversible reaction in the presence and in the absence of catalyst, though the catalysed reaction will attain this equilibrium much more quickly at the same temperature. An intriguing feature of catalysis is the specificity of many catalysts which may cause a dramatic acceleration in the rate of one reaction but have no perceptible effect on the rates of what apparently are other very similar reactions.

We cannot here discuss all forms of catalysis, and we must therefore ignore some important catalytic mechanisms such as that found in some free radical reactions where the catalyst aids in initiating and propagating chain reactions. Instead, we will consider the mode of action of those catalysts that so change the mechanism of a reaction that it proceeds to its normal destination by a new and speedier route in which the catalyst participates as a regenerated reactant. What is observed as the catalysed reaction is likely to be the one reaction proceeding simultaneously by two different mechanisms, namely (1) the slower uncatalysed mechanism, together with (2) the faster catalyst-involving mechanism. Only when the catalysed mechanism greatly predominates can one ignore the kinetic contribution of the uncatalysed mechanism. It follows from the involvement of the catalyst as a regenerated reactant, that the rate of a catalysed reaction will depend on the concentration of the catalyst that is supplied (see p. 260). This suggests as the definition of this type of catalyst, 'a substance whose concentration appears in the rate equation, but which is not represented in the stoichiometric equation for the net reaction'.

For a reaction to be accelerated by catalysis, the velocity of the rate-limiting step in the catalysed mechanism must obviously be greater than the velocity of the uncatalysed reaction. According to the Arrhenius equation ($k = Ae^{-E/RT}$), the greater velocity of the catalysed reaction could be due to its rate-limiting step possessing *either* (i) a lower energy of activation, *or* (ii) a larger value for the pre-exponential term, A, than the slowest step in the uncatalysed reaction. In *most* cases the energy of activation of the catalysed mechanism is indeed lower than that of the uncatalysed mechanism, by an amount characteristic of the catalyst but independent of its concentration; e.g. E for the decomposition of H_2O_2 (uncatalysed) is about 71 to 75 kJ mol^{-1}, but it is only 46 to 50 kJ mol^{-1} when the reaction is catalysed by platinum, and is still less (21 to 25 kJ mol^{-1}) when the decomposition is catalysed by the enzyme catalase. On the other hand, the acid-catalysed hydrolysis of *p*-methoxybiphenylyl benzoate has a higher energy of activation than the uncatalysed hydrolysis of this substance, though in moderately acid solution its hydrolysis proceeds virtually entirely by the catalysed route. In this case, the greater rate of the catalysed reaction is due to its possessing the much larger value of A.

Some of the main features of this type of catalysis can be illustrated by the following example. Suppose that the uncatalysed, one-step reaction $A \rightarrow P+Q$ proceeds slowly at a given temperature due to its high energy of activation (E_u). The addition of a little catalyst C accelerates the reaction by participating in an alternative two-step mechanism whose slower step has an energy of activation which is less than E_u. Two feasible mechanisms for the catalysed reaction would be:

(A)

$$\begin{array}{lrcl} \text{Step 1:} & A+C & \xrightarrow{\text{slow}} & AC \\ \text{Step 2:} & AC & \xrightarrow{\text{fast}} & C+P+Q \\ \hline \text{Net reaction:} & A & \longrightarrow & P+Q \\ \hline \end{array}$$

(B)

$$\begin{array}{lrcl} \text{Step 1:} & A+C & \underset{}{\overset{\text{fast}}{\rightleftharpoons}} & AC \\ \text{Step 2:} & AC & \xrightarrow{\text{slow}} & C+P+Q \\ \hline \text{Net reaction:} & A & \longrightarrow & P+Q \\ \hline \end{array}$$

In either case:

(a) the net, catalysed reaction is the exact thermodynamic and stoichiometric replica of the overall uncatalysed reaction, though its mechanism, and hence its rate equation, are quite different;
(b) a minute amount of catalyst could be very effective because it is continuously regenerated;
(c) the specificity of action of C, if it acts as a catalyst only of this reaction and no other, is explained by the fact that it enters into specific combination with reactant A;
(d) if the uncatalysed reaction proceeds infinitely slowly, the addition of C might appear to initiate the reaction, since only the catalysed reaction

proceeds at a measurable rate. Yet C has only accelerated a reaction that is thermodynamically feasible.

Acid-base catalysis

Homogeneous catalysis (performed entirely in a single phase) is limited to gaseous or liquid systems and generally involves a specific chemical reaction between reactant (substrate) and catalyst. Acid-base catalysis deserves special mention in this context since it is possibly the most commonly encountered type of homogeneous catalysis in aqueous systems.

In acid catalysis, the substrate acts as a base and accepts a proton from the acid catalyst to form a protonated intermediate. Breakdown of this acid intermediate to yield the products of the reaction involves transference of the proton to an available base and so, directly or indirectly, leads to the reformation of the acid catalyst. In *specific acid catalysis* (e.g. the hydrolysis of acetals) only H_3O^+ can act as the catalytic proton donor. In *general acid catalysis* (e.g. keto-enol transformations, ester hydrolyses) any acid can act as catalyst, though its catalytic effectiveness will be a function of its acid dissociation constant (K_a). This means that general acid catalysis can take place under circumstances in which no H_3O^+ can be present, for example in non-aqueous solvents such as benzene. Thus general acid catalysis of the reaction $X \rightleftharpoons Y$ might proceed by the following mechanism (where the acid catalyst is HC)

$$\begin{aligned} X+HC &\rightleftharpoons XH^+ + C^- \\ XH^+ + C^- &\rightleftharpoons Y+HC \\ \hline \text{Net reaction:} \quad X &\rightleftharpoons Y \end{aligned}$$

In *general base catalysis*, the substrate acts as an acid and donates a proton to any available base. This catalytic base is then regenerated as a consequence of the proton uptake that takes place during the decomposition of the substrate-derived intermediate to yield the products of the reaction. The catalytic effectiveness of the base is a function of its basic dissociation constant (K_b), and it is usually found that OH^- is the best catalyst of a reaction susceptible to general base catalysis (e.g. the hydrolysis of nitramide). Therefore, if the reaction $YH \rightleftharpoons Z$ is base catalysed, it might proceed by the following mechanism

$$\begin{aligned} YH+OH^- &\rightleftharpoons Y^- + H_2O \\ Y^- + H_2O &\rightleftharpoons Z+OH^- \\ \hline \text{Net reaction:} \quad YH &\rightleftharpoons Z \end{aligned}$$

Since water is capable of acting as an acid and as a base, it is likely that reactions susceptible to acid-base catalysis will proceed in neutral, aqueous solution at rates greater than their true uncatalysed rates.

Heterogeneous catalysis

Heterogeneous catalysis is frequently encountered in the case of reactions which proceed in the gaseous or liquid phase but are catalysed by the presence of certain solids. It seems that these reactions are speeded by adsorption of reactant molecules onto the surface of the solid—hence many of these catalysts are most effective in finely divided form. The kinetics of physical adsorption of the reactant molecules onto the catalytic surface must therefore enter into any assessment of the kinetics of heterogeneously catalysed reactions. Assuming that the degree of adsorption can be measured at various temperatures, and its specificity explained by the chemical properties of adsorbent and reactants, we have still to explain how the rate of decomposition of the adsorbed molecules is greater than that of unadsorbed molecules. Among several possible explanations, two appear to be most reasonable, viz.

(i) the interaction between adsorbed reactant and catalyst could be such that those chemical bonds which are subsequently to be broken during the reaction are weakened (in the sense that the energy of activation of this reaction is lessened);

(ii) reaction between adsorbed molecules might be facilitated since (a) as a consequence of their close proximity, the chance of their 'meeting' will be greater than by random collision of free molecules, and (b) they could be held in the steric configuration most favourable for interaction.

Summary: Why study the kinetics of chemical reactions?

Aims

These are twofold:

1. to be able to predict how the rate of a reaction will be affected by changes in reaction conditions;
2. to aid the determination of the mechanism of the reaction, that is, to identify the sequence of reaction events that intervene between reactants and terminal products.

Methods

Much information of value can be obtained from quite elementary experiments. From simple rate measurements, it should prove possible to determine an experimental rate equation for the overall reaction and a value for its energy of activation, though much more skill is needed to interpret such results than to obtain them. The investigator, using an experimentally derived rate equation as a basis for 'inspired guesswork', would devise plausible mechanisms for the reaction, calculate their theoretical rate equations, and discard those in disagreement with the experimental equation. He would then use all the kinetic information that he could gather, such as response of velocity to changes in temperature, pH or the addition of various catalysts or inhibitors, to eliminate mechanisms that were incompatible with these findings until, eventually, only one survived. Until new inspiration or contrary evidence suggested a more probable alternative, this would be considered the 'established' mechanism of the reaction.

It will be obvious that this 'classical' kinetic analysis of reaction mechanisms will be successful in proportion to the resource and experience of the interpreter. The simplicity of the experimental methods carries with it the danger that too much conjecture might be based on too little fact, or that the investigator might be misled by his findings, either because his kinetic experiments were poorly designed, or because he was unaware of the inherent limitations of his findings. The answer to such criticism is not to be found in the proliferation of hypotheses and the construction of elaborate equations to explain the same minimal data; rather, it requires the use of improved experimental methods designed to supplement these simple velocity studies and, if possible, to follow more closely the progress of the sequential reaction events. Thus sophisticated kinetic techniques, such as rapid flow procedures coupled with very sensitive physical methods of product analysis, might enable one to identify transient intermediates, and to follow the sequential rise and fall in their concentrations during the course of reaction. Such methods make it possible in many cases to obtain the values of the individual rate constants of component steps in a complex reaction, and could enable one to identify with certainty its rate-limiting step.

Limitations

With particular reference to the topics discussed in this chapter, two limitations of their usefulness should be mentioned:

1 Simple velocity studies of a reaction could tell us a great deal about its rate-limiting step, e.g. the value of its rate constant and its enthalpy of activation, yet they are not able to define the chemical structures of the reactants in this step, or the structure of its transition complex.
2 There is no alternative to the experimental determination of a rate constant. Even the most sophisticated analysis of its determinants by the transition state theory, has not yielded a means of predicting the rate of a reaction in solution from thermodynamic or concentration data alone.

Successes

Kinetic studies have in many instances suggested mechanisms of reaction that have, by analogy, made related chemical processes intelligible. In so doing, they have aided the search for analogous reactions, the design of specific catalysts and inhibitors, and the selection of reaction conditions that favour a desired reaction. The application of kinetic methods to biological systems has particularly contributed to the recognition that biochemical events are not 'unique physiological processes, powered by vital force and directed by living ferments', but are in reality chemical reactions whose feasibility, extent and rate are determined by normal thermodynamic and kinetic considerations, even when they are enzyme-catalysed and occur very rapidly at room temperature.

PROBLEMS

1. When the decomposition of compound A in 1 mol dm^{-3} aqueous solution was followed at 303 K, it was found that its concentration decreased by 20% in 10 minutes. Calculate the rate constant of the reaction assuming it obeys (a) zero order, (b) 1st order, and (c) 2nd order kinetics with respect to A.

2. A unidirectional reaction between two compounds, A and B, proceeds with 2nd order kinetics overall (rate constant at 310 K$=5\times10^{-2}$ dm^3 mol^{-1} s^{-1}). Calculate the concentrations of A and B that will remain after 30 s of reaction, if initially there were present 0.2 mol dm^{-3} A and 0.1 mol dm^{-3} B.

3. Compounds A and B are interconvertible by a chemically reversible reaction whose 1st order rate constants at 310 K in the forward and back directions are respectively, $k_1=2.5\times10^{-6}$ s^{-1}, and $k_{-1}=5\times10^{-4}$ s^{-1}. If the reaction was started with 20 mmol dm^{-3} A and no B present, calculate the concentration of B at equilibrium.

4. The following values were obtained for the 2nd order rate constant of a bimolecular reaction at the temperatures indicated:

Temperature/K	Rate constant/dm^3 mol^{-1} s^{-1}
285	1.07×10^{-2}
290	2.82×10^{-2}
298	0.126
306	0.525

Calculate (a) the energy of activation, and (b) the enthalpy of activation of this reaction at 303 K.

5. A reaction which was 1st order with respect to A had an initial velocity of 0.5 μmol cm^{-3} s^{-1} when 0.1 mol dm^{-3} A was supplied at 293 K. If the energy of activation of the reaction is 33.47 kJ mol^{-1}, what will be the initial rate of utilization of 0.1 mol dm^{-3} A at 310 K?

6. A metabolite decomposes with 1st order kinetics in aqueous acid solution at 298 K. The value of the 1st order rate constant varies with the pH of the solution as shown:

pH	Rate constant/s^{-1}
3.0	8.50×10^{-4}
3.1	7.17×10^{-4}
3.3	5.26×10^{-4}
3.7	3.30×10^{-4}
5.0	2.07×10^{-4}
7.0	2.00×10^{-4}

Do these findings suggest acid (proton) catalysis of the decomposition?

11

The Kinetics of Enzyme-Catalysed Reactions

The proteins that specifically and very effectively catalyse the chemical reactions of metabolism are called enzymes. Even though some are catalytically active only when associated with special cofactors of small molecular weight, the explanation of the specificity of action of every enzyme must be sought in its protein structure. Yet this is so complex that although some features of enzymic behaviour, e.g. heat lability, pH-sensitivity, can in part be explained by analogy with known properties of other proteins, we are able, as yet, to describe very few enzymes in terms of their molecular structures. Meanwhile, each enzyme is identified and characterized by its catalytic effect on one or more chemical reactions.

Many aspects of metabolic regulation and cellular differentiation are explicable in terms of mechanisms that either control the variety and quantity of enzymes synthesized by a living cell, or that supervise their subsequent activity. Thus every biologist should be prepared to employ kinetic methods to characterize and assay enzymes on the basis of their catalytic activity, and to identify those conditions and compounds that can modify this activity.

Why one should measure the initial velocity of an enzyme-catalysed reaction

Consider an enzyme which catalyses the conversion of a reactant S (here called the substrate) into a product P, i.e. $S \rightarrow P$. The rate of the catalysed reaction can be measured by following the progressive utilization of substrate, or formation of product (Fig. 11.1). The 'progress curve' so

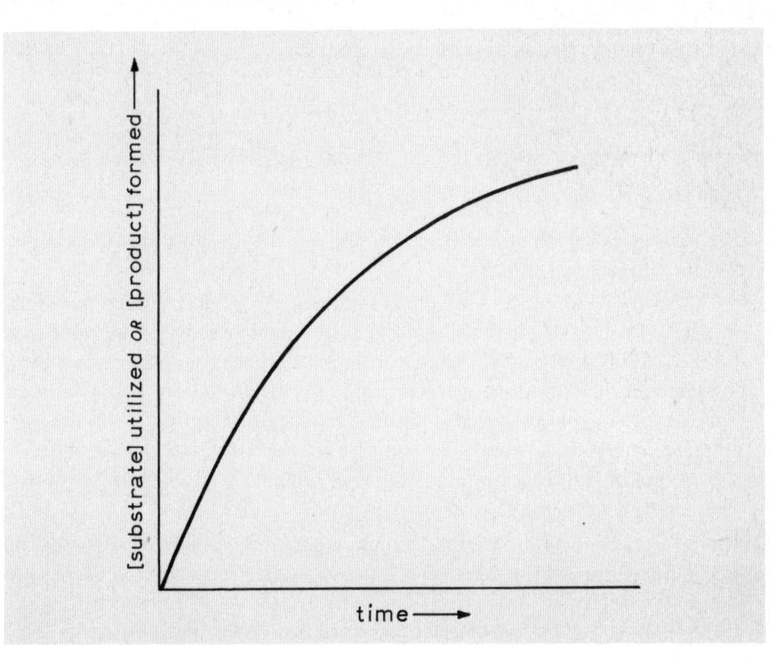

Fig. 11.1 Progress curve of an enzyme-catalysed reaction. The rate of the reaction (measured at any instant as the slope of the tangent to the curve) decreases as the reaction proceeds.

obtained generally shows that the velocity of the reaction decreases with time. There may be several contributory causes; e.g.

(a) If the reaction is significantly reversible, the velocity of the 'back reaction' will increase as the concentration of the product rises. The net rate of transformation of S will therefore decrease.
(b) If the substrate is not provided in sufficient excess, its concentration will decrease significantly during the course of the reaction so causing a progressive 'falling off' in rate.
(c) The enzyme may be labile, in which case its activity might rapidly decrease with time under the chosen reaction conditions.
(d) The product of the reaction may inhibit the activity of the enzyme.

To obtain an accurate measure of the activity of an enzyme, it is therefore necessary to measure the *initial velocity* of the enzyme-catalysed reaction so that interference by all these factors will be minimal. If the rate of the reaction decreases very quickly, then the initial velocity would be measured as the slope of the tangent to the progress curve at the earliest possible time. In practice, progress curves are often effectively linear at early times, so that the construction of tangents is not usually necessary.

HOW TO ASSAY THE CATALYTIC ACTIVITY OF AN ENZYME

As for any other chemical reaction, the velocity of an enzyme-catalysed reaction can be measured by:

(i) a sampling technique (discontinuous assay), whereby portions of the reaction mixture are withdrawn at intervals, the reaction stopped, and the samples analysed for their contents of substrate and/or product;

(ii) a continuous monitoring technique, in which use is made of some distinctive, quantitatively measurable, physical property of the substrate or product which can be assayed without interfering with the progress of the reaction (e.g. when the formation of NADH is followed spectrophotometrically at 340 nm).

In general, a continuous method of assay is preferable since it ensures that the true initial rate of the reaction is taken. Because it is essential to measure this initial velocity, it is also desirable, when possible, to follow the reaction by assaying the formation of product, since newly formed product is likely to be more accurately measured than the disappearance of a small proportion of the supplied substrate.

The rate of an enzyme-catalysed reaction, like that of any chemical reaction, will depend on the reaction conditions, e.g. temperature, pH; these must be kept constant and must be reported. Furthermore, as the concentration of secondary reactants and cofactors will affect the reaction rate (see later), these must be present in excess if a measure of the enzyme's full catalytic potential is to be obtained. By varying one reaction condition whilst keeping the others constant, optimum conditions for the enzyme's action can eventually be discovered.

The activity of an enzyme cannot be judged solely by the rate of the enzyme-catalysed reaction unless it is also known that the rate of reaction is negligible in the absence of the enzyme. Now enzymes, like all other proteins, are denatured by heating, and 'boiled enzyme' (kept at 100°C for 10 to 15 min) is generally catalytically inactive. Therefore, in order to be able to attribute the catalytic activity of a crude cell extract to an enzyme, it is advisable to measure the 'basal' rate of the reaction in the presence of the boiled cell extract as well as in the complete absence of added extract. In this way it is possible to detect whether the extract also contains heat-stable (non-enzymic) activators or inhibitors of the reaction.

In summary, to measure the maximum catalytic activity of a quantity of enzyme:

1 Measure initial reaction rates—preferably by a method which continuously monitors the appearance of product.
2 Provide reactants, other than the enzyme, in adequate excess.
3 Maintain and report optimal reaction conditions.
4 Determine the rate of the non-enzymic reaction (employing boiled enzyme).
5 Measure the quantity of enzyme present (usually one can measure only the protein content of the reaction mixture).
6 As a final confirmation of the validity of the assay:
   (a) check that the initial rate is directly proportional to the enzyme concentration;
   (b) confirm that the initial rate is unchanged when the concentrations of substrates are further increased;
   (c) if the assay is discontinuous, check that the rate is linear with time during the period wherein those samples are taken from which the initial velocity is calculated.

The catalytic activity of an enzyme is measured in 'units of enzyme', 1 unit of enzyme being the quantity of enzyme which produces a certain rate of reaction under defined conditions of reaction. The usual unit of activity is 'that quantity of enzyme which catalyses the utilization of 1 micromole of substrate per minute under certain, specified conditions'. Generally speaking, it is not possible in an impure preparation to measure the amount of enzymically active protein that is responsible for the measured number of units of activity; instead, one must measure the concentration of the enzyme in a given solution as units/cm^3, or its specific activity in the preparation as units/mg *total* protein. The unit in which to express the specific activity of an enzyme is therefore μmol substrate transformed min^{-1} mg^{-1} protein, the temperature, pH and any other relevant reaction conditions being reported.

It is important to realize that the specific activity of an enzyme is not a measure of the activity of the enzyme, but is an indirect measure of the fraction of the total protein of a given preparation which the enzyme comprises. It is therefore a somewhat cryptic measure of the purity of the enzyme in the given preparation, i.e.

Purity of enzyme in a given sample (as the fraction of the total protein that is provided by the enzyme)

$$= \frac{\text{Specific activity of enzyme in the sample}}{\text{Specific activity of } \textit{pure enzyme}}.$$

KINETIC STUDIES OF ENZYME-CATALYSED REACTIONS

As was the case with the simpler chemical reactions considered in Chapter 10, studies of the kinetic behaviour of an enzyme-catalysed reaction are usually directed to the elucidation of its mechanism; in particular, it is hoped that they will disclose in what manner and sequence the enzyme combines with its substrates and releases the various products of the reaction. The measurement of initial velocities of enzyme-catalysed processes under different reaction conditions is only one aspect of this kinetic study. Examination of isotope exchange between substrates and products, and highly sophisticated, ultra-rapid measuring techniques, sometimes enable one to follow the course of events during the pre-steady state phase of interaction of an enzyme with its substrate(s), and also during the phase of relaxation which ensues when the established steady state is disturbed by a sudden change in reaction conditions (see references[12, 17, 20, 22, 31]).

In this chapter we shall consider what information we can hope to derive from simple initial velocity studies alone. Not only are such studies 'classical', in the sense that they were the first to provide evidence in support of the notion that an enzyme enters into specific chemical combination with its substrates, but they are easily performed with the minimum of special equipment and can yield information concerning the mechanism of an enzyme-catalysed process when only relatively impure preparations of the enzyme are available.

Effect of changes in substrate concentration on the initial velocity of an enzyme-catalysed reaction

If we measure the initial velocity (v_0) of a simple reaction $S \rightarrow P$ when this is catalysed by a given concentration of enzyme (e_0) under constant reaction conditions, we find that v_0 varies with the concentration of supplied substrate [S]. When the measured values of v_0 are plotted against the corresponding values of [S] as in Fig. 11.2, a characteristic, rectangular hyperbolic plot is usually obtained, demonstrating that at low values of [S], the initial velocity of the reaction is directly proportional to the substrate concentration, but that at larger values of [S], the initial velocity is maximal, and its value (V_{max}) is independent of the actual concentration of the substrate so long as this is 'saturating'.†

This means that:

(a) at very low concentrations of substrate, the reaction is 1st order with respect to substrate,

$$v_0 = k[S]$$

(b) at high concentrations of substrate, v_0 has a maximum value (V_{max}), and the reaction is zero order with respect to substrate,

$$V_{max} \text{ is independent of } [S]$$

(so long as [S] is sufficiently large).

† Although the hyperbolic plot between v_0 and [S] in Fig. 11.2 is for a single substrate reaction, a similar plot is obtained for a multi-substrate reaction so long as the concentration of only one substrate (the variable substrate) is changed, while the concentrations of all other substrates (the fixed substrates) are kept constant.

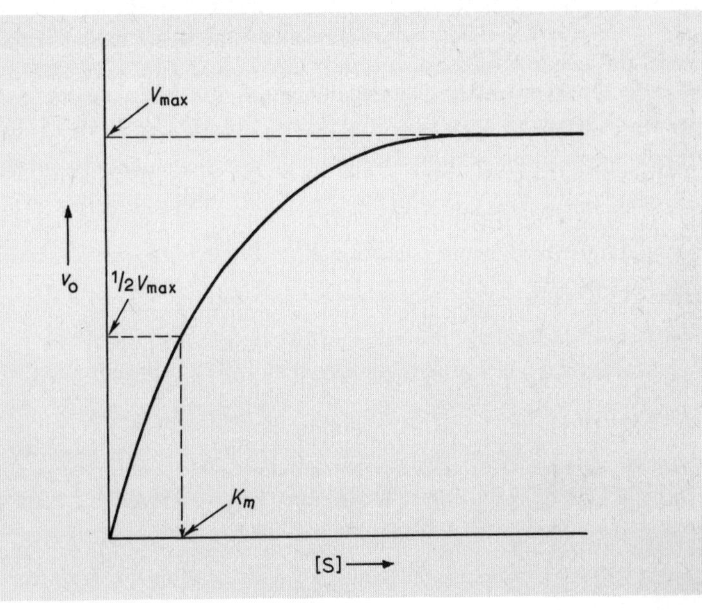

Fig. 11.2 Plot of initial velocity v_o versus concentration of substrate [S], for a typical enzyme-catalysed reaction.

These simple, initial velocity equations only define the behaviour of the reaction at the extremes of the range of substrate concentrations employed in Fig. 11.2. But in Fig. 11.2, we see that when we plot measured values of v_o against [S] (over the complete range of values of [S]), we obtain a rectangular hyperbola. This means that the relationship between v_o and [S] must be definable in terms of the equation of a rectangular hyperbola which for this purpose we can write in the form,

$$v_o = \frac{a[S]}{[S]+b} \quad \text{where } a \text{ and } b \text{ are constants.}$$

Applying this equation to the experimental plot, we find that a equals the maximum velocity V_{max}, and that b is the substrate concentration at which the reaction proceeds at half its maximum initial velocity (this is generally symbolized by K_m). Thus the experimentally derived equation which relates v_o and [S] is normally written as,

$$v_o = \frac{V_{max}[S]}{[S]+K_m}$$

This is known as the ***Michaelis equation*** and its denominator constant, K_m, is called the ***Michaelis constant***. It should be noted that this experimentally derived equation makes no assumptions concerning the mechanism of the enzymic reaction; it merely describes how v_0 is *observed* to vary with [S], in terms of the two experimentally measurable constants V_{max} and K_m. These constants (or kinetic parameters) may be defined as follows:

V_{max} is the maximum initial velocity of the enzyme-catalysed reaction under the given conditions, and is the limiting value that v_0 approaches as $[S] \to \infty$. The value of V_{max} is measured in units of quantity of substrate transformed/unit time for a given concentration of enzyme.

K_m is the Michaelis constant. It can be defined only in experimental terms and equals the value of [S] at which v_0 equals $\frac{1}{2}V_{max}$. The value of K_m is therefore measured in units of concentration, e.g. for most single substrate enzymic reactions K_m has a value between 10^{-5} mol dm^{-3} and 10^{-2} mol dm^{-3}.

That K_m is equal to that substrate concentration at which the reaction proceeds at half its maximal initial rate, can be shown by substituting $\frac{1}{2}V_{max}$ for v_0 in the Michaelis equation. Then,

$$\frac{V_{max}}{2} = \frac{V_{max}[S]}{[S]+K_m}$$

or

$$\frac{V_{max}}{2}([S]+K_m) = V_{max}[S]$$

Dividing throughout by $V_{max}/2$,

$$[S]+K_m = 2[S] \qquad \text{whence, } K_m = [S].$$

Since V_{max} and K_m define the quantitative relationship between v_0 and [S] for a simple enzyme-catalysed reaction, their values are distinctive for each reaction proceeding under given conditions (reflecting, as we shall see later, the magnitudes of rate constants involved in the reaction mechanism). It is therefore necessary that we should be able to measure the values of these parameters.

Experimental determination of values of K_m and V_{max}

Obviously the values of V_{max} and K_m could be read off the hyperbolic plot of v_0 versus [S], as shown in Fig. 11.2. However, it could prove difficult in practice to draw an accurate hyperbola if the experimental points of this graph were appreciably scattered. Furthermore, it might be difficult to supply sufficiently high concentrations of *S* to ensure the attainment of V_{max}, or it could prove difficult to measure the low initial velocities obtained with very small concentrations of *S*. All of these difficulties are circumvented by the Lineweaver and Burk method of determining the values of V_{max} and K_m. This makes use of the fact that the reciprocal of the equation of a rectangular hyperbola is the equation of a straight line. By taking the reciprocal of the Michaelis equation, $v_o = \frac{V_{max}[S]}{[S]+K_m}$, we obtain the following equation,

$$\frac{1}{v_o} = \frac{[S]}{V_{max}[S]} + \frac{K_m}{V_{max}[S]}$$

Rearranged, this becomes the Lineweaver–Burk equation,

$$\frac{1}{v_o} = \frac{1}{V_{max}} + \frac{K_m}{V_{max}} \times \frac{1}{[S]}$$

If $1/v_o$ is plotted against $1/[S]$, a straight line is obtained whose slope is K_m/V_{max}, whose intercept on the $1/v_o$ ordinate equals $1/V_{max}$, and whose intercept on the negative side of the $1/[S]$ abscissa equals $-1/K_m$† (see Fig. 11.3). This straight line, 'double reciprocal' plot is to be preferred to the hyperbolic plot as a means of determining V_{max} and K_m, since a straight line is more accurately drawn and extrapolated 'freehand' than is any curve, while should there be any doubt about the 'best' straight line, its slope and intercepts can quite easily be statistically determined from the experimental data.

EXAMPLE

The enzyme isocitrate lyase, present in Escherichia coli *grown aerobically on acetate as sole source of carbon, catalyses the cleavage of* L_s*-isocitrate to glyoxylate and succinate,*

$$\textit{isocitrate} \longrightarrow \textit{glyoxylate} + \textit{succinate}.$$

The activity of the enzyme can be measured by following the production of glyoxylate at pH 6·8 and 30°C.

The following data were obtained when the initial velocity of the reaction, catalysed by a fixed concentration of isocitrate lyase, was measured over a range of concentrations of isocitrate.

Concentration of L_s*-isocitrate* / μmol dm^{-3}	*Initial velocity* / nmol min^{-1}
18	4·08
24	4·64
30	5·10
40	5·62
100	6·90

† That the intercept on the $1/[S]$ axis equals $-1/K_m$ can be shown by substituting 0 for $1/v_o$ in the Lineweaver–Burk equation, when it is found that $1/[S] = -1/K_m$.

Calculate the value of K_m for the enzyme-catalysed reaction under these assay conditions.

Values of $1/[S]$ and $1/v_0$ are calculated, and are plotted as shown in Fig. 11.3.

$\frac{[S]}{\text{mmol dm}^{-3}}$	$\frac{1}{[S]} \times \text{mmol dm}^{-3}$	$\frac{v_0}{\text{nmol min}^{-1}}$	$\frac{10}{v_0} \times \text{nmol min}^{-1}$
0.018	55.6	4.08	2.45
0.024	41.6	4.65	2.15
0.03	33.3	5.10	1.96
0.04	25.0	5.62	1.78
0.10	10.0	6.90	1.45

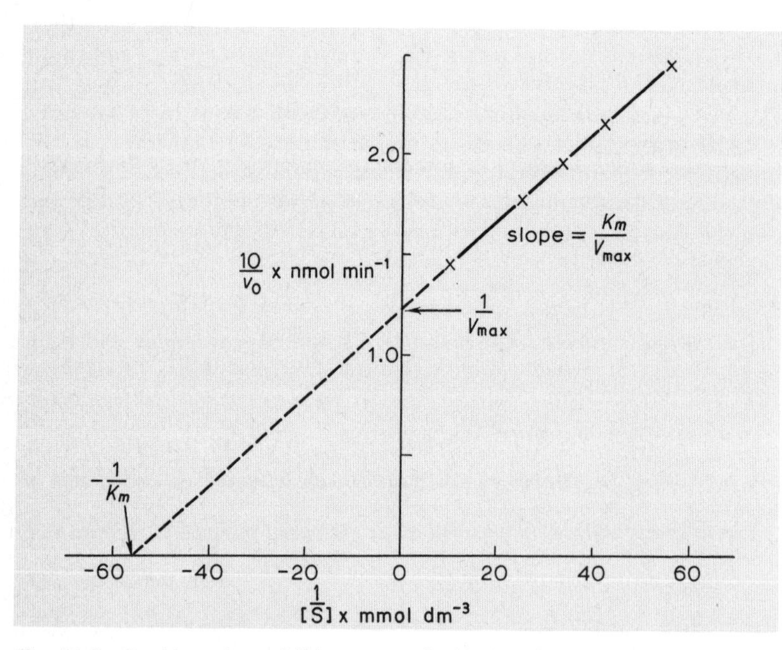

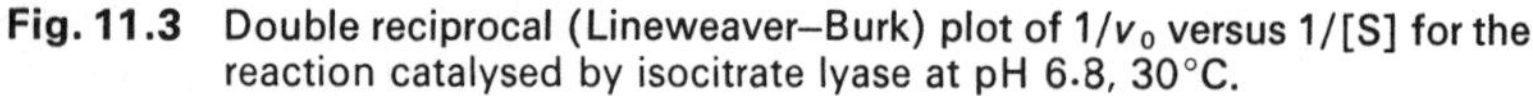
Fig. 11.3 Double reciprocal (Lineweaver–Burk) plot of $1/v_0$ versus $1/[S]$ for the reaction catalysed by isocitrate lyase at pH 6.8, 30°C.

When the linear reciprocal plot is extrapolated, it intersects the negative portion of the abscissa at -56.0 dm^3 mmol^{-1}.

$\therefore \quad -56.0 = 1/K_m,$

when K_m has the same units as [S], i.e. mmol dm^{-3}

$$\therefore \qquad K_m = 0.018 \text{ mmol dm}^{-3}$$

Thus K_m for isocitrate lyase, under the reaction conditions employed, equals 1.8×10^{-5} mol dm^{-3}.

Though the double-reciprocal (i.e. Lineweaver–Burk) plot is used extensively in kinetic studies with enzymes, it tends to give undue weighting to the least accurate points, i.e. those obtained at low concentrations of substrate. Alternative methods of rearranging the Michaelis–Menten equation have been proposed, which distribute the points more evenly; these are illustrated in Fig. 11.4.

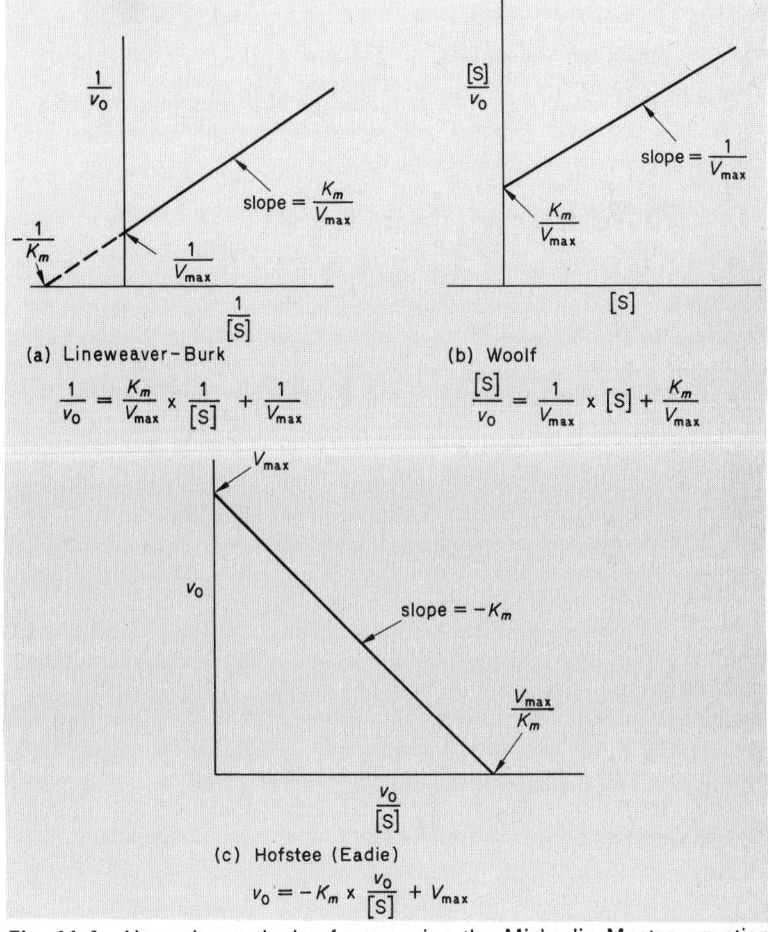

Fig. 11.4 Alternative methods of rearranging the Michaelis–Menten equation to yield linear plots from which values of K_m and V_{max} are easily obtained.

The significance of V_{max} and K_m in terms of the mechanism of an enzyme-catalysed reaction

The simplest explanation of the enzymic catalysis of an irreversible, single substrate reaction, supposes that the catalysed reaction involves an initial combination between the enzyme and its substrate to form an enzyme-substrate complex (ES). The enzyme-catalysed reaction would then proceed by a minimally two-step mechanism in which the enzyme participates as a regenerated reactant, i.e.

$$\text{Uncatalysed mechanism:} \qquad S \longrightarrow P$$

$$\text{Catalysed mechanism:} \qquad E+S \longrightarrow ES \longrightarrow E+P$$

If it is further assumed that the combination of the enzyme with its substrate is a reversible process, the mechanism of the enzyme-catalysed reaction should be written as follows,

$$E+S \underset{k_{-1}}{\overset{k_1}{\rightleftharpoons}} ES \overset{k_2}{\longrightarrow} E+P$$

On the basis of this mechanism, the hyperbolic relationship between v_o and [S] is explained by supposing that it is the second step (i.e. the decomposition of $ES \rightarrow E+P$) that is rate-limiting. Under these circumstances the initial velocity of the reaction is the velocity of this second step, which in turn is proportional to the concentration of ES, i.e.

$$v_o = k_2[\mathrm{ES}]$$

Given a fixed concentration of enzyme (e_o), the concentration of ES increases to a maximum as the concentration of substrate is increased. Therefore, according to the two-step hypothesis, because v_o is proportional to [ES], it is the form of the dependence of [ES] upon [S] that determines the form of the relationship between v_o and [S].

It follows that if only we can derive the value of [ES] as a function of [S] (involving only known or measurable parameters of the assumed mechanism), we can use the equation $v_o = k_2[\mathrm{ES}]$ to relate v_o to [S], and so obtain a theoretical, initial velocity equation for the reaction. We could then test the validity of our assumed mechanism by comparing the theoretically derived equation with the experimental Michaelis equation.

We shall consider two methods of deriving an initial velocity equation for the irreversible, single substrate reaction with the mechanism,

$$E+S \underset{k_{-1}}{\overset{k_1}{\rightleftharpoons}} ES \overset{k_2}{\longrightarrow} E+P$$

These treatments will be considered separately since they make quite different assumptions.

1. *The steady state assumption* (Briggs and Haldane, 1925)

The concentration of substrate ([S]) is usually much greater than the concentration of enzyme (e_o) that is used in any reaction, since the very large molecular weight of the enzyme ensures that it is generally present in very low molar concentration. In these circumstances (i.e. $[S] \gg e_o$), if *ES* is formed very rapidly and thereafter decreases in concentration relatively slowly, then during the reasonably short time interval over which the initial velocity is measured and when the concentration of substrate remains virtually unchanged, the concentration of *ES* also remains constant. In fact, a 'steady state' exists wherein the concentration of *ES* remains constant due to its being produced and destroyed at equal rates.

We will employ the following symbols to represent the concentrations of participants in the reaction:

e_o = initial (and therefore total) concentration of enzyme,
c = steady state concentration of the enzyme-substrate *complex*, i.e. [ES] under steady state conditions,
[S] = concentration of supplied substrate (where $[S] \gg e_o$).

Then,

Rate of formation of $ES = k_1(e_o - c) \times [S]$
Rate of destruction of ES,
(i) by dissociation $= k_{-1}c$
(ii) by decomposition $= k_2c$

In the steady state,

$$\text{Rate of formation of } ES = \text{Rate of destruction of } ES$$

$$\therefore \quad k_1(e_o - c)[S] = k_{-1}c + k_2c$$

$$\text{or} \quad k_1e_o[S] - k_1c[S] = k_{-1}c + k_2c$$

Dividing throughout by k_1c,

$$\frac{e_o[S]}{c} - [S] = \frac{k_{-1} + k_2}{k_1}$$

The right-hand side of this equation, being composed of three rate constants, is itself a constant, i.e.

$$\frac{k_{-1} + k_2}{k_1} = K$$

$$\therefore \quad \frac{e_o[S]}{c} = K + [S]$$

Therefore, c (which is the steady state concentration of ES) $= \dfrac{e_o[S]}{[S] + K}$

Since, according to the two-step hypothesis, $v_o = k_2 c$,

$$v_o = \frac{k_2 e_o [S]}{[S]+K}\dagger$$

But $k_2 e_o = V_{max}$ (since the concentration of ES (i.e. c) equals e_o at the saturating concentrations of substrate at which $v_o = V_{max}$),

$$\therefore \qquad v_o = \frac{V_{max}[S]}{[S]+K}$$

Here we have an equation which is identical in form to the Michaelis equation.

2. *The rapid equilibrium assumption* (suggested by Henri, 1902, but popularized by Michaelis and Menten, 1913)

In this view of the two-step mechanism, it is assumed that the decomposition of the enzyme-substrate complex ($ES \xrightarrow{k2} E+P$) proceeds so much more slowly than its dissociation ($ES \xrightarrow{k_{-1}} E+S$), that the complex will be in equilibrium with enzyme and substrate.

On the basis of this assumption of rapid attainment of equilibrium in the reversible first step, the following initial velocity equation is obtained,

$$v_o = \frac{V_{max}[S]}{[S]+K_s}$$

where K_s is the dissociation constant of the enzyme-substrate complex (equals k_{-1}/k_1).

This equation also has the same form as the experimental Michaelis equation, but it differs from the similar equation derived by the steady state treatment in its interpretation of the meaning of the denominator constant.

What then is the significance of V_{max} and K_m?

V_{max} According to both interpretations of the two-step mechanism offered above, V_{max} has a defined meaning reflecting the fact that the decomposition of the ES complex is rate-limiting, i.e.

$$V_{max} = k_2 e_o$$

(where k_2 is the rate constant of the rate-limiting step). It is therefore

† This equation, $v_o = \frac{k_2 e_o [S]}{[S]+K}$ explains why, at any given substrate concentration, there is a stoichiometric relationship between the initial velocity of the reaction and the concentration of enzyme used ($v_o \propto e_o$).

feasible that by following changes in the value of V_{max} under different reaction conditions we might obtain information concerning the kinetic properties of the rate-limiting step in the decomposition of ES.

K_m According to the Briggs and Haldane steady state assumption, $K_m = \frac{k_{-1}+k_2}{k_1}$, but by the equilibrium hypothesis of Michaelis and Menten $K_m = k_{-1}/k_1 = K_s$ (the dissociation constant of ES). It would appear, therefore, that in the Briggs and Haldane view the situation envisaged by Michaelis and Menten is only a special case in which k_2 is so much smaller than k_{-1} that its value is negligible in the ratio $\frac{k_{-1}+k_2}{k_1}$,

i.e. $$K_m = K_s \quad \text{when} \quad k_2 \ll k_{-1}$$

Evidently, although K_m is an experimentally measurable kinetic parameter, and an 'operationally' definable feature of an enzymic process, its value can only be interpreted in terms of contributory rate constants if the mechanism of the reaction is known.

Confirmation of the existence of enzyme-substrate complexes

Recent advances in experimental techniques (including use of rapid flow methods in conjunction with sensitive spectrophotometry) have enabled the initiation and progress of several enzyme-catalysed reactions to be studied in such detail that the formation and fate of transient intermediates have been followed.

In one of the earliest attempts, Britton Chance (1943) used a pure preparation of the enzyme peroxidase to catalyse the oxidation by H_2O_2 of a colourless reduced dye (AH_2) to a coloured product (A). Chance discovered that this reaction was accomplished via the preliminary formation of a peroxidase-H_2O_2 complex detectable by its characteristic spectrum. By sensitive and rapid spectrophotometry, the concentrations of the peroxidase-H_2O_2 complex, and of the coloured dye product, could be simultaneously assayed throughout the reaction when substrate-like concentrations of the enzyme were used (though now the reaction was completed in less than 2 s !). The rate at which the complex decomposed to yield product could be controlled in these experiments by varying the supplied concentration of the reduced dye AH_2, and the formation of the complex could be studied in the absence of this product-yielding decomposition by omitting AH_2 altogether. In this way Chance was able to observe a rapid initial increase in the concentration of the peroxidase-H_2O_2

complex which then decreased more slowly as the complex was utilized in the oxidation of AH_2. The rate of formation of the coloured dye A gave the velocity of the overall reaction, and he found that, at any instant, this was directly proportional to the current concentration of the peroxidase-H_2O_2 complex.

These findings confirmed that at least two of the assumptions of the two-step hypothesis were probably justified in the case of this peroxidase system, viz.

(a) that a specific enzyme-substrate complex is an intermediate in the reaction;

(b) that the overall velocity of the enzyme-catalysed reaction at any time is determined by the concentration of this enzyme-substrate complex.

Yet several difficulties of interpretation were posed by Chance's findings. Assuming the following mechanism for the reaction,

$$\text{Peroxidase} + H_2O_2 \underset{k_{-1}}{\overset{k_1}{\rightleftharpoons}} \text{Peroxidase} - H_2O_2 \xrightarrow[AH_2 \;\; A]{k_2} \text{Peroxidase} + 2H_2O$$

he obtained the following values for the rate constants of the contributory steps when a certain, quite large concentration of AH_2 was supplied: $k_1 = 10^7$ dm^3 mol^{-1} s^{-1}, $k_{-1} = 0.2$ s^{-1} and $k_2 = 4.2$ s^{-1}. The Michaelis constant K_m of the reaction under these conditions was approximately 4×10^{-7} mol dm^{-3}. From the values of the individual rate constants, the dissociation constant K_s of the peroxidase-H_2O_2 complex (equals k_{-1}/k_1) is 2×10^{-8} mol dm^{-3}. This means that the Michaelis and Menten assumption that K_m equals K_s was not valid in this case, where K_m was evidently a kinetic (steady state) constant rather than a thermodynamic (equilibrium) constant (since its value was more nearly equal to the ratio of the rate constants of the 'forward' steps ($k_2/k_1 = 4.2 \times 10^{-7}$ mol dm^{-3}) than to the value of K_s). This illustrates how unwise it is, in the absence of other evidence, to assume that K_m equals K_s and that the value of K_m reflects the affinity of the enzyme for its substrate.

In the peroxidase-catalysed reaction, in which both H_2O_2 and AH_2 are reactants, we have incidentally encountered an example of a multisubstrate reaction. The special problems relating to the kinetic behaviour of such reactions will be mentioned later (p. 309); meanwhile, two important lessons can be learnt from the peroxidase system.

The first of these concerns the fact that initial velocity studies and determination of $K_m^{H_2O_2}$, for the peroxidase-catalysed reaction may be carried out, as for a single substrate reaction, by keeping the concentration of AH_2 constant and varying the concentration of H_2O_2 (though the value of $K_m^{H_2O_2}$ so obtained would be an 'apparent K_m'; see p. 312). This is why we were careful to point out that the values for the rate constants given above were obtained using a certain fixed concentration of AH_2. By decreasing the concentration of AH_2, the rate of decomposition of the peroxidase-H_2O_2 complex is decreased (the observed value of k_2 is lessened). Indeed, by sufficiently lowering the concentration of AH_2, the apparent value of k_2 can be so adjusted that it is as small as the value of k_{-1}, when K_m becomes approximately equal to K_s. Thus in the one system, the Michaelis constant can, under certain conditions, approximately equal the dissociation constant of ES, whilst under different conditions (in this instance, when a higher concentration of fixed substrate is supplied) its value is very different from the value of K_s. The same moral emerges—never assume in the absence of corroboratory evidence that the Michaelis constant reflects the affinity of an enzyme for its substrate. Indeed, there are now many more examples demonstrating that K_m and K_s for the same enzymic reaction may have very different values; e.g. it was found that for human serum cholinesterase using *o*-nitrophenylbutyrate as substrate K_m was approximately twice the value of K_s, while for a succinic oxidase system K_m was some fifteen times greater than K_s.

The second lesson taught by Chance's work with the peroxidase system is that in enzyme kinetic studies (as in reaction kinetics generally) one should always be aware of the assumptions made in proposing a reaction mechanism, and must be prepared to modify one's view in the light of new evidence. Thus despite the fact that he had been able to 'explain' the peroxidase-catalysed oxidation of AH_2 by H_2O_2 by a very reasonable two-step mechanism and had been able to calculate plausible values for the rate constants of all three contributory reactions, Chance later obtained fresh evidence that the reaction actually proceeds in four steps. Although the peroxidase-H_2O_2 complex whose formation and concentration he had originally studied was in fact not the *first* product of interaction of the enzyme with H_2O_2, the rate-limiting step in the revised four-step mechanism was still the reaction of this complex with AH_2. Thus Chance's conclusion that the overall reaction velocity is determined by the concentration of *an* enzyme-substrate complex still stands, even though it has turned out that more than one such complex is an intermediate in this reaction.

Pre-steady, steady and equilibrium states during an enzyme-catalysed reaction

Computer analysis of proposed reaction mechanisms, with plausible values for kinetic constants being introduced until the 'best fit' is achieved with the experimentally determined rates of appearance and disappearance of reactants, intermediates and products, has already proved a most potent analytical tool in the hands of the enzyme kineticist. If we assume that a single substrate enzyme-catalysed reaction has the mechanism,

$$E+S \underset{k_{-1}}{\overset{k_1}{\rightleftharpoons}} ES \xrightarrow{k_2} E+P$$

and then assume (i) that the three rate constants (k_1, k_{-1} and k_2) all have the same value (1.0), and (ii) that the initial substrate concentration $[S]_0$ is just ten times the total enzyme concentration [E], the computer predicts that the reaction will follow the course shown in Fig. 11.5. Three quite distinctive phases are recognizable in this reaction, each of which may be profitably exploited for purposes of kinetic analysis. These are:

1. *The pre-steady state*, wherein the change in [S] is negligible compared with the proportional changes in [E] or [ES]. 'Stopped flow' and other rapid reaction techniques can be used to investigate this phase.
2. *The steady state*, during which there is no change in [ES] (i.e. $d[ES]/dt=0$) and the change in [S] is much greater than the change in residual [E]. This is the phase routinely examined in simple kinetic studies of the type described in this chapter.
3. *The equilibrium state*, wherein the reactants E and S have achieved their equilibrium concentrations and there is no net change in their concentrations (although, of course, flux continues in both directions). This phase may be studied either by isotopic exchange, or by techniques by which the equilibrium is displaced and measurements are made during relaxation to the equilibrium state, e.g. the original Eigen techniques of T-jump, P-jump etc.

INHIBITION OF ENZYME-CATALYSED REACTIONS

Though inhibitors may act by combining with substrates, cofactors or various forms of the one enzyme, we shall only consider those inhibitors which combine directly with the enzyme. Some of these compounds are limited in their inhibitory effect to a single enzyme; others affect a large number of enzymes that serve very different catalytic functions. Although it is possible to classify such inhibitors in several different ways, it is necessary at the outset to distinguish between those that act irreversibly, and those that act reversibly.

An ***irreversible inhibitor*** reacts with an enzyme in such a way that subsequent dialysis does not restore its catalytic activity, i.e. a

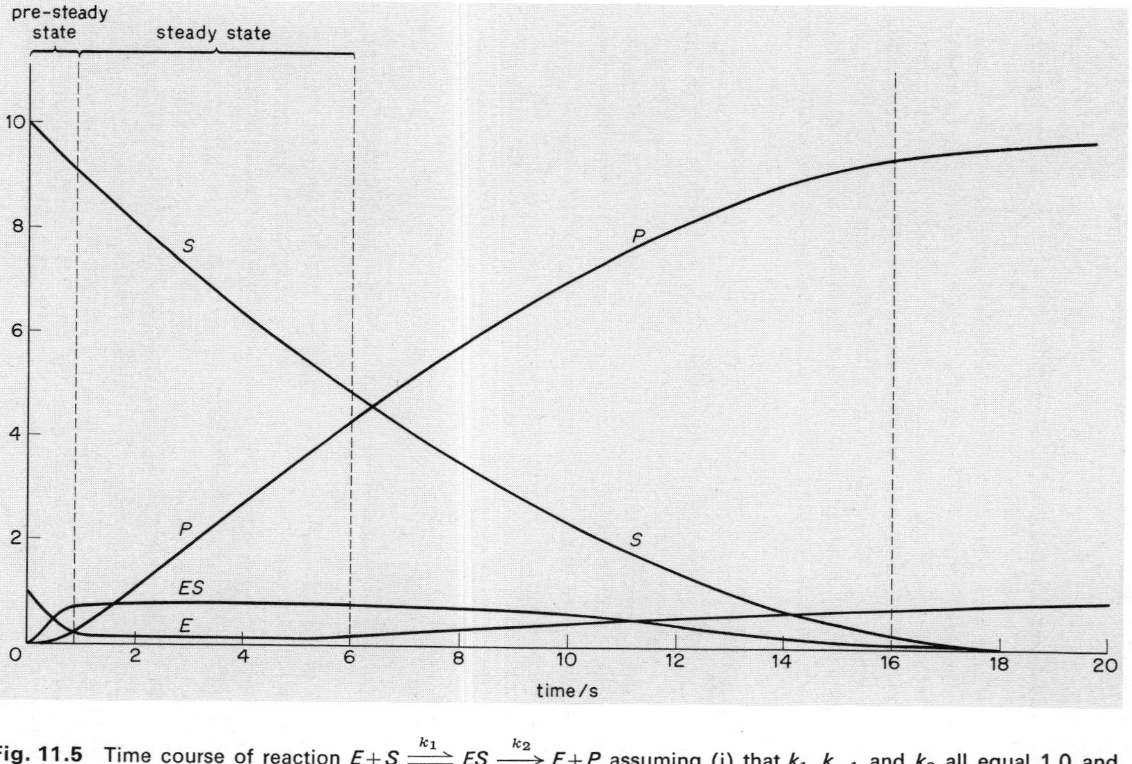

Fig. 11.5 Time course of reaction $E+S \underset{k_{-1}}{\overset{k_1}{\rightleftharpoons}} ES \xrightarrow{k_2} E+P$ assuming (i) that k_1, k_{-1} and k_2 all equal 1.0 and (ii) that the initial substrate concentration $[S]_0$ is $10 \times$ total enzyme concentration [E].

stoichiometric reaction of enzyme with inhibitor produces an undissociable complex and so inactivates the enzyme

$$E + I \longrightarrow EI$$

Reaction between the enzyme and the inhibitor is therefore progressive, taking place at a rate defined by the rate constant of this irreversible binding reaction. This means that irreversible inhibition of catalytic activity (which is what is generally measured) is itself progressive, increasing in extent until, when excess inhibitor is supplied, all of the binding sites on the enzyme capable of reacting with the inhibitor have done so, and are substituted. Inhibition by iodoacetamide is of this type, since it is the result of irreversible reaction between the inhibitor and essential -SH groups of the enzyme (e.g. essential cysteine side chains), leading to the creation of covalent (S-alkyl) bonds,

$$\text{Enz-SH} + \text{ICH}_2\text{CONH}_2 \longrightarrow \text{Enz-S-CH}_2\text{CONH}_2 + \text{HI}.$$

Similarly, certain organophosphorus compounds irreversibly inhibit enzymes such as trypsin or cholinesterase by irreversibly reacting with the -OH groups of essential serine side chains, to yield stable serine phosphate esters. Such irreversible inhibition must be distinguished from destructive inactivation of the enzyme molecule brought about by (a) general or localized (active site) denaturation, or (b) chemical degradation.

In contrast, the reaction between a ***reversible inhibitor*** and an enzyme (or ES complex) is freely reversible, e.g.

$$E + I \rightleftharpoons EI$$

where E represents the form of the enzyme capable of binding the inhibitor. This means that reversible inhibition is *not* progressive, in the sense that it does not proceed until all of the susceptible sites on the enzyme bind the inhibitor all of the time. Instead, the extent of inhibition is determined by the composition of the equilibrium that is attained between E, I and EI. This has two important consequences:

1. the extent of inhibition of a given amount of enzyme by a reversible inhibitor will be related to the concentration of the inhibitor [I] and to the equilibrium constant of the inhibitor-binding reaction $E + I \rightleftharpoons EI$ (this K_{eq} equals $1/K_i$ where K_i is the dissociation constant of EI);
2. the reversibly-inhibited enzyme should have its catalytic activity restored by dialysis (since, by removing 'free' I as it is released, the complete dissociation of EI is eventually accomplished).

We have seen that some information about the creation and decomposition of enzyme-substrate complexes can be obtained from initial velocity studies of enzyme-catalysed reactions. We have noted in particular the 'classical' behaviour of a single substrate reaction, in which the initial velocity varies hyperbolically with the substrate concentration—behaviour that is summarized in the Michaelis equation and quantitatively defined by the values of V_{max} and K_m. By repeating initial velocity studies in the presence of different fixed concentrations of a reversible inhibitor and describing the changed kinetic behaviour by a suitably modified initial velocity equation, it is likely that we could discover whether that inhibitor interferes predominantly with the binding of substrate, or with the subsequent decomposition of the *ES* complex. In turn this could inform us of the likely nature of the reaction of the inhibitor with the enzyme and might also extend our knowledge of the mechanism of the uninhibited process (see p. 311).

In fact, three types of reversible inhibitor can readily be distinguished by their characteristic effects on the linear double reciprocal plot of $1/v_0$ against $1/[S]$ (i.e. the Lineweaver–Burk plot) for an enzyme-catalysed reaction.

1 *Competitive inhibitors* alter its slope but do not change its intercept on the $1/v_0$ axis.
2 *Non-competitive inhibitors* alter both its slope and its intercept on the $1/v_0$ axis.
3 *Uncompetitive inhibitors* alter its intercept on the $1/v_0$ axis but do not change its slope.

We shall briefly consider the three groups separately to see what these kinetic findings tell us about their mode of action. (N.B. *All* are reversible inhibitors.)

1. Competitive inhibition

Figure 11.6 shows the effect of a fixed concentration of a truly competitive inhibitor on the plots of (a) v_0 versus [S], and (b) $1/v_0$ against $1/[S]$ for an enzyme-catalysed reaction. From Fig. 11.6a we see that this competitive inhibition is relieved by addition of more substrate until, at sufficiently high concentrations of substrate, the same maximum velocity is displayed as is exhibited by the uninhibited reaction at smaller concentrations of substrate. This would seem to indicate that the competitive inhibitor interferes with the reaction by diminishing the ease of access of substrate to the active site of the enzyme. The most reasonable explanation of this is that the competitive inhibitor itself combines reversibly with the enzyme at its substrate-binding site.

Interaction of enzyme and competitive inhibitor at the substrate-binding site produces an *EI* complex which can only dissociate to re-form *E* plus *I*. Interaction of substrate with the same site produces the 'normal'

ES complex which can either dissociate or decompose, i.e. $E+S \rightleftharpoons ES \rightarrow E+P$. So long as any of the substrate-binding, catalytic sites on the enzyme are binding I in place of S then the concentration of ES must be smaller than in the absence of the inhibitor. It follows that the rate of the overall reaction (which is proportional to [ES]) is decreased. Yet because inhibitor and substrate are in competition for the limited number of catalytic sites provided by a fixed amount of enzyme, the degree of inhibition will depend on the *ratio* of the concentrations in which inhibitor and substrate are supplied. Competitive inhibition is therefore overcome by increasing the concentration of the substrate and thus raising the [S]/[I] ratio.

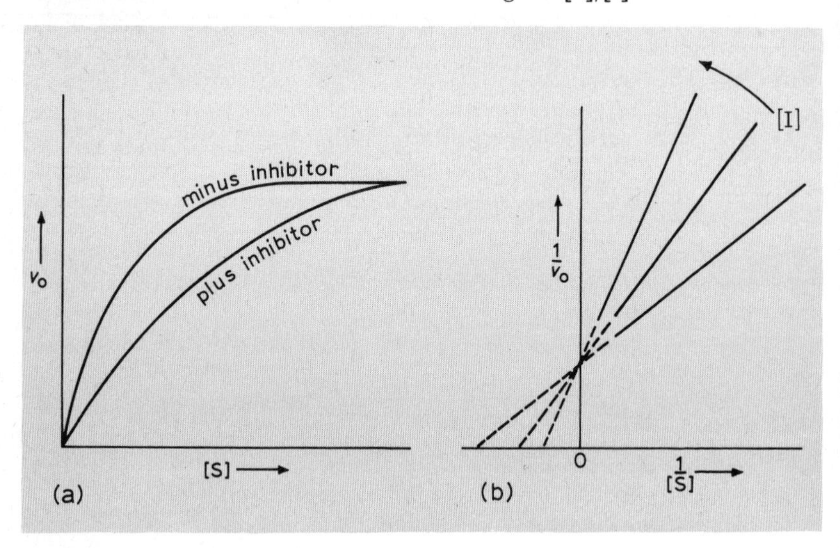

Fig. 11.6 Effect of a competitive inhibitor on the shape of (a) the hyperbolic plot of v_o versus [S], and (b) the double reciprocal plot of $1/v_o$ versus $1/[S]$ for an enzyme-catalysed reaction.

This explanation qualitatively accounts for the changed shape of the plot of v_o versus [S] in the presence of a fixed concentration of a competitive inhibitor (Fig. 11.6a). Quantitatively treated, it also shows why a competitive inhibitor alters only the slope of the linear reciprocal plot of $1/v_0$ versus $1/[S]$ (Fig. 11.6b), for if we apply the Briggs and Haldane steady state treatment to the situation in which inhibitor is supplied at concentration [I], we obtain the following modified Michaelis equation,

$$v_o = \frac{V_{max}[S]}{[S]+K_m\left(1+\frac{[I]}{K_i}\right)}$$

in which K_i is the dissociation constant of the EI complex. If we compare this equation with the Michaelis equation for the uninhibited reaction, i.e. $v_0 = \frac{V_{max}[S]}{[S]+K_m}$, it is obvious that the competitive inhibitor has altered the value of the denominator constant so that it has the new value $K_m\left(1+\frac{[I]}{K_i}\right)$. The maximum velocity is unchanged, so that the intercept of the reciprocal plot on the $1/v_0$ axis is also unchanged (equals $1/V_{max}$). However the slope of the reciprocal plot is increased in the presence of the inhibitor $\left(\text{since slope}=K_m/V_{max}, \text{ the slope is increased by the factor } 1+\frac{[I]}{K_i}\right)$. Since,

$$\text{slope} = \frac{K_m}{V_{max}}\left(1+\frac{[I]}{K_i}\right)$$

by rearranging this equation we obtain,

$$\text{slope} = \frac{K_m}{V_{max}K_i}\times[I]+\frac{K_m}{V_{max}}$$

Thus, if we plot the slopes of the lines obtained in the (primary) double reciprocal plot of $1/v_0$ vs. $1/[S]$, against the [I] at which each was obtained, the result is a linear (secondary) plot of slope vs. [I] which intercepts the [I] axis at a point corresponding to K_i (Fig. 11.7). It is for this

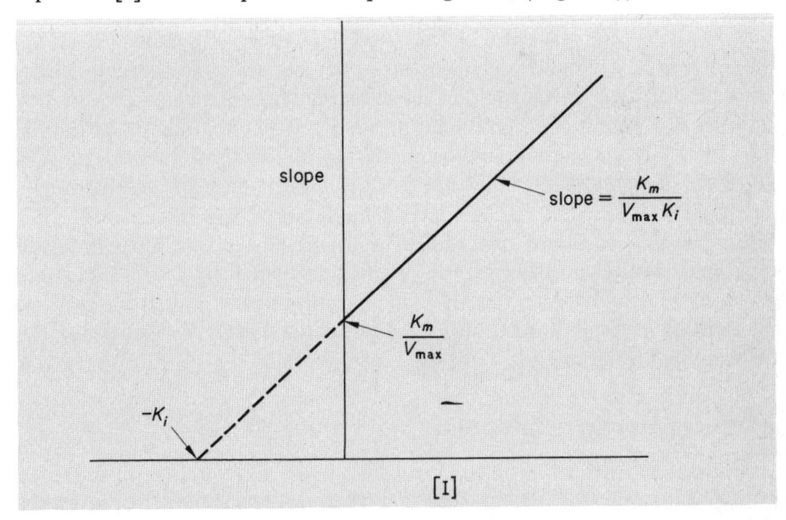

Fig. 11.7 Secondary plot of slope (of lines obtained in a primary, double reciprocal plot) versus [I], indicating linear competitive inhibition.

reason that this simple form of competitive inhibition is also known as *linear competitive inhibition.* In linear competitive inhibition, I and S cannot be present on the enzyme at the same time, i.e. *either* a molecule of E reacts with I to yield EI, *or* it reacts with S to give ES.

The same total exclusion of S occurs in the more complex situation wherein reaction of one molecule of I at the catalytic site can lead to the reaction of a second molecule of I so that both are involved in the exclusion of S. In this case the secondary plot of slope *vs.* [I] is a parabola, and the phenomenon is termed *parabolic competitive inhibition.* Yet another sub-type, namely *hyperbolic competitive inhibition*, is found when I and S can be present simultaneously in combination with the enzyme but the IES complex breaks down at the same rate as the ES complex.

My purpose in mentioning these non-linear forms of competitive inhibition is merely to indicate that still more information concerning the mechanism of the process can be derived from the primary double reciprocal plot merely by undertaking the secondary plot of slope *vs.* [I].

Examples of competitive inhibition of enzymic reactions are not uncommon. Understandably, since substrate and competitive inhibitor compete for the same specific binding site, they usually possess similar molecular structures; e.g. malonate is a competitive inhibitor of succinic dehydrogenase, and sulphanilamide competitively inhibits the enzymic utilization of the structurally similar compound *para*-aminobenzoate for the biosynthesis of the folic acid group of coenzymes. In such cases, hopes are raised that the structure of the 'complementary' catalytic site on the enzyme might be deducible from the structures and stereochemical configuration of those groups which the substrate analogue must possess if it is to act as a potent competitive inhibitor. Even so, one must not presume the identity of the sites on an enzyme that bind a substrate with those sites that bind its competitive inhibitor until the mechanism of their competition has been elucidated. As Cleland has pointed out, these substances could possibly combine with different forms of the enzyme which are separated from each other in the reaction sequence by freely reversible steps. Each would then affect the concentration of the alternative form of the enzyme available for binding its competitor merely by displacing the intervening equilibria.

2. Non-competitive inhibition

A non-competitive inhibitor of an enzymic reaction alters both the slope of the Lineweaver–Burk plot and its intercept on the $1/v_0$ ordinate. Using a series of fixed concentrations of the inhibitor, a family of straight lines is obtained when $1/v_0$ is plotted against $1/[S]$. As the inhibitor concentration is increased, so the magnitudes of the slopes and intercepts of these lines increase, thus ensuring that, when extrapolated, they intersect somewhere to the left of the ordinate. This point may be *above*, *below*, or (as shown in Fig. 11.8) *on* the $1/[S]$ abscissa.

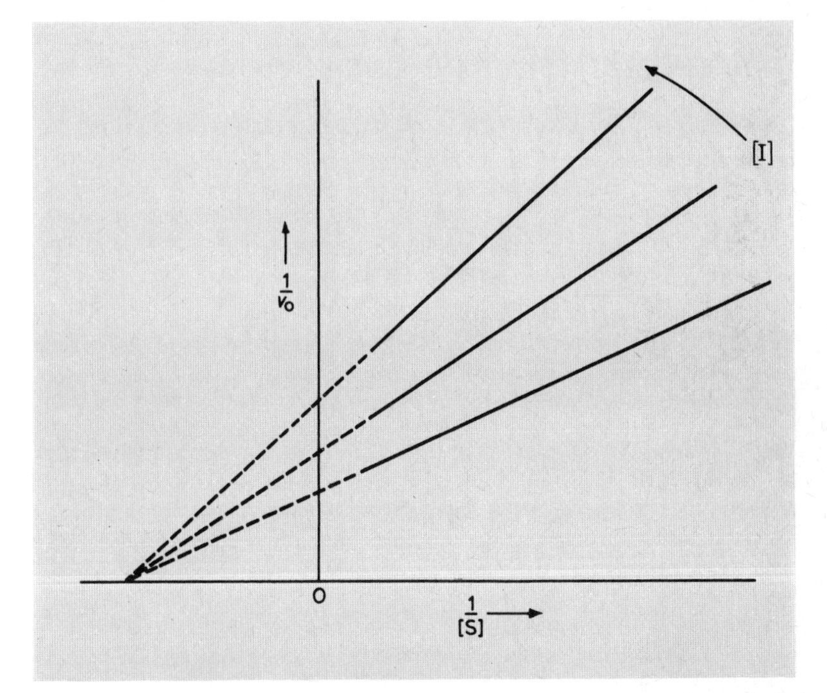

Fig. 11.8 Pure non-competitive inhibition diagnosed by the effect of the inhibitor on the double reciprocal plot of $1/v_0$ versus $1/[S]$ for an enzyme-catalysed reaction.

If, using values obtained from the primary, double reciprocal plot, we construct secondary plots of (i) slope *versus* [I], and (ii) intercept on the $1/v_0$ axis *versus* [I], each of these could turn out to be linear, parabolic or hyperbolic, so that in all nine classes of non-competitive inhibitor can be distinguished by this simple test.

The simplest *linear, non-competitive inhibitors* combine with various forms of the enzyme at a site other than its catalytic site; thus besides E and ES, enzyme will also be present in the form of IE and IES complexes, neither of which can however break down to yield products (i.e. IE and IES are 'dead-end' complexes). Since I is a reversible inhibitor, the IE complex can dissociate into I and E in a reaction whose equilibrium constant is K_i

$$IE \rightleftharpoons I+E \quad K_{eq} = K_i$$

But the IES complex can dissociate in two ways,

$$IES \rightleftharpoons I+ES \quad K_{eq} = K_I$$

and,

$$IES \rightleftharpoons IE+S$$

In its double reciprocal form the equation for non-competitive inhibition is,

$$\frac{1}{v_0} = \frac{K_m}{V_{max}}\left(1+\frac{[I]}{K_i}\right) \times \frac{1}{[S]} + \frac{1}{V_{max}}\left(1+\frac{[I]}{K_I}\right)$$

As mentioned above, the double reciprocal plots of $1/v_0$ vs. $1/[S]$ can intersect above, below or on the $1/[S]$ axis; from the above equation it can be shown that only when K_i happens to equal K_I will the intersection occur at $-1/K_m$ (as shown in Fig. 11.8).

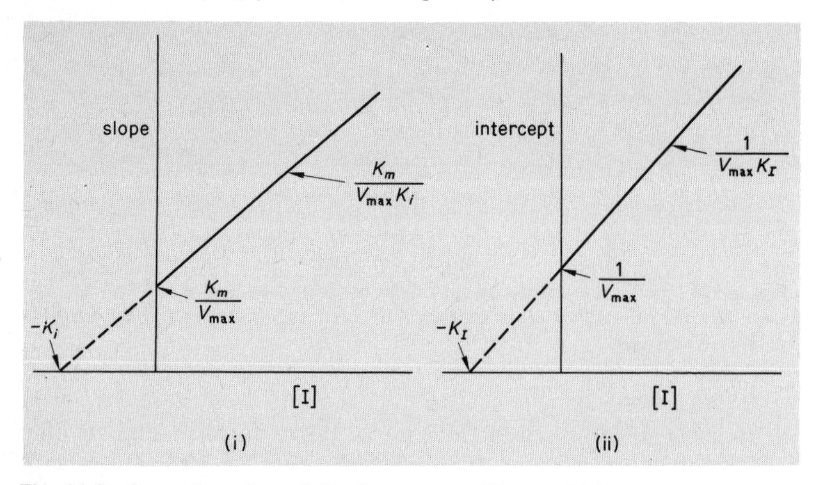

Fig. 11.9 Secondary plots of (i) slope versus [I], and (ii) intercept versus [I] for linear, non competitive inhibition.

You might like to use the equation of the primary double reciprocal plot given above, to verify the fact that when the secondary plots of (i) slope *vs.* [I] and (ii) intercept *vs.* [I] are prepared for linear competitive inhibitors they will themselves have the slopes and intercepts shown in Fig. 11.9, which allow the values of K_i and K_I to be determined.

3. Uncompetitive inhibition

According to Cleland, the uncompetitive inhibitor combines with forms of the enzyme (notably *ES*) that do not themselves combine with substrate. The *IES* so created is a dead-end complex that can only dissociate to yield *I* and *ES* (equilibrium constant $= K_i$). Increasing the supply of *S* will neither prevent binding of the inhibitor nor cause a release of substrate-available enzyme so that inhibition is not relieved.

The uncompetitive inhibitor increases the intercept on the $1/v_0$ ordinate of the double reciprocal plot without changing its slope (Fig. 11.10). The new intercept has the value $\left(\frac{1+[I]/K_i}{V_{max}}\right)$, and since the slope $(=K_m/V_{max})$ is unchanged, the inhibitor must cause the numerator in this ratio to change by the same factor. Thus in the presence of uncompetitive inhibitor the Michaelis equation is:

$$v_0 = \frac{\dfrac{V_{max}}{(1+[I]/K_i)} \times [S]}{[S] + \left(\dfrac{K_m}{1+[I]/K_i}\right)}$$

which in the double reciprocal form is,

$$\frac{1}{v_0} = \frac{K_m}{V_{max}} \times \frac{1}{[S]} + \frac{1}{V_{max}}\left(1+\frac{[I]}{K_i}\right)$$

The secondary plot of intercept *versus* [I] will be linear and will readily yield the value of K_i (as in Fig. 11.9(ii)).

Classical uncompetitive inhibition of this type is not common in the case of single substrate reactions, but is frequently observed with multi-substrate reactions in whose mechanisms more than one enzyme-reactant complex participate.

Before concluding this section, it might be pointed out that although, for simplicity, we have considered the effects of reversible inhibitors on single substrate reactions, their action on multisubstrate systems can be examined in a similar fashion. The mode of action of a given inhibitor (i.e. whether competitive, uncompetitive or non-competitive) would be tested using the various reactants in turn as the variable substrate whilst maintaining the rest at fixed concentrations. Indeed this is a most profitable kinetic study when the products of a multisubstrate reaction are tested in turn as reversible inhibitors of the 'forward' reaction. Thus, one product may act as a competitive inhibitor when substrate *A* is made the variable substrate, but as an uncompetitive inhibitor when the second substrate *B* is made the variable substrate. In the case of another bisubstrate reaction of different

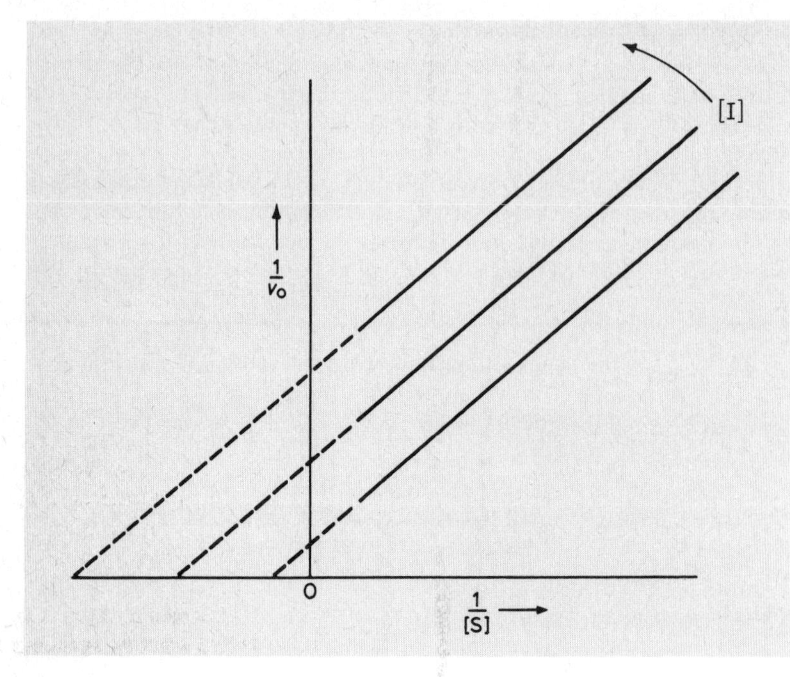

Fig. 11.10 Uncompetitive inhibition of an enzyme-catalysed reaction; characteristic appearance of double reciprocal plots of $1/v_0$ versus $1/[S]$.

mechanism, a product might act as a competitive inhibitor when A' is the variable substrate, but as a non-competitive inhibitor when B' is the variable substrate. From such patterns of product inhibition one is often able to distinguish between alternative sequential mechanisms for bisubstrate reactions which cannot be differentiated by 'straightforward' initial velocity studies (see p. 311 and references[12, 31]).

EXAMPLE

The activity of L-*aspartate 4-carboxylase (aspartate β-decarboxylase) can be assayed manometrically by following the evolution of CO_2 from* L-*aspartate.*

$$\text{L-}aspartate \longrightarrow \text{L-}alanine + CO_2$$

The following results were obtained in an experiment designed to assess the potency of threo-β-*hydroxyaspartate as an inhibitor, at 303 K and pH 5, of a preparation of aspartate 4-carboxylase from a microbial source.*

Concentration of L-*aspartate* / *μmol dm*$^{-3}$	*Initial velocity* / *μmol* CO_2 *liberated min*$^{-1}$ *mg protein*$^{-1}$	
	No inhibitor	*Plus 0.02 mol dm*$^{-3}$ DL-threo-β-OH*aspartate*
25	17.0	—
33.3	21.3	5.7
50	27.8	8.1
100	41.7	14.7
200	52.6	25.0

Calculate the value of K_m *of the uninhibited reaction. Does* threo-β-*hydroxy-aspartate act as a competitive or non-competitive inhibitor?*

Values of $1/[S]$ and $1/v_0$ are calculated and plotted as shown in Fig. 11.11.

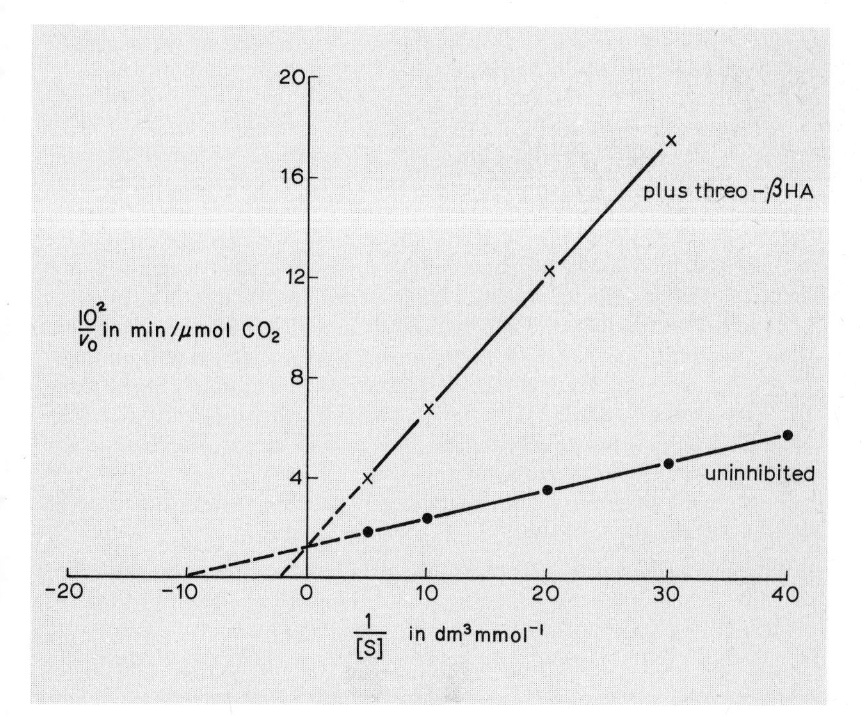

Fig. 11.11 Competitive inhibition of aspartate 4-carboxylase by *threo*-β-hydroxy-aspartate.

[S]/μmol dm^{-3}	$\frac{1}{[S]}\times$ mmol dm^{-3}	$\frac{10^2}{v^0}\times$ μmol CO_2 min^{-1}	
		Uninhibited	Inhibited
25	40	5.88	—
33.3	30	4.70	17.54
50	20	3.60	12.34
100	10	2.40	6.80
200	5	1.90	4.00

The reciprocal plot for the uninhibited reaction, when extrapolated backwards, intersects the $1/[S]$ axis at -10.1 dm^3 mmol^{-1}.

$$\therefore \qquad -1/K_m = -10.1 \text{ dm}^3 \text{ mmol}^{-1}$$

and

$$K_m = 9.9\times10^{-5} \text{ mol dm}^{-3}$$

The reciprocal plot for the reaction in the presence of *threo*-β-hydroxyaspartate, has a different slope but intersects the $1/v_0$ ordinate at the same point as does the reciprocal plot of the uninhibited reaction. Therefore, *threo*-β-hydroxyaspartate acts as a *competitive* inhibitor of the aspartate 4-carboxylase-catalysed decarboxylation of L-aspartate.

ALLOSTERIC EFFECTS

Investigation of the regulation and coordination of enzymic processes in living cells has revealed the existence of several enzymes whose kinetic behaviour is not always classical, and whose catalytic activity is affected by the binding of 'modifying compounds' to specific sites which are not identical with its catalytic (active) sites. Depending on whether the catalytic activity of the enzyme is enhanced or diminished by their presence, these 'modifiers' behave respectively as reversible activators or inhibitors. To distinguish them from reversible effectors that act directly at the catalytic site (isosteric effectors), these 'second site' activators and inhibitors are called ***allosteric effectors***, and the specific sites at which they are bound are called ***allosteric sites***.†

Features of enzymes whose catalytic activities are subject to allosteric control include the following:

(a) Kinetic behaviour that suggests the presence in the enzyme molecule of multiple, interacting binding sites. Frequently the plot of v_0 against [S] is distinctly S-shaped (sigmoidal), and the reciprocal plot of $1/v_0$ against $1/[S]$ is consequently non-linear. Such behaviour suggests the

† According to this nomenclature, the classical competitive inhibitor that combines with the enzyme at its substrate-binding site, is an isosteric inhibitor; the pure non-competitive inhibitor that attaches to the enzyme at a site that is different from its substrate binding site is an allosteric inhibitor.

existence of multiple substrate-binding sites and of a cooperative effect whereby the binding of substrate at one site facilitates the utilization of substrate at another site. (The interaction of sites binding identical molecules, is termed a *homotropic effect* and is always cooperative.)

(b) Altered catalytic activity as a consequence of the binding at distinct allosteric sites of effectors which are structurally dissimilar to the substrate. Interactions between sites binding structurally different molecules are called *heterotropic effects*; they can either be cooperative or antagonistic. With any enzyme that exhibits heterotropic effects, homotropic effects are likely to be demonstrated by one or other type of bound molecules (substrate or activator or inhibitor).

(c) Release of catalytic activity from allosteric control by treatments (e.g. pH, heat, heavy metal cations, urea, high ionic strength) which can alter the conformation of the enzyme protein (possibly without damaging the effector-binding sites).

(d) Tendency for the enzyme protein to disaggregate into catalytically inactive subunits (sometimes making for unusual lability during purification).

Several theories have been advanced to explain this behaviour, and in particular to explain how events at catalytic sites can be affected by the binding of small molecular weight effectors at distinct allosteric sites. The most promising of these theories proposes that enzymes susceptible to allosteric control are 'oligomers', i.e. symmetrical aggregates of a unit protein structure (the protomer). According to this model, allosteric interactions are the consequence of a reversible, effector-induced alteration in the quaternary structure of the oligomer which involves changes in inter-protomer bonding. It is thought likely that the oligomer can exist in at least two states of differing catalytic activity and that the binding of the effector serves to displace the equilibria between these states.

The kinetic test of any such hypothesis consists, as always, of deriving initial velocity equations based on the assumptions of the theory and then attempting, by statistical methods, to match actual data obtained from initial velocity studies to these theoretical equations. If the theory invokes conformational changes, then evidence of induced change in the size or shape of the enzyme molecule coincident with changes in susceptibility to allosteric control may be sought by ultracentrifugal, electrophoretic or optical rotatory dispersion methods, and where possible by 'direct' examination of the appearance of the molecules by electron microscopy. Explanations of allosteric effects in terms of subunit interactions have thus been encouraged by the finding that several enzymes which demonstrate these effects actually possess polymeric structures and, furthermore, that allosteric control is lost when these are caused to disaggregate.

Allosteric control of the activity of a key enzyme has proved to be the means whereby a controlled flow of metabolites is maintained through many a metabolic route (e.g. feedback inhibition of an early enzyme of a biosynthetic pathway by the end product of that pathway). Yet the discovery of allosteric phenomena has not only disclosed an important regulatory mechanism, it has been a salutary reminder to the enzymologist that he must be concerned not only with the chemical groups at the active site of an enzyme, but also with its three-dimensional configuration and with the quaternary structure of the whole molecule.

ENZYMIC CATALYSIS OF READILY REVERSIBLE REACTIONS

In deriving the Michaelis equation (p. 288) the assumption was made that the enzyme catalyses an irreversible reaction by reversibly forming an ES complex which irreversibly decomposes to yield product in a rate-limiting reaction. A different mechanism must be proposed if the reaction is significantly reversible, i.e. $S \rightleftharpoons P$. The simplest mechanism involving ES would be the following two-step process,

$$E+S \underset{k_{-1}}{\overset{k_1}{\rightleftharpoons}} ES \underset{k_{-2}}{\overset{k_2}{\rightleftharpoons}} E+P$$

Yet it would seem more likely that if in the forward reaction S is utilized via the creation of any enzyme-substrate complex, then in the reverse reaction when P is utilized, an enzyme-product complex EP is first formed. This would mean that at least two different enzyme-reactant complexes must participate in a minimally three-step mechanism,

$$E+S \underset{k_{-1}}{\overset{k_1}{\rightleftharpoons}} ES \underset{k_{-2}}{\overset{k_2}{\rightleftharpoons}} EP \underset{k_{-3}}{\overset{k_3}{\rightleftharpoons}} E+P$$

Provided that initial velocities are measured (i.e. of $S \rightarrow P$ when $[P]=0$, and of $P \rightarrow S$ when $[S]=0$), the rate equations for the forward and back reactions reduce to the form of the classical Michaelis equation.† Thus values of $V_{\max}$ and K_m can be measured for the forward reaction (V_f and K_f), and other values for the back reaction (V_r and K_r). It can be shown that the values of these parameters are related to the equilibrium constant of the overall reaction by the equation $K_{eq}=V_f K_r / V_r K_f$ (the Haldane relationship).

† Relatively simple techniques exist for calculating both general and initial rate equations for reversible reactions of quite complex mechanism, using a graphic method whose rules can be applied without necessarily understanding *why* they work[24].

One consequence of the ability of the product to combine with the enzyme to form an EP complex in steady state equilibrium with the ES complex, is that the product must act as a competitive inhibitor of the forward ($S \rightarrow P$) reaction. Therefore although it is unnecessary in *initial* velocity studies to worry about the accumulation of product, one should be aware of the possibility of product inhibition of enzymic reactions (even of those that are virtually irreversible), and of the likelihood that in all enzymic processes an EP complex is formed prior to the release of 'free' product.

ENZYMIC CATALYSIS OF REACTIONS INVOLVING TWO SUBSTRATES

In many enzymic reactions of metabolic importance, two substrates are utilized (one of which may be a coenzyme), and two products are formed, i.e.

$$A + B \rightleftharpoons C + D$$

Such reactions include those in which certain groups are transferred from a donor molecule to an acceptor (e.g. reactions catalysed by transaminases, dehydrogenases and phosphotransferases). Even hydrolytic enzymes (e.g. esterases) catalyse two-substrate reactions, though because water is always present in excess we tend to overlook its role as second substrate.

Evidently the mechanism of a two-substrate reaction is likely to be more complicated than that of a single-substrate enzymic process. In particular, we must now discover whether the substrates are randomly bound by the enzyme or whether they are bound, and their products released, in some obligatory sequence.

To illustrate the major types of mechanism that can be expected, we can use the convention introduced by Cleland as a shorthand way of representing the course of events in which the enzyme is a participant. According to this convention, a horizontal line represents the enzyme, and reading from left to right along this line we follow in sequence the creation and fate of various enzyme complexes (indicated under the line). Vertical arrows indicate the binding of substrates or the release of products.

(a) Sequential mechanisms

These are characterized by the binding of both substrates prior to the release of either product.

(i) *Ordered mechanisms*

Here the substrates (and products) are bound in an obligatory order. A good example of this type of mechanism is provided by several NAD^+-

dependent dehydrogenases which first bind NAD^+ and are only then able to bind the oxidizable substrate, i.e.

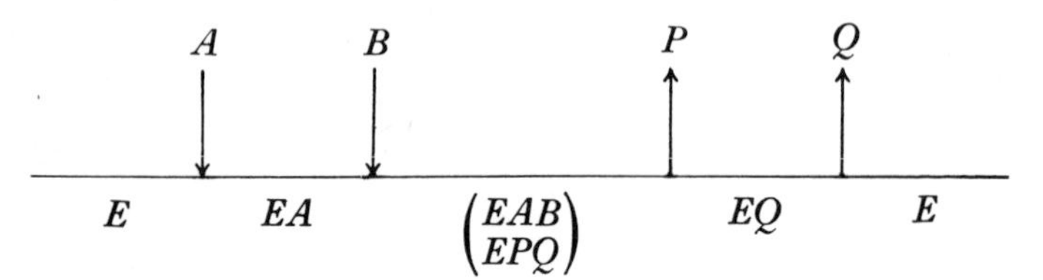

(ii) *Random mechanisms*

In these cases one assumes that there must be two distinct binding sites on the 'free' enzyme, one for each substrate (or product). Since the substrates are independently bound, either may be the first to react with the enzyme (and thus be the 'leading substrate').

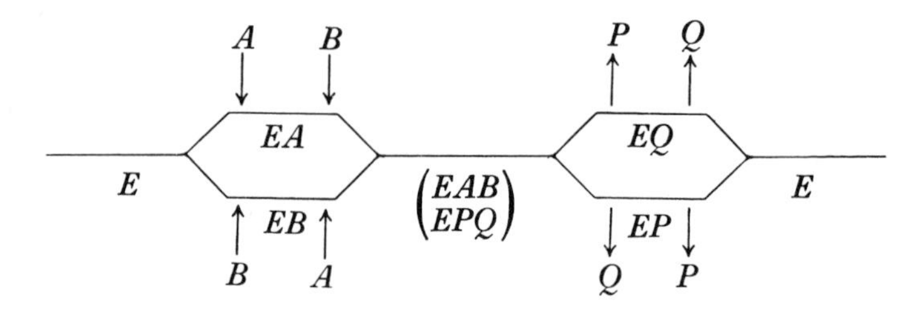

Various phosphotransferases (e.g. creatine kinase, yeast hexokinase) undertake a form of this mechanism in which the central complexes ($EAB \rightleftharpoons EPQ$) are very rapidly interconverted.

(b) Non-sequential mechanisms (ping-pong mechanisms)

These are characterized by the fact that one product is formed and released before the second substrate is bound.

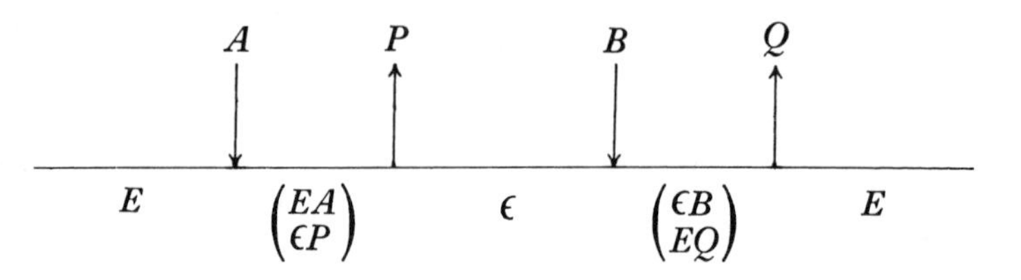

The enzyme binds A to create an EA complex which is converted into a form ϵP which breaks down to yield one product P, and a second stable form of the enzyme (ϵ) capable of binding the second substrate B. Conversion of ϵB to EQ is followed by decomposition of this to release the second product Q and regenerate the primary stable form of the enzyme (E). The enzyme thus exists in two stable forms, one of which is capable of reacting with substrate A to produce the alternative form which is capable of reacting with the second substrate B to regenerate the A-binding form. It is this shuttling back and forth of the enzyme between these forms that Cleland emphasized when he called this a *ping-pong mechanism*. Probably the most common group of enzymes to operate by this mechanism are the pyridoxal phosphate-dependent transaminases.

To perform initial velocity studies with these two-substrate systems, values of v_0 are measured, using different concentrations of one substrate in the presence of a series of fixed concentrations of the second. In this way, a family of plots is obtained of $1/v_0$ against $1/[B|$ using different fixed concentrations of A. The appearance of these primary reciprocal plots will show whether the mechanism is sequential or non-sequential (ping-pong). If sequential, the plots intersect, since their slopes increase and their intercepts (on the $1/v_0$ axis) decrease as the concentration of the fixed substrate is increased. In contrast, if the mechanism is ping-pong, the plots do not intersect, since their slopes are identical irrespective of the concentration of fixed substrate (i.e. the plots are parallel at all concentrations of fixed substrate).

Though we can very simply discover whether a sequential mechanism is involved, these initial velocity studies cannot by themselves distinguish between various alternative 'subtypes' such as the ordered and random (rapid equilibrium) mechanisms illustrated above. This is because these, and other types of ordered sequential mechanism, all behave according to the same type of initial velocity equation and yield linear primary reciprocal plots of similar appearance. However, by extending these kinetic studies to include an investigation of the types of inhibition caused by the products of the reaction (tested individually with both substrates used in turn as the variable substrate), it is possible to identify which type of sequential mechanism is involved (p. 303). Together with isotope exchange methods and 'binding studies', these product-inhibition patterns can elucidate which is the leading substrate in the mechanism. Details of these methods are given in the admirably explicit review by J. F. Morrison.[31]

The quantitative definition of the kinetic behaviour of a two-substrate enzymic process involves more parameters than the two needed to define the single-substrate irreversible reaction. Thus four parameters must be determined for a reaction proceeding by a sequential mechanism, viz.

$$V_{max}, K_m{}^A, K_m{}^B \text{ and } K_s{}^A$$

where V_{max} = limiting maximum velocity
$K_m{}^A$ = limiting Michaelis constant for A
$K_m{}^B$ = limiting Michaelis constant for B
$K_s{}^A$ = dissociation constant for A

These *limiting* values of the kinetic parameters cannot be determined from primary reciprocal plots obtained at arbitrary concentrations of fixed substrate (even if this is *thought* to be saturating). For example $K_m{}^A$ is the value that this parameter approaches as the concentration of the fixed substrate B approaches infinity. Therefore these limiting values are obtained from secondary plots of (i) the intercepts, and (ii) the slopes of the primary reciprocal plots against the reciprocal of the concentration of the fixed substrate, as explained in reference [12].

EFFECTS OF TEMPERATURE ON ENZYMIC REACTIONS

Over a relatively small range of temperatures (0°C to 50°C i.e. 273 K to 323 K) the velocity of an enzyme-catalysed reaction usually first increases, and then decreases, as the temperature is raised, making it appear as if there is an optimum temperature at which the enzyme is most active (Fig. 11.12).

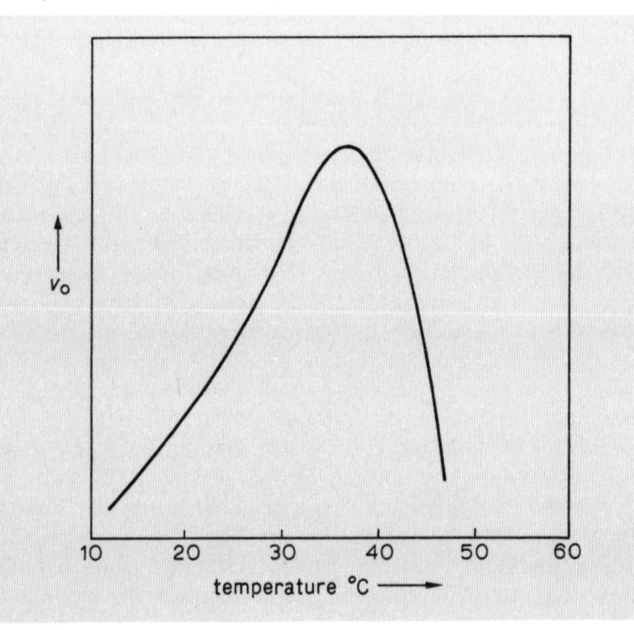

Fig. 11.12 Effect of temperature on the activity of a 'representative' enzyme.

This behaviour is the outcome of two concurrent events:

1. a genuine increase in rate of the catalysed reaction in response to increased temperature (cf. p. 264);
2. a progressive inactivation (denaturation) of the enzyme at a rate which increases as the temperature rises.

Most proteins (including enzymes) are notoriously liable to irreversible denaturation at temperatures above about 40° to 50°C, and the very high temperature coefficient of denaturation usual for proteins could explain the dramatic loss of catalytic activity suffered by many enzymes at these relatively mild temperatures. The few enzymes that retain considerable activity at higher temperatures (e.g. 60°C to 100°C) are therefore of unusual stability.† At any temperature, as the fraction of the total enzyme that is denatured increases with time, so will its catalytic effectiveness diminish. Thus the 'optimum temperature' of an enzyme is a meaningless term until the time of its exposure to that temperature is also recorded (as well as the composition of the assay medium, e.g. pH, ionic strength). The thermal stability of the enzyme (and of its substrates) must therefore be determined before the effect of temperature on the rate of the catalysed reaction can properly be investigated. (The thermal stability of an enzyme is easily tested by maintaining it for varying times at temperatures subsequently used in the kinetic studies, and assaying its previous and subsequent activity at one favourable temperature.)

Temperature-dependence of the initial velocity of an enzyme-catalysed reaction

We have seen previously (p. 264) that the extent by which the rate of a single-step reaction is affected by a given change in temperature is determined by the magnitude of its energy of activation. As the temperature increases from T_1 K to T_2 K, the rate constant changes from k' to k'' as predicted by the Arrhenius equation, which in its integrated form is $\log k = \text{constant} - E/2{\cdot}303\, R \times 1/T$, so that

$$\log k'' - \log k' = \frac{E^*(T_2 - T_1)}{19.14\, T_1 T_2} \quad \text{(p. 268)}$$

The value of the energy of activation (E^*) is determined experimentally by plotting $\log k$ against $1/T$, when the slope of the resultant straight line equals $-E^*/19.14$. Furthermore, E^* equals $(\Delta H^\ddagger + RT)$ where $\Delta H^\ddagger$ is the enthalpy of activation of the reaction, and is the difference between the enthalpy of the reactants and the enthalpy of the transition state.

† A very few enzymes have been shown to be 'cold sensitive', decreasing in activity more rapidly at low temperatures (0°C to 10°C) than at normal room temperatures.

Turning now to the simplest mechanism proposed for a single-substrate, irreversible enzymic process, we see that a change in temperature will affect the magnitudes of all three rate constants (k_1, k_{-1}, and k_2), since

$$E+S \underset{k_{-1}}{\overset{k_1}{\rightleftharpoons}} ES \xrightarrow{k_2} E+P$$

This means that the change in temperature affects the value of K_s (equals k_{-1}/k_1) and thus alters the steady state concentration of ES at a 'non-saturating' concentration of substrate. Meanwhile the temperature also affects the rate of decomposition of ES by causing a change in the value of k_2. It is possible to study specifically the effect of temperature changes on the velocity of this second step since at saturating concentrations of substrate, when [ES] equals e_o, $v_o = V_{max} = k_2 e_o$ (see p. 290). By obtaining linear reciprocal plots of $1/v_o$ against $1/[S]$ at different temperatures one can follow changes in the value of V_{max} with temperature.

Suppose that at temperature T_1 K, the maximum velocity is V'_{max} and the rate constant of the rate-limiting step in the decomposition of ES is k'_2. If at temperature T_2 K these parameters have the values V''_{max} and k''_2, then since $V_{max} = k_2 e_o$,

$$\log \frac{V''_{max}}{V'_{max}} = \log \frac{k''_2 e_o}{k'_2 e_o} = \log \frac{k''_2}{k'_2} = \frac{E^*(T_2 - T_1)}{19.14\, T_1 T_2}$$

An approximate value for E^* (in J mol^{-1}) can thus be calculated from the values of V_{max} at two temperatures, though a more exact value is obtained by plotting $\log V_{max}$ against $1/T$ and measuring the slope of the straight line obtained (equals $-E^*/19.14$). It is therefore a relatively simple matter to obtain a value for the energy of activation of an enzyme-catalysed reaction, but the mechanism of the reaction must be known before this value can be attributed to a specific rate-limiting step (whose $\Delta H^{\ddagger}$ will equal $E^* - RT$).

Suppose therefore that a single-substrate, reversible enzymic reaction is known to possess the following mechanism in which k_2 is much smaller than either k_1 or k_3, and k_{-2} is similarly much smaller than k_{-3} or k_{-1},

$$E+S \underset{k_{-1}}{\overset{k_1}{\rightleftharpoons}} ES \underset{k_{-2}}{\overset{k_2}{\rightleftharpoons}} EP \underset{k_{-3}}{\overset{k_3}{\rightleftharpoons}} E+P$$

The energies of activation for the forward reaction (E^*_f) and for the reverse reaction (E^*_r), can be determined by initial velocity studies (determining values of V_{max} at various temperatures from Lineweaver–Burk plots). Since the rate-limiting step in the forward reaction is $ES \xrightarrow{k_2} EP$, the kinetically measured value of E^*_f evidently equals $(\Delta H^{\ddagger}_2 + RT)$; similarly, E^*_r equals $(\Delta H^{\ddagger}_{-2} + RT)$. If the enthalpy profile was constructed for this reaction, it might well have the appearance shown in Fig. 11.13, where $EX_2^{\ddagger}$ represents the transition state between ES and EP.

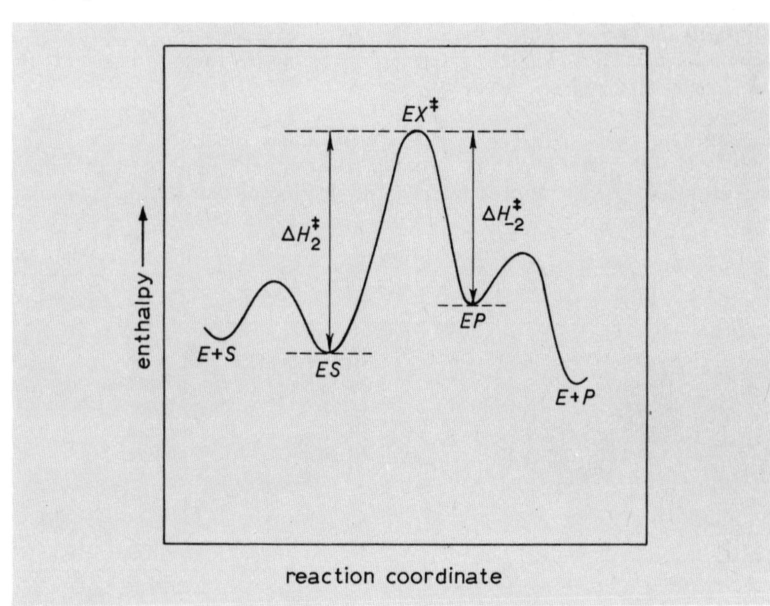

Fig. 11.13 Possible enthalpy profile for the enzyme-catalysed reaction

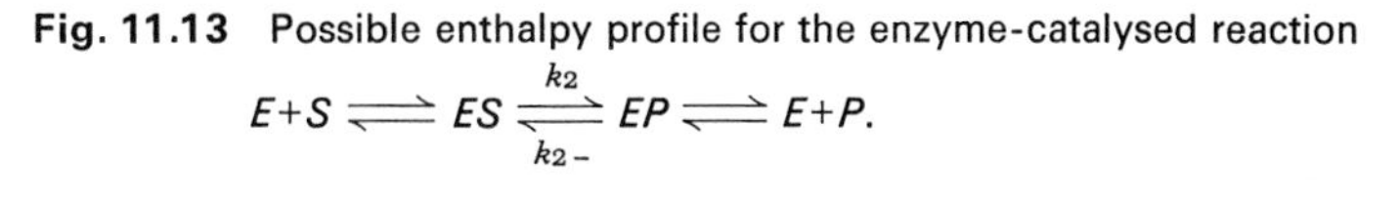

$$E+S \rightleftharpoons ES \underset{k_{2-}}{\overset{k_2}{\rightleftharpoons}} EP \rightleftharpoons E+P.$$

Though it is not possible to comprehend the full significance of the energy of activation of a reaction if its mechanism is not known, the experimental finding that its value is smaller for an enzyme-catalysed reaction than for the same but uncatalysed reaction at least explains why that reaction is more rapid in the presence of the enzyme. Thus the observation that the decomposition of hydrogen peroxide ($2H_2O_2 \rightarrow 2H_2O + O_2$) is accelerated under otherwise identical reaction conditions by Fe^{2+} ions, and even more markedly by the enzyme catalase, can be correlated with the finding that the values for the energy of activation of the reaction are as follows (kJ mol^{-1}): (i) uncatalysed, 70 to 75; (ii) in the presence of Fe^{2+}, 38 to 42; (iii) in the presence of catalase, 23.

EXAMPLE

The following values of V_{max} were determined for an enzyme-catalysed reaction at pH 7.6 and at the various temperatures listed:

Temperature/K	285	293	303	308	313
V_{max}/μmol s^{-1}	16.40	25.12	41.70	52.73	66.10

Calculate the energy of activation of the reaction. Assuming the existence of a single rate-limiting step in the decomposition of the ES complex, calculate its enthalpy of activation at 298 K.

Assuming that, over the comparatively small range of temperatures employed, the value of E^* is virtually constant, E^* may be determined from the slope of the approximately straight line plot of log V_{max} against $1/T$.

T/K	$\frac{10^3}{T} \times$ K	V_{max}/μmol s^{-1}	log V_{max}
285	3.51	16.40	1.2148
293	3.41	25.12	1.4000
303	3.30	41.70	1.6201
308	3.25	52.73	1.7220
313	3.20	66.10	1.8202

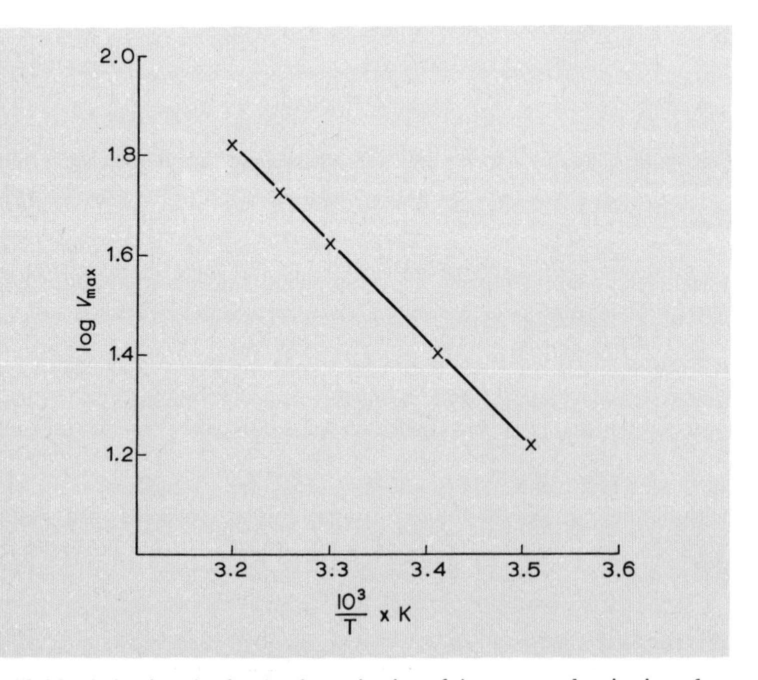

Fig. 11.14 Arrhenius plot for the determination of the energy of activation of an enzyme-catalysed reaction.

The slope of the straight line plot in Fig. 11.14 is $-1950 = -E^*/19.14$

$\therefore$ energy of activation $E^* = 19.14 \times 1950 = \underline{37.35 \text{ kJ mol}^{-1}}$

Assuming that the decomposition of the enzyme-substrate complex involves a single rate-limiting step whose rate constant k equals V_{max}/e_o, then the enthalpy of activation of this step in the reaction equals $E^* - RT$

$\therefore$ at 298 K $\Delta H^\ddagger$ of the rate-limiting step $= 37\,350 - (8.314 \times 298)$ J mol^{-1}

i.e. $\Delta H^\ddagger = (37.35 - 2.50) \text{ kJ mol}^{-1} = \underline{34.85 \text{ kJ mol}^{-1}}$

You should always be very careful in your use of the term 'energy of activation' with respect to an enzymic reaction; be aware of any assumptions you make in your determination of a value for E^*, remembering that, ideally, you would wish to know the value of each rate constant involved in the mechanism of the reaction, and the effect of changes in temperature on each of these, before reaching a conclusion concerning the true 'energy profile' of the reaction.

EFFECTS OF pH ON THE RATES OF ENZYME-CATALYSED REACTIONS

An enzyme is catalytically active over only a restricted pH range and usually has a quite marked optimum pH. This is reflected in the appearance of the pH-activity curve (Fig. 11.15) which is obtained by plotting the initial velocity of the enzyme-catalysed reaction against the pH of the assay medium. This optimum is generally near pH 7, though a few enzymes are remarkable in possessing abnormally low or high pH optima, e.g. pepsin (pH 2), arginase (pH 10).

The decreased catalytic activity on both sides of the optimum pH (Fig. 11.15), could be due in part to irreversible denaturation of the enzyme protein. In this case, unless the activity is measured over a very short period, an 'apparent' pH optimum is measured whose value would very likely depend on the length of time for which the enzyme was exposed to the unfavourable pH. To eliminate this complication, the pH stability of the enzyme (and of its substrates) should be predetermined (cf. determination of the thermal stability of the enzyme, p. 313).

The remaining reversible effects of pH change on the rate of a single-substrate enzyme-catalysed reaction are largely attributable to the pH-dependence of the ionization of those acidic and basic groups in the enzyme and its substrate that are involved in the formation of the ES complex, and of other dissociable groups that are involved in the decomposition of the complex. One might attempt to distinguish between the effects of pH change on (a) the combination of the enzyme with its substrate, and (b) the decomposition of the ES complex, by studying the pH-dependence of the values of

K_s and V_{max} respectively. The interpretation of such findings is fraught with dangers—especially when the mechanism of the reaction is quite unknown.[20] Even so, in the hands of experts, such studies have sometimes been the first to implicate groups (with distinctive pK_a values) whose involvement in the enzymic process has then been substantiated by other

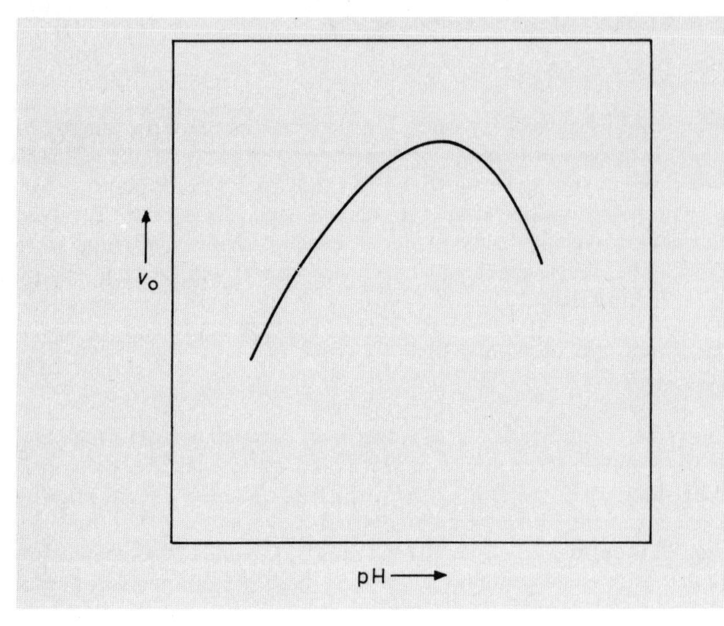

Fig. 11.15 pH-Activity curve for a 'representative' enzyme.

methods. For example, the presence of an essential -SH group at the active site of papain, and the involvement of a histidine residue of chymotrypsin in its catalysis of polypeptide hydrolysis, were both first suggested by the influence of pH changes on the kinetic behaviour of the reactions catalysed by these enzymes.

Finally it should be mentioned that, because the pK_a of a weakly dissociable group changes in value with temperature (in the manner determined by its heat of ionization), the extent of its ionization at any pH will, in part, be determined by the temperature. It follows from this that the shape of the pH-activity curve of any enzyme is likely to be temperature-dependent.

A cautionary tale

Every biologist who deals with enzymes must possess a sound working knowledge of the methods of enzyme kinetics and an understanding of the principles that underlie them. Only when he is aware of the factors that determine the rate of an enzyme-catalysed reaction, can he so design his assay methods that they yield a true measure of enzymic activity. Inevitably, the apparently optimal pH and temperature will be found, and those concentrations of substrates and cofactors that are 'saturating' determined. Values of V_{max} and K_m will be measured under different conditions. This much is virtually routine, yet the interpretation that is put upon such data is frequently suspect (particularly when it relates to multisubstrate reactions). Two equally mistaken views are, alas, all too prevalent:

1. that once an enzyme has been purified to maximum specific activity and 'apparent' values of K_m and V_{max} have been determined under some arbitrary set of reaction conditions, the mission of enzyme kinetics has been fulfilled and the enzyme has been 'characterized';
2. that if the very minimum of kinetic data can somehow be moulded to fit a simplifying hypothesis, it will reveal the chemical structures of intermediate enzyme-substrate complexes and of the enzyme's catalytic site.

The beginner would do well to remember that the more expert the enzymologist, the more cautiously will he interpret simple kinetic data. This is not because the values themselves are inexact, but rather because he recognizes that their meaning can sometimes be ambiguous; hence his interest in confirming or disproving suggested mechanisms by the use of methods designed to identify and follow the fate of transient intermediates.

PROBLEMS

1. The enzyme β-hydroxyaspartate aldolase catalyses the cleavage reaction whereby L-*erythro*-β-methyl-β-hydroxyaspartate is converted into pyruvate and glycine, i.e.

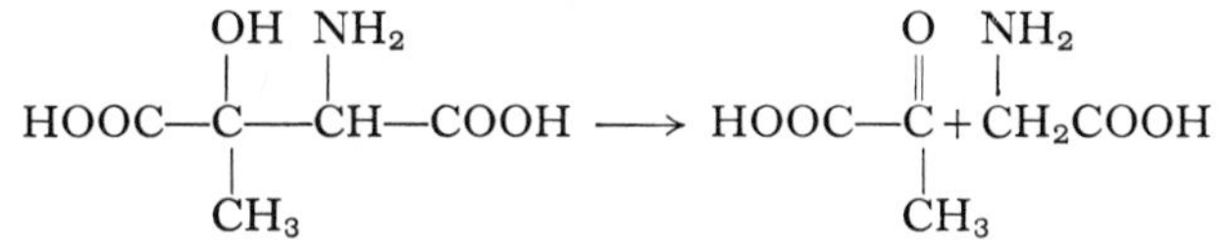

The following results were obtained when the initial velocity of the enzyme-catalysed reaction was measured at pH 8 and 30°C, using various concentrations of the hydroxyamino acid as substrate and a fixed concentration of a bacterial extract containing the aldolase. (Using boiled extract, no pyruvate was produced from the substrate during the assay times employed.)

Concentration of L-*erythro*-β-CH_3-β-OHaspartate / mmol dm^{-3}	Initial velocity / μmol pyruvate formed min^{-1}
4.0	2.50
3.0	2.24
2.25	1.95
1.50	1.55

Calculate the value of K_m (the Michaelis constant) of the reaction.

2. The enzyme-catalysed conversion of $A \rightarrow P$ proceeds by the following mechanism,

$$E+A \rightleftharpoons EA \longrightarrow E+P$$

and the initial velocity of the reaction is decreased by increasing the ionic strength of the reaction mixture. Initial velocity studies using various concentrations of A in media differing only in their ionic strengths, showed that the values of K_m and V_{max} of the reaction were proportionately altered by a given change in ionic strength (i.e. K_m is linearly related to V_{max} at various ionic strengths).

Does the ionic strength of the medium affect (i) the rate of formation of EA, or (ii) the rate of dissociation of EA, or (iii) the rate at which EA decomposes to yield P?

3. The enzyme trehalase specifically catalyses the hydrolysis of the disaccharide trehalose, i.e.

$$\text{trehalose} + H_2O \rightleftharpoons 2 \text{ glucose}$$

If, under certain reaction conditions (at pH 7 and 30°C), a sample of trehalase exhibited a V_{max} of 1.5 μmol glucose formed min^{-1} mg^{-1} protein, what is the value of V_{max} for the same preparation of the enzyme under these conditions, but in the presence of 5 mmol dm^{-3} trehalose 6-phosphate which acts as a competitive inhibitor ($K_i = 2$ mmol dm^{-3})?

4. Acetate-grown *Pseudomonas aeruginosa* contains an enzyme capable of hydrolysing propionamide,

$$CH_3CH_2CONH_2 + H_2O \rightleftharpoons CH_3CH_2COOH + NH_3$$

Initial velocity studies, in which rates of production of ammonia from propionamide were measured at pH 7.2 and 37°C, demonstrated that urea inhibited this enzyme-catalysed reaction. In one such experiment in which the effects of two different concentrations of urea were examined over a range of concentrations of propionamide, the following results were obtained:

Propionamide / mmol dm^{-3}	Initial velocity / μmol NH_3 liberated min^{-1} mg^{-1} protein: No inhibitor	Urea (1 mmol dm^{-3})	Urea (2 mmol dm^{-3})
5.0	160	111	76
6.67	194	140	95
10.0	263	183	128
20.0	400	279	188
50.0	576	400	277

Determine the value of the apparent K_m of the uninhibited reaction. Does urea act as a competitive, uncompetitive or non-competitive inhibitor of the reaction?

5. Ca^{2+} ions were found to be inhibitory to the activity of an enzyme which showed a requirement for Mg^{2+} ions. When initial velocity studies were performed at 293 K and pH 7 using a range of concentrations of Mg^{2+}

ions in the presence of different fixed concentrations of Ca^{2+}, the following results were obtained:

Concentration of Mg^{2+} / mmol dm^{-3}	Initial velocity / μmol product min^{-1} [Ca^{2+}]: Nil	2×10^{-5} mol dm^{-3}	4×10^{-5} mol dm^{-3}
0.33	20.4	9.26	5.99
0.5	25.65	12.66	8.47
1.0	33.3	20.0	14.28
2.0	40.0	28.57	22.23

What type of inhibition is displayed by the Ca^{2+} ions? Calculate the value of K_i for Ca^{2+}, and compare this with the dissociation constant of the Mg^{2+}-enzyme complex which, under the conditions of the experiment, was 4.2×10^{-4} mol dm^{-3}.

6. An enzyme of microbial origin catalysed the synthesis of glycine from glyoxylate according to the following stoichiometric equation:

$$\text{glyoxylate} + \text{L-aspartate} \rightleftharpoons \text{glycine} + \text{oxaloacetate}$$

Initial velocity studies were performed on the 'forward' reaction, measuring the initial rate of production of oxaloacetate when a sample of the enzyme was incubated at 298 K and pH 7.1 with a range of concentrations of L-aspartate (as variable substrate), and different concentrations of glyoxylate (as fixed substrate). The following results were obtained:

[L-aspartate] / mmol dm^{-3}	Initial velocity of forward reaction / μmol oxaloacetate formed min^{-1} mg^{-1} protein [glyoxylate]: 10 mmol dm^{-3}	5 mmol dm^{-3}	2 mmol dm^{-3}
1.0	7.70	6.25	4.55
1.5	10.75	8.00	5.47
2.0	13.31	9.52	6.06
3.0	17.25	11.36	6.90
5.0	22.23	13.32	7.58

Does the reaction proceed by a sequential or by a non-sequential mechanism?

7. A pyrophosphatase present in potatoes catalyses the hydrolysis of inorganic pyrophosphate at pH 5.3. The following values of V_{max} were determined at different temperatures with a partially purified preparation of this enzyme:

Temperature/K	288	298	308	313
V_{max}/μmol min^{-1}	6.53	10.47	16.79	20.65

Calculate the value of the energy of activation of the enzyme-catalysed reaction and compare with the corresponding value of the energy of activation for the uncatalysed hydrolysis of pyrophosphate (equals 121.3 kJ mol^{-1})

12

Oxidation and Reduction

There are at least two reactants in any oxidation-reduction reaction: (1) a reducing agent which, by losing electrons, is oxidized; and (2) an oxidizing agent which gains these electrons and is reduced. One can in fact view the oxidation-reduction reaction as consisting of the transfer of electrons from a reducing agent to an oxidizing agent, so that there can be no oxidation without a concomitant reduction.

When a reducing agent is oxidized, it is transformed into its oxidized form and, since the reaction is reversible, the reduced and oxidized forms of the compound constitute a conjugate pair which is called a ***reduction-oxidation couple*** (or ***redox couple***). If the reduced member of this redox couple is designated ***the reductant***, then the oxidized member can be considered to be its ***conjugate oxidant***. Thus,

$$\text{reductant} \rightleftharpoons \text{conjugate oxidant}^{n+} + ne$$

where n equals the number of electrons (e) liberated from one molecule of the reductant. Note that the relationship between a reductant (electron donor) and its conjugate oxidant (electron acceptor) is analogous to the relationship between an acid (proton donor) and its conjugate base (proton acceptor) that is described by the Brönsted and Lowry theory (p. 102).

Just as the oxidation of a reductant yields its conjugate oxidant, so reduction of an oxidant produces its conjugate reductant,

$$\text{oxidant}^{n+} + ne \rightleftharpoons \text{conjugate reductant}$$

It follows that any oxidation-reduction reaction must involve two redox couples that differ in their affinity for electrons. The couple that demonstrates the greater electron affinity assumes the oxidizing role, and its oxidant will be reduced. The other couple, with the lesser affinity for electrons, plays the reducing role and donates electrons to the oxidizing couple; as a consequence, the reductant of the reducing couple is oxidized.

Consider, for example, a typical oxidation-reduction reaction in which $\mathbf{red_1}$ is oxidized when it donates one electron per molecule to $\mathbf{ox_2}$. The overall reaction can be considered to be the resultant of two 'half reactions' as follows:

$$\text{Oxidation:} \qquad \text{red}_1 \rightleftharpoons \text{conjugate ox}_1 + e$$
$$\text{Reduction:} \qquad \text{ox}_2 + e \rightleftharpoons \text{conjugate red}_2$$

$$\text{Overall oxidation-reduction:} \quad \text{red}_1 + \text{ox}_2 \rightleftharpoons \text{conjugate ox}_1 + \text{conjugate red}_2$$

If, by the isothermal transfer of electrons from $\mathbf{red_1}$ to $\mathbf{ox_2}$, the Gibbs free energy of the system is decreased, the overall reaction is thermodynamically spontaneous from left to right as written (p. 191). When the Gibbs free energy content of the system reaches its minimum value, net reaction ceases and the system is at equilibrium (p. 214). Since the oxidation and reduction involves electron transfer from $\mathbf{red_1}$ to $\mathbf{ox_2}$, the reaction could perform electrical work; this would be equal to its $-\Delta G$ if the reaction could be carried out at constant temperature and pressure in a thermodynamically *reversible* manner (p. 188). As we shall see, this affords a unique opportunity to measure, experimentally, the value of $-\Delta G$ for an oxidation-reduction reaction.

The oxidation-reduction reaction as a generator of electricity

In the apparatus illustrated in Fig. 12.1, vessel A contains a solution of a reducing agent while vessel B contains a solution of an oxidizing agent. Electron-transmitting, but otherwise inert, metal electrodes dip into these solutions. Assume that the oxidizing redox couple in vessel B demonstrates a greater affinity for electrons than does the reducing redox couple in A; the tendency will then be for electrons to pass from A to B. The intensity of this tendency is measured as the ***electromotive force*** of that electrical cell which consists of the reducing half cell (A), with the oxidizing half cell (B). No electrons can pass from A to B until the electrodes are connected by an 'electrical conductor' (which could be a metal wire). Even then, the impelling electromotive force cannot bring about the net transference of electrons from the reducing half cell to the oxidizing half cell until some means is provided for maintaining the electrical neutrality of both half cells (that is, for keeping equal numbers of positive and negative charges in each half cell). In the situation illustrated in Fig. 12.1, for electrons to be transferred from A to B there must either be an equivalent, accompanying flow of positively charged ions (A → B), or an equivalent, but contrary, flow of negatively charged ions (B → A). Unlike electrons, such compensatory ions cannot travel through the external metallic circuit, and an ion-

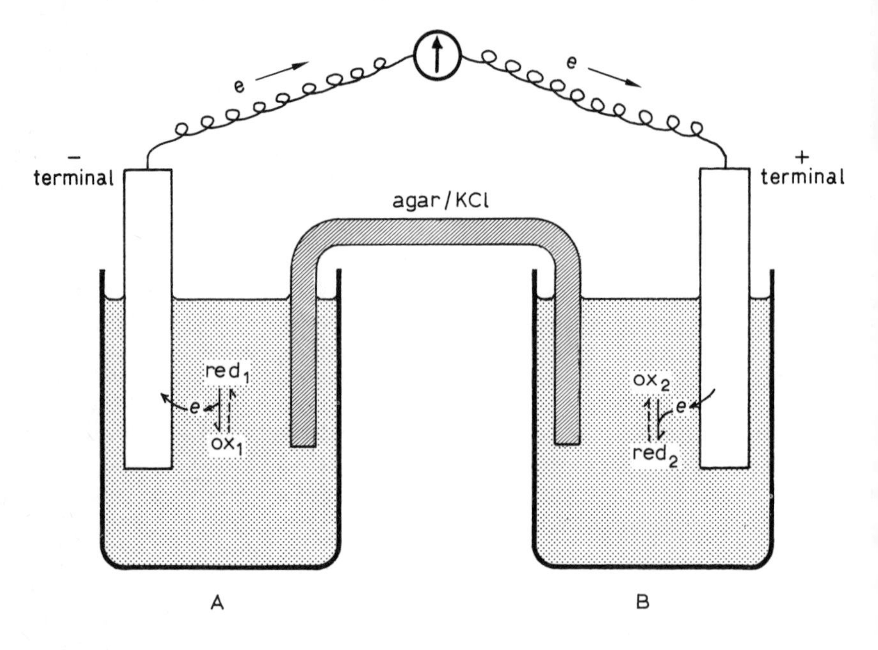

Fig. 12.1 An electrical cell constructed from two redox 'half cells'

conducting bridge must be constructed between the solutions in A and B in order to permit ion transfer whilst maintaining the separation of the interacting redox couples. A tube, containing a high concentration of KCl set in agar gel, serves as such a bridge and completes the electrical circuit; the spontaneous flow of electrons from A to B may now take place with consequent oxidation of the reductant in A, and reduction of the oxidant in B.

Electrode potentials

The potential difference of an electrical cell expresses the difference that exists between its half cells in their affinity for electrons. In this chapter we will adopt the convention employed by all biochemists whereby the affinity that a half cell (redox couple) demonstrates for electrons is termed its *electrode potential.* The greater is a half cell's affinity for electrons the larger (more positive) is the value of its electrode potential. The electrode potential is therefore a measure of the tendency of a redox couple to undergo reduction

$$\text{oxidized form} + ne \rightarrow \text{reduced form}$$

Accordingly, this type of electrode potential is often called the *reduction electrode potential* of a half cell, and redox couples are written in 'shorthand' as oxidized form|reduced form to emphasize that the reported electrode potential refers to oxidized form → reduced form.

If two half cells of differing electrode potentials are suitably joined to construct an electrochemical cell, discharge of this cell (i.e. the spontaneous cell reaction) will consist of a reduction reaction in the oxidizing half cell of more positive electrode potential, coupled with an oxidation reaction in the other reducing half cell of smaller (less positive, more negative) electrode potential.

The magnitude of the electrode potential of any half cell is determined by its chemical composition and the prevailing temperature. Therefore whenever the composition of an electrochemical cell is described, it is necessary to report the concentrations (strictly speaking, activities) of those of its components that are in solution. It has become conventional to summarize the composition of an electrical cell in the form of a cell diagram in which the constituents of the half cells are displayed on either side of a pair of vertical, parallel lines that represents the bridge between them. The convention further directs that the components are so ordered (from left to right as written) that the cell reaction under consideration is also indicated. Thus in the cell,

$$A, B \parallel C, D$$

the cell reaction is $A + C \rightarrow B + D$.

The potential of the cell (ΔE) measures the difference in electrode potentials of its component half cells

i.e. ΔE = (Electrode potential of the half cell on the *right* of the bridge) *minus* (Electrode potential of the half cell on the left of the bridge).

Thus, in the above example of cell $A, B \parallel C, D$,

$$\Delta E = E(C \mid D \text{ couple}) - E(B \mid A \text{ couple})$$

If the electrode potential of the $C \mid D$ couple were more positive than the electrode potential of the $B \mid A$ couple, then,

(i) ΔE would have a positive value;
(ii) the cell reaction $A+C \rightarrow B+D$ would have a negative value of ΔG; (since $\Delta G = -nF\Delta E$; see p. 198), and would therefore be spontaneous;
(iii) spontaneous discharge of the cell would be accomplished by left to right flow of electrons in the external circuit.

However, if the electrode potential of the $C|D$ couple were smaller (less positive, more negative) than that of the $B|A$ couple, then,

(a) ΔE would have a negative value;
(b) the cell reaction (for the cell as written), i.e. $A+C \rightarrow B+D$, would have a positive value of ΔG. The spontaneous reaction with a negative ΔG would therefore be the reverse reaction viz: $D+B \rightarrow C+A$;
(c) spontaneous discharge of the cell would be accomplished by right to left flow of electrons in the external circuit.

The biologist is mainly concerned with assessing the feasibility of reactions between redox couples of known electrode potentials. Thus if, when considering such processes in electrochemical terms, you purposely place the redox couple of smaller electrode potential on the left hand (of the bridge), you will ensure that the cell reaction as written will be the spontaneous process proceeding by a flow of electrons from left to right in the external circuit, impelled by a positive cell potential difference. For example, given the two redox couples of known electrode potentials fumarate (0.1 mol dm^{-3})|succinate (0.1 mol dm^{-3}), $E = 0.03\ V$, and FAD (10^{-3} mol dm^{-3})|$FADH_2$ (10^{-3} mol dm^{-3}), $E = -0.22\ V$, the cell created by placing the couple of lesser electrode potential on the left of the bridge, is:

$^-$Pt, $FADH_2$ (10^{-3} mol dm^{-3}), FAD (10^{-3} mol dm^{-3}) ‖ fumarate (0.1 mol dm^{-3}), succinate (0.1 mol dm^{-3}), Pt$^+$

The potential difference of this cell would be positive,

$$\begin{aligned} \Delta E &= E \text{ (right hand couple)} - E \text{ (left hand couple)} \\ &= +0.03\ V - (-0.22\ V) \\ &= +0.25\ V \end{aligned}$$

and the cell reaction, $FADH_2$ + fumarate $\rightleftharpoons$ FAD + succinate would be the spontaneous process, having a negative value of ΔG.

To emphasize the direction of spontaneous electron flow, positive and negative signs may be assigned to the electrodes (forming the + and − terminals of the cell). The use of these symbols is illustrated in Fig. 12.1; the $-ve$ sign is given to the electrode from which electrons are spontaneously 'driven' into the external circuit and the +*ve* sign is given to the electrode which accepts electrons from this circuit.

The maximum potential difference of an electrochemical cell (which is what we have hitherto merely called its potential difference) is measurable in volts as the electromotive force of the cell†; 1 V being defined as that e.m.f. which promotes a standard rate of flow of electrons (1 A) through a medium which exhibits a standard degree of resistance to their passage (1 Ω). Because the e.m.f. of a cell measures the difference between the electrode potentials of its component half cells, these electrode potentials are also measured in volts.

MEASUREMENT OF ELECTRODE POTENTIALS

A half cell displays its affinity for electrons only when it is suitably coupled with another donor or acceptor of electrons. There is no practical way of measuring the actual magnitude of the electrode potential of an isolated half cell. It is, however, possible to measure the extent by which the electrode potential of one half cell differs from that of any other. In this way the magnitudes of the electrode potentials of various half cells can be assessed relative to one another and can be assigned values on a scale whose zero is defined as being equal to the actual electrode potential of a reference half cell. This device should not present any difficulty since, after all, intensities of warmth are represented in a similar manner on a scale of temperatures. For example, it is arbitrarily decreed that, for the purpose of the centigrade scale of temperatures, freezing water at standard atmospheric pressure shall be taken as the reference standard, whose temperature shall be designated as zero (0°C). Systems whose temperatures are greater than that of freezing water may now be assigned positive values in degrees centigrade in proportion to the extent that their warmth exceeds that of this standard. Conversely, systems colder than freezing water are assigned negative values in degrees centigrade in proportion to the difference between their heat intensities and that of freezing water. Not only is the position of the zero decreed, but the scale's unit of measurement is also quite arbitrarily defined.

The scale on which electrode potentials are measured is called ***the hydrogen scale*** since it takes as its zero value the electrode potential of the standard hydrogen half cell (p. 329). Its scale division is the ***volt*** (see above). On this hydrogen scale of electrode potentials:

† The e.m.f. of a cell may be defined in several ways, e.g. (i) as the limiting value of the electric potential difference for zero flow of current through the cell; (ii) as being equal in sign and magnitude to the electric potential of the terminal on the right (of the cell diagram) when that of the similar terminal on the left is taken as zero, the circuit being open.

(a) the standard hydrogen half cell (standard hydrogen electrode) possesses at all temperatures an electrode potential of 0.00 V;
(b) a half cell which is more reducing than the standard hydrogen half cell is credited with a negative electrode potential (E_h is $-$ve) whose magnitude measures the extent by which its electron affinity is less than that of the standard hydrogen electrode;
(c) a half cell which is more highly oxidizing than the standard hydrogen half cell is credited with a positive electrode potential (E_h is $+$ve) whose magnitude measures the extent by which its electron affinity exceeds that of the standard hydrogen electrode.

This means that the (reduction) electrode potential of any redox couple ([ox] | [red]) is the e.m.f. in volts of the cell that is formed when that couple acting as a half cell is joined to a standard hydrogen electrode in the following manner:

$$\text{Pt, } H_2 \text{ (1 atm.), H } (a = 1) \parallel \text{[ox], [red], Pt}$$

It is the value of its electrode potential on the hydrogen scale (E_h) that is termed the *redox potential* of a redox couple.

EXAMPLE

The following cells were produced by joining a standard hydrogen electrode to half cells containing either redox couple $A|AH_2$ or $B|BH_2$ at 298 K and pH 8:

$^-$*redox couple $A|AH_2$ ‖ standard hydrogen electrode*$^+$ *e.m.f.* $= +0.3$ *V*
$^-$*standard hydrogen electrode ‖ redox couple $B|BH_2$*$^+$ *e.m.f.* $= +0.55$ *V*

Calculate the redox potential of each couple under the conditions that existed in these half cells. If, under identical conditions, a cell was created by joining redox couple $A|AH_2$ in one half cell to redox couple $B|BH_2$ in another, which would form its reducing half cell, and what is the maximum e.m.f. it could generate?

In the cell,

$$^-AH_2, A \parallel \text{standard hydrogen electrode}^+$$

the standard hydrogen electrode demonstrates the greater electron affinity, and so forms the oxidizing half cell.

$$\text{e.m.f.} = \Delta E_h = E_h(\text{right}) - E_h(\text{left}) = E_h(\text{S.H.E.}) - E_h(\text{A}|\text{AH}_2)$$

By definition, $E_h(\text{S.H.E.}) = 0, \quad \therefore 0.3 = 0 - E_h(\text{A}|\text{AH}_2)$

whence, $E_h(\text{A}|\text{AH}_2) = \underline{-0.3 \text{ V}}$

In the cell,

$$^{-}\text{standard hydrogen electrode} \parallel \text{B}|\text{BH}_2{}^{+}$$

the standard hydrogen electrode forms the reducing half cell, and the redox potential of the $B|BH_2$ couple must have a positive value,

$$\text{e.m.f.} = \Delta E_h = E_h(\text{B}|\text{BH}_2) - E_h(\text{S.H.E.})$$

$$\therefore \quad 0.55 = E_h(\text{B}|\text{BH}_2) - 0, \quad \text{whence } E_h \text{ of B}|\text{BH}_2 = \underline{+0.55\ \text{V}}$$

Since $E_h(B|BH_2)$ is more positive than $E_h(A|AH_2)$, in the following cell,

$$^{-}\text{AH}_2, \text{A} \parallel \text{B}, \text{BH}_2{}^{+}$$

$$E_h = -0.3\ \text{V} \quad E_h = +0.55\ \text{V}$$

The e.m.f. generated by this cell $\Delta E_h = (+0{\cdot}55-(-0{\cdot}3)) = \underline{0.85\ \text{V}}$

Reference half cells

(a) *The standard hydrogen electrode*

This consists of H_2 gas at standard atmospheric pressure (101 325 Pa) in equilibrium with H^+ ions in solution at unit activity, i.e.

$$\underset{\text{(oxidant)}}{\text{H}^+ + e} \rightleftharpoons \underset{\text{(conjugate reductant)}}{\tfrac{1}{2}\text{H}_2}$$

Electrons can be transmitted to and from this redox couple via an inert platinum electrode, so that the composition of the half cell can be represented as H_2 (1 atmosphere), H^+ ($a=1$), Pt.

A solution of HCl of pH 0 could provide the unit activity of H^+ ions, and H_2 gas at 1 atmosphere could be bubbled through this solution and adsorbed onto platinum black on the surface of the platinum foil electrode (so as to ensure that the gas is equilibrated with the H^+ ions in solution).

(b) *The saturated calomel electrode*

Though the standard hydrogen electrode is the ultimate reference half cell, great care must be taken in its construction. More convenient half cells, whose values of E_h on the hydrogen scale are accurately known over a wide range of temperatures, are more often employed as practical standards. One of the most popular of these is the saturated calomel electrode, composed of an electrode of mercury in contact with a paste of mercurous chloride (calomel) in saturated aqueous potassium chloride, i.e.

$$\text{KCl (sat.)}, \text{Hg}_2\text{Cl}_2 \text{ (solid)} \mid \text{Hg}$$

The electrode potential of this half cell is attributable to the redox couple,

$$Hg_2Cl_2\ (s) + 2e \rightleftharpoons 2Hg + 2Cl^-$$

and at 298 K, E_h equals +0.244 V. This means that the cell composed of a saturated calomel electrode joined to a standard hydrogen electrode, generates an e.m.f. of 0.244 V at 298 K,

$$^-Pt,\ H_2\ (1\ atm)\ H^+\ (pH\ 0)\ \|\ KCl\ (sat.),\ Hg_2Cl_2\ (solid)\ |\ Hg^+$$

The redox potential of any couple can easily be calculated from the e.m.f. of the cell that it forms with a saturated calomel electrode. Returning to the analogy of the temperature scale, it is as though a temperature is required in °C, when it is known by how many degrees centigrade it differs from 244°C.

EXAMPLE

(a) *Calculate the values of E_h for the redox couples that form the following cells when joined to a saturated calomel electrode at 298 K:*

$^+$*S.C.E.* ‖ *redox couple (1)*$^-$ *e.m.f.* = *0.256 V*
$^-$*S.C.E.* ‖ *redox couple (2)*$^+$ *e.m.f.* = *+0.256 V*

(b) *Calculate E_h of the redox couple $X|XH_2$, if at 298 K its electrode potential on the 'saturated calomel electrode scale' ($E_{S.C.E.}$) equals −0.3 V (E_h of the saturated calomel electrode at 298 K = +0.244 V)*

(a) Since redox couple (1) spontaneously donates electrons to the saturated calomel electrode it must possess the lesser value of E_h at 298 K,

$$\text{e.m.f.} = \Delta E_h = E_h\ \text{(right half cell)} - E_h\ \text{(left half cell)}$$
$$= E_h\ \text{(redox couple 1)} - E_h\ \text{(S.C.E.)}$$
$$\therefore\quad -0.256 = E_h\ \text{(redox couple 1)} - (+0.244)\ \text{V}$$
$$\therefore\quad E_h\ \text{(redox couple 1) at 298 K} = 0.244 - 0.256$$
$$= \underline{-0.012\ \text{V}}$$

Since redox couple (2) accepts electrons from the saturated calomel electrode it must possess the larger value of E_h at 298 K,

$$\text{e.m.f.} = E_h\ \text{(redox couple 2)} - E_h\text{(S.C.E.)}$$
$$\therefore\quad 0.256 = E_h\ \text{(redox couple 2)} - 0.244$$
$$\therefore\quad E_h\ \text{(redox couple 2) at 298 K} = 0.256 + 0.244 = \underline{+0.50\ \text{V}}$$

(b) If the electrode potential of the saturated calomel electrode is assumed to be zero, the redox potential of the $X|XH_2$ couple is −0.3 V. In fact, on the hydrogen scale, the electrode potential of the S.C.E. at 298 K is +0.244 V,

$$\therefore \quad E_h \text{ of redox couple } X|XH_2 \text{ at 298 K} = +0.244 - 0.30$$
$$= -0.056 \text{ V}$$

Potentiometric measurement of electromotive force

An electrochemical cell manifests its true e.m.f. only when it produces no current, and operates in a thermodynamically reversible manner (p. 188). The 'null point' potentiometric method of e.m.f. measurement measures the e.m.f. of a cell as a multiple of the standard e.m.f. generated by a stable reference cell. (A favourite reference cell is the 'unsaturated' Weston cell with an e.m.f. of 1.019 V at room temperature.)

In diagrams of electrical circuits, an electrical cell is symbolically represented as two parallel, vertical lines of unequal height, ı|. The shorter of these lines designates the reducing half cell (−) of smaller E_h, the taller line represents the oxidizing half cell (+) of larger E_h. This means that the spontaneous flow of electrons will be as shown in Fig. 12.2a, namely, from the 'shorter to the taller' of these lines in the circuit diagram. If two cells are incorporated into the same electrical circuit in such a way that they tend to impel the flow of electrons in opposite directions (Fig. 12.2b), they are said to be opposed. Whether electrons then flow in the common circuit and, if so, in which direction this flow takes place, will be determined by the relative magnitudes of the opposed e.m.fs.

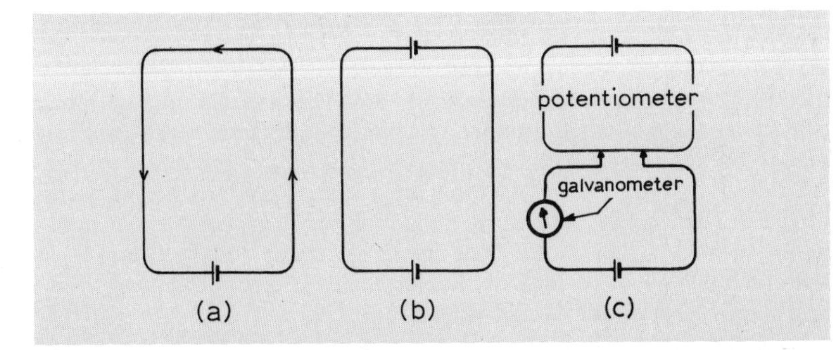

Fig. 12.2 Potentiometric measurement of the e.m.f. of an electrical cell. (a) Electron flow from test cell alone in circuit; (b) nil electron flow in circuit when test cell is opposed by cell of equal potential; (c) potentiometer supplying a measurable and adjustable potential to oppose the potential of the test cell.

Only if the opposed e.m.fs. are exactly equal will no current flow; should they differ, electron flow will occur in the direction dictated by the polarity of the cell with the larger e.m.f. The rate of flow of electrons (the current) then depends on the magnitude of the difference between the opposed e.m.fs.

This forms the basis of the potentiometric, null point method for measuring e.m.fs. The test cell is opposed by a source of e.m.f. of variable but measured magnitude, supplied by an instrument known as a *potentiometer.* A current-measuring instrument (galvanometer) is incorporated into the circuit to indicate the presence of current and the direction of electron flow (Fig. 12.2c).

The e.m.f. supplied by the potentiometer can be adjusted very rapidly until it exactly equals the e.m.f. of the test cell. This point, at which the opposed e.m.fs. are perfectly balanced, is indicated by there being no current in the circuit (this is the nullifying or null point). The e.m.f. supplied by the potentiometer at the null point is 'read off' the instrument from a scale that has been pre-calibrated using a standard cell of known e.m.f. Since no current flows in the circuit at the null point, the e.m.f. supplied by the potentiometer here equals the *true e.m.f.* of the test cell, which at this null point theoretically 'operates' in a thermodynamically reversible manner.

The quantity of electricity that can be obtained from an oxidation-reduction reaction

A reducing agent undergoes univalent oxidation by 'losing' one electron per molecule; i.e.

$$\text{reductant} \rightleftharpoons \text{oxidant} + e$$

Since one mol of any compound consists of 6.023×10^{23} molecules (p. 18) the same number of electrons are derived from the univalent oxidation of 1 mol of any reductant. The charge carried by this 1 mol of electrons (i.e. by 6.023×10^{23} electrons) is 96487 coulombs of electricity, so that univalent oxidation of 1 mol of any reductant yields 96487 C; conversely, univalent reduction of any oxidant consumes 96487 C mol^{-1}. This molar quantity of electricity (96487 C mol^{-1}) is therefore a convenient unit of measurement in electrochemical work; it is termed the Faraday constant (symbol, F).

The electrical work that can be performed by an isothermal oxidation-reduction reaction

Consider the electrical cell,

$$^{-}\text{Pt, red}_1\text{, ox}_1 \parallel \text{ox}_2\text{, red}_2\text{, Pt}^{+}$$

which is composed of the two redox couples,

$$\text{ox}_1 + ne \rightleftharpoons \text{red}_1 \quad E_h(1)$$

$$\text{ox}_2 + ne \rightleftharpoons \text{red}_2 \quad E_h(2).$$

The spontaneous cell reaction is $red_1 + ox_2 \rightleftharpoons ox_1 + red_2$, and the positive e.m.f. of this cell at its potentiometric null point will be $\Delta E_h = E_h(2) - E_h(1)$. The oxidation of 1 mol of red_1 yields $n \times 96487$ C of electricity which is employed to reduce an equivalent amount of ox_2. This means that at its null point this cell can in theory perform an amount of electrical work equal to $-n \times 96487 \Delta E_h$ C V mol^{-1} (since 1 C V = 1 J, this equals $-n \times 96\,487 \Delta E_h$J mol^{-1} = $-nF\Delta E_h$ J mol^{-1}).

It has been emphasized that an electrochemical cell at its potentiometric null point is operating in a thermodynamically reversible manner. Furthermore, we have seen that the work (other than pressure-volume work) performed by an isothermal, thermodynamically reversible reaction at constant pressure, is equal to the concurrent decrease in the Gibbs free energy of the system (p. 188). Therefore, when an oxidation-reduction reaction occurs isothermally in an electrochemical cell at its potentiometric null point, the electrical work performed per mole of reductant oxidized equals the decrease in Gibbs free energy per mole, i.e. $-\Delta G = nF\Delta E_h$ J mol^{-1} of reductant oxidized. This affords a means of measuring the value of ΔG for any oxidation-reduction reaction.

In summary, for any isothermal oxidation-reduction reaction,

$$-\Delta G = nF\Delta E_h \text{ J mol}^{-1}$$

or

$$\Delta G = -nF\Delta E_h \text{ J mol}^{-1}$$

where ΔG = increase in Gibbs free energy of the system in J mol^{-1}

n = number of electrons 'lost' per molecule of reducing agent oxidized (or quantity of electricity flowing through the cell for the reaction, measured in F),

F = Faraday constant = 96487 C mol^{-1} = 96487 J V^{-1} mol^{-1}

ΔE_h = e.m.f. of the thermodynamically reversible oxidation-reduction cell (in V).

EXAMPLE

Calculate the value of ΔG for the spontaneous oxidation-reduction reaction in a solution in which E_h of the $A|AH_2$ redox couple is -0.32 V and E_h of the $B|BH_2$ redox couple is $+0.25$ V. (Faraday constant = 96487 J V^{-1} mol^{-1}.)

To obtain the required value of ΔE_h you may employ any of the following methods.

Method (1) Draw the cell diagram for that electrochemical cell which will have a positive potential difference and whose cell reaction will therefore be the spontaneous oxidation-reduction process. This involves placing on the right of the bridge the couple with the more positive electrode potential.

$$^{-}\text{Pt, AH}_2\text{, A} \parallel \text{B, BH}_2\text{, Pt}^{+}$$

then,
$$\Delta E_h = E_h(\text{right}) - E_h(\text{left})$$
$$= +0.25-(-0.32)\ \text{V} = \underline{+0.57\ \text{V}}$$

Method (2) Consider what will constitute the spontaneous half (cell) reactions under the prevailing reaction conditions. Since the $B|BH_2$ couple has the more positive redox potential this will constitute the electron acquiring half cell in which the spontaneous reaction will be a reduction. The spontaneous reaction in the other half cell will consist of oxidation of the reduced component of the $A|AH_2$ couple. Thus,

Reducing couple: $AH_2 \rightleftharpoons A+2H^+ +2e$

Oxidizing couple: $B+2H^+ +2e \rightleftharpoons BH_2$

Net reaction: $AH_2+B \rightleftharpoons A+BH_2$

Since E_h ($=-0.32$ V) of the $A|AH_2$ couple is its reduction potential (applicable to the reaction $A+2H^+ +2e \rightarrow AH_2$), the oxidation of AH_2 that is the reverse of this process will have an *oxidation potential* equal to $-E_h$ ($= +0.32$ V). The E_h of the $B|BH_2$ couple acting as an electron accepting half cell is $+0.25$ V. Thus the ΔE_h of the complete cell is the sum of the reduction potential of the oxidizing couple and the oxidizing potential of the reducing couple.

$$\Delta E_h = +0.25+(+0.32)\ \text{V} = \underline{+0.57\ \text{V}}$$

Method (3)—*recommended.* While Method (2) is pedantically correct in its reasoning, you can arrive at the same conclusions without becoming involved with translation of the reduction potential of one couple into its oxidation potential of opposite sign.

Once again, you first identify the electron accepting (oxidizing) couple as that with the more positive E_h, and to check that mass and charge balances are maintained, you should write out the contributory reactions of the oxidizing and reducing couples as in Method (2).

Electron accepting process: $B+2H^+ +2e \rightleftharpoons BH_2$
(oxidizing couple)

Electron donating process: $AH_2 \rightleftharpoons A+2H^+ +2e$
(reducing couple)

Net spontaneous reaction: $AH_2+B \rightleftharpoons A+BH_2$

$$\Delta E_h = E_h(\text{oxidizing couple}) - E_h(\text{reducing couple})$$
$$= +0.25-(-0.32)\ \text{V} = \underline{+0.57\ \text{V}}$$

[*Note* that this approach is formally equivalent to that used in Method (1) but has the advantage of emphasizing the component half reactions and so reminding one that these must be balanced. If you like, you could in Method (3) draw the conventional cell diagram to reassure yourself that your identification of the roles of the couples is correct.]

Having obtained the value of ΔE_h you can now proceed to calculate ΔG of the reaction from the equation,

$$\Delta G = -nF\Delta E_h$$

$$\text{where}\begin{cases} n = 2 \\ F = 96\,487\ \text{C mol}^{-1} = 96\,487\ \text{J V}^{-1}\ \text{mol}^{-1} \\ \Delta E_h = +0.57\ \text{V} \end{cases}$$

$$\therefore \quad \Delta G = -2\times 96\,487\times(+0.57)\ \text{J mol}^{-1} = \underline{-120\ \text{kJ mol}^{-1}}$$

REDOX POTENTIALS

Standard redox potentials

If the components of a redox couple are together present in their standard states, at a constant temperature and at standard atmospheric pressure, the redox potential is the standard redox potential of the couple at that temperature ($E^{\ominus}$). As before (p. 192), all solids are considered to be in their standard states, the standard state for a gas is taken to be the gas at a partial pressure of 1 atmosphere (i.e. 101 325 Pa), and the standard state of a solute is unit activity (p. 65).

Table 12.1

All redox potentials are measured in volts on the hydrogen scale at a constant temperature (298 K if not otherwise reported).

E_h = redox potential of a redox couple of an arbitrary but reported composition

$E^{\ominus}$ = true, standard redox potential at pH 0 with all components in their standard states (unit *activity* for solutes)

$E_{\ominus}$ = practical, standard redox potential at pH 0 but with other solute components present in 1 mol dm^{-3} *concentration*. Usually measured as the 'midpoint redox potential at pH 0,' its value is assumed to equal $E^{\ominus}$ for redox couples in *dilute*, aqueous solution

$E'_{\ominus}$ = $E_{\ominus}$ at a pH other than 0. If the pH is not stated, it is assumed to equal pH 7. The value of $E'_{\ominus}$ is usually measured as the 'midpoint redox potential at the reported pH'

To conform with our previous use of the symbol $G^{\ominus}$ to represent the standard Gibbs free energy of a system, we shall use the symbol $E^{\ominus}$ to represent the true standard redox potential of a couple whose solute components are all present in unit activity. In most studies concerned with redox couples in dilute aqueous solution, the activity coefficients of the components are ignored, the assumption being made that the ratio of the activity coefficients of the oxidized and reduced members of a redox couple will approximately equal 1. The chief exception to this simplifying assumption concerns H^+ ions, whose *activity* in any solution is always reported (generally as the pH value of the solution). We shall here employ the symbol $E_{\ominus}$ to represent the practical, standard redox potential of a couple whose solute components are present in 1 mol dm^{-3} concentration at pH 0. Should the pH be other than 0, the modified value of the practical standard potential is represented by the symbol $E'_{\ominus}$ and the pH is recorded. If the temperature is not stated, it is assumed that it is 298 K.

To help you to distinguish between these symbols, they are listed and defined in Table 12.1; the meaning of the term 'midpoint redox potential' is explained later (p. 341).

The relationship between the standard e.m.f. and the equilibrium constant of an oxidation-reduction reaction

The electrochemical cell that is composed of half cells containing redox couples in their standard states, generates a unique e.m.f. at its null point. This e.m.f. represents the difference between the standard redox potentials of the oxidizing and reducing couples, and can be termed the standard potential difference of the cell ($\Delta E^{\ominus}$). Its value is related to the standard Gibbs free energy change associated with the oxidation-reduction reaction in the cell, for $\Delta G^{\ominus} = -nF\Delta E^{\ominus}$. Furthermore, since $\Delta G^{\ominus} = -RT \ln K_{eq}$ (p. 196), the value of $\Delta E^{\ominus}$ is related to the magnitude of the thermodynamic equilibrium constant K_{eq} of the oxidation-reduction reaction,

$$-nF\Delta E^{\ominus} = -RT \ln K_{eq} \quad \text{or} \quad \Delta E^{\ominus} = \frac{RT}{nF} \ln K_{eq}$$

Since we shall consider the components of redox couples in terms of their concentrations and not in terms of their activities, their standard (midpoint) redox potentials will be reported as $E_{\ominus}$ or $E'_{\ominus}$ values. Thus, in this chapter, we shall also use the symbol $\Delta E_{\ominus}$ to designate the limiting practical standard potential difference of an electrochemical cell (measured as the practical, standard e.m.f.). In so far as we are justified in assuming that when the oxidized and reduced components of a redox couple are present at equal concentrations in aqueous solution the ratio of their activity coefficients also equals 1, we may also assume that $-nF\Delta E_\ominus$ approximately equals $\Delta G^\ominus$ and that $nF\Delta E_\ominus/2.303RT$ approximately equals $\log K_{eq}$.

EXAMPLE

If at 298 K and pH 7, $E'_\ominus$ of the $NAD^+ \mid NADH + H^+$ redox couple is -0.32 V, and $E'_\ominus$ of the oxaloacetate^{2-} | malate^{2-} redox couple is -0.166 V, calculate the equilibrium constant at this temperature and pH of the oxidation of malate by NAD^+ (catalysed in living organisms by the enzyme malate dehydrogenase).

$$\textit{malate}^{2-} + NAD^+ \rightleftharpoons \textit{oxaloacetate}^{2-} + NADH + H^+$$

In the oxidation of malate by NAD^+, since NAD^+ is the electron acceptor the $NAD^+ \mid NADH + H^+$ redox couple is the oxidizing couple, and the oxaloacetate^{2-} | malate^{2-} redox couple is the reducing couple.

$$\therefore \quad \Delta E'_\ominus \text{ of this reaction} = E'_\ominus(\text{NAD}^+ \mid \text{NADH} + \text{H}^+) - E'_\ominus(\text{oxaloacetate}^{2-} \mid \text{malate}^{2-})$$

$$= -0.32 - (-0.166) \text{ V}$$

$$= \underline{-0.154 \text{ V}}$$

Since $\quad -nF\Delta E'_\ominus = \Delta G^{\ominus\prime} = -2.303RT \log K'_{eq}$

$$\therefore \quad \log K'_{eq} = \frac{nF\Delta E'_\ominus}{2.303RT}$$

where for the present reaction,

$$\begin{cases} n = 2 \\ F = 96\,487 \text{ C mol}^{-1} = 96\,487 \text{ J V}^{-1} \text{ mol}^{-1} \\ \Delta E'_\ominus = -0.154 \text{ V} \\ R = 8.314 \text{ J mol}^{-1} \text{ K}^{-1} \\ T = 298 \text{ K} \end{cases}$$

$$\therefore \quad \log K'_{eq} = -\frac{2 \times 96\,487 \times 0.154}{2.303 \times 8.314 \times 298} = -5.21 = \bar{6}.79$$

$$\therefore \quad K'_{eq} = \text{antilog } \bar{6}.79 = \underline{6.17 \times 10^{-6}}$$

The relationship between E_h and $\Delta G_{\frac{1}{2}}$

The e.m.f. of an electrochemical cell at its null point (ΔE_h) is related to the free energy change (ΔG) of its oxidation-reduction reaction by the equation,

$$\Delta G = -nF\Delta E_h$$

We have already seen that ΔE_h can be regarded as the difference between two contributory half cell potentials, so that,

$$\Delta E_h = E_h \text{ (right)} - E_h \text{ (left)}$$

In a similar way, it is often convenient for calculation purposes to regard ΔG as the difference between two free energy terms—one for each contributory redox couple. These terms are symbolized as $\Delta G_{\frac{1}{2}}$ and their values are calculated from the equation,

$$\Delta G_{\frac{1}{2}} = -nFE_h$$

where $\Delta G_{\frac{1}{2}}$ is attributed to the redox couple whose potential on the hydrogen scale is E_h.

Since E_h is the e.m.f. of the cell which is formed by joining the redox couple (on the right of the bridge) to a standard hydrogen half cell (on the left of the bridge), it follows that $\Delta G_{\frac{1}{2}}$ can be thought of as the change in Gibbs free energy that is associated with the operation of this cell at its null point. By defining the redox potential of the standard hydrogen half cell as zero, we have in effect decreed that donation or acceptance of electrons by the standard hydrogen half cell involves no change in Gibbs free energy. Thus $\Delta G_{\frac{1}{2}}$ is entirely attributable to the half reaction of the other half cell; that is,

$$\text{oxidant} + ne \rightleftharpoons \text{reductant}$$

for the redox couple whose potential is E_h.

Values of $\Delta G^{\ominus}$ are additive (p. 194). Thus $\Delta G^{\ominus}$ for any oxidation-reduction reaction is equal to the algebraic sum of the values of $\Delta G_{\frac{1}{2}}^{\ominus}$ of its component half reactions (suitably multiplied when necessary, as illustrated in the following example).

EXAMPLE

Calculate the value of $\Delta G^{\ominus\prime}$ for the oxidation of lactate by oxidized cytochrome c at 298 K and pH 7, if

$E'_{\ominus} = +0.25$ *volt for* $cyt.c_{ox}(Fe^{3+}) + e \rightleftharpoons cyt.c_{red}(Fe^{2+})$

$E'_{\ominus} = -0.19$ *volt for* $pyruvate + 2H^{+} + 2e \rightleftharpoons lactate.$

(Faraday constant = 96 487 J V^{-1} mol^{-1}.)

The oxidation-reduction reaction between lactate (a 2-electron donor) and oxidized cytochrome c (here assumed to be a 1-electron acceptor), proceeds as shown,

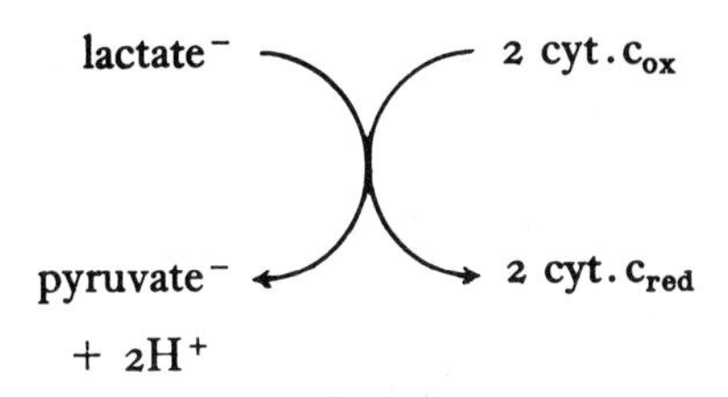

1. *$\Delta G^{\ominus\prime}_{\frac{1}{2}}$ of lactate oxidation*

$$\text{pyruvate}^- + 2\text{H}^+ + 2e \rightleftharpoons \text{lactate}^-$$

$\Delta G^{\ominus\prime}_{\frac{1}{2}}$ for this reduction of pyruvate $= -nFE'_0$

$= -2 \times 96\,487 \times (-0.19)$ J mol^{-1}

$= +36.66$ kJ mol^{-1}

It follows that $\Delta G^{\ominus\prime}_{\frac{1}{2}}$ for the reverse half reaction (the oxidation of lactate$^-$) equals -36.66 kJ mol^{-1}

2. *$\Delta G^{\ominus\prime}_{\frac{1}{2}}$ of reduction of oxidized cytochrome c*

$$\text{cyt.c}_{\text{ox}}(\text{Fe}^{3+}) + e \rightleftharpoons \text{cyt.c}_{\text{red}}(\text{Fe}^{2+})$$

$\Delta G^{\ominus\prime}_{\frac{1}{2}}$ for this reduction of cyt.c_{ox} $= -nFE'_0$

$= -1 \times 96\,487 \times 0.25$ J mol^{-1}

$= -24.12$ kJ mol^{-1}

3. *$\Delta G^{\ominus\prime}$ for the oxidation of lactate by oxidized cytochrome c*

The net reaction is,

$$\text{lactate}^- + 2\text{cyt.c}_{\text{ox}} \rightleftharpoons \text{pyruvate}^- + 2\text{H}^+ + 2\text{cyt.c}_{\text{red}}$$

It is thermodynamically equal to the sum of the following components,

$$\text{lactate}^- \rightleftharpoons \text{pyruvate}^- + 2\text{H}^+ + 2e$$

$$2\text{cyt.c}_{\text{ox}} + 2e \rightleftharpoons 2\text{cyt.c}_{\text{red}}$$

The modified, standard free energy change associated with the oxidation of 1 mol of lactate at 298 K and pH 7 is -36.66 kJ. The same term for the reduction of 1 mol of cyt.c_{ox} is -24.12 kJ; so for the reduction of 2 mol of cyt.c_{ox} it will equal -48.24 kJ.

Therefore, $\Delta G^{\ominus\prime}$ for the oxidation of 1 mol of lactate with the concurrent reduction of 2 mol of cyt.c_{ox}

$= \Delta G^{\ominus\prime}_{\frac{1}{2}}$(lactate to pyruvate) $+ 2\Delta G^{\ominus\prime}_{\frac{1}{2}}$(cyt.$c_{ox}$ to cyt.c_{red})

$= -36.66 + (-48.24) = -84.9$ kJ mol^{-1}

The redox potential of any couple depends upon its degree of oxidation

The value of ΔG for a reaction $A+B \rightleftharpoons C+D$ is related to the magnitude of its standard free energy change (ΔG°) in the following manner (see p. 204).†

$$\Delta G = \Delta G^{\circ} + RT \ln \frac{(C)(D)}{(A)(B)}$$

(where the bracketed terms represent the activities of the reaction components). This equation can be used to relate the values of $\Delta G_{\frac{1}{2}}$ and $\Delta G^{\circ}_{\frac{1}{2}}$ for the reduction of a redox couple by a standard hydrogen electrode. If the oxidizing couple were multivalent, the net reaction would be

$$\frac{n}{2}H_2 + \text{ox} \rightleftharpoons \text{red} + nH^+$$

and,
$$\Delta G_{\frac{1}{2}} = \Delta G^{\circ}_{\frac{1}{2}} + RT \ln \frac{(\text{red})(H^+)^n}{(\text{ox})(H_2)^{n/2}}$$

where the bracketed terms again refer to the activities of the reaction components as they exist throughout the reaction. However, both (H^+) and (H_2) equal 1, since in the standard hydrogen electrode the pH is zero, and the hydrogen gas is present at standard atmospheric pressure (unit activity);

$$\therefore \qquad \Delta G_{\frac{1}{2}} = \Delta G^{\circ}_{\frac{1}{2}} + RT \ln \frac{(\text{red})}{(\text{ox})}$$

Since $\Delta G_{\frac{1}{2}}$ equals $-nFE_h$, and $\Delta G^{\circ}_{\frac{1}{2}}$ equals $-nFE^{\circ}$ (p. 338),

then,
$$-nFE_h = -nFE^{\circ} + RT \ln \frac{(\text{red})}{(\text{ox})}$$

Reversing the signs and dividing throughout by nF, we obtain the following equation,

$$E_h = E^{\circ} - \frac{RT}{nF} \ln \frac{(\text{red})}{(\text{ox})}$$

Since $-\ln \frac{a}{b}$ is equal to $+\ln \frac{b}{a}$ (see p. 8),

$$E_h = E^{\circ} + \frac{RT}{nF} \ln \frac{(\text{ox})}{(\text{red})}$$

† For a definition of ΔG and ΔG°, see Table 7.1, p. 193.

If we use concentration terms instead of activities, this equation becomes,

$$E_h = E_{\ominus}+\frac{RT}{nF}\ln\frac{[\text{ox}]}{[\text{red}]} = E_{\ominus}+2.303\frac{RT}{nF}\log\frac{[\text{oxidant}]}{[\text{conjugate reductant}]}$$

At 298 K, the factor $2.303RT/nF$ is approximately equal to $0.059/n$; at 303 K, it equals $0.06/n$. This equation (known as the ***Nernst equation***) quantitatively expresses the self-evident fact that the more of a redox couple that is present in its oxidized form, the greater will be the oxidizing tendency of the couple (the more positive its value of E_h).

Whenever the oxidant and reductant of a redox couple are present in equal concentration, the term $\log\frac{[\text{ox}]}{[\text{red}]}$ equals zero, and E_h is equal to $E_{\ominus}$. This explains why, in practice, the standard redox potential ($E_{\ominus}$) of a redox couple is assumed to be equal to the midpoint redox potential which it displays when its oxidized and reduced forms are present in equal concentrations (not necessarily 1 mol dm^{-3}). The identity of these terms is sometimes emphasized by representing the value of $E_{\ominus}$ by the symbol E_m. When the midpoint potential is measured in a medium whose pH is not zero, its value is designated as $E'_{\ominus}$ and the pH is recorded (Table 12.1).

EXAMPLE

Nicotinamide adenine dinucleotide (NAD) participates in many metabolic oxidation-reduction reactions as a bivalent redox couple according to the half reaction,

$$NAD^+ + 2H^+ + 2e \rightleftharpoons NADH + H^+$$

If $E'_{\ominus}$ of this couple is -0.32 V at 298 K and pH 7, calculate its redox potential at this temperature and pH when in solution: (a) 90% in its reduced form, and (b) 10% in its reduced form.

To obtain the values of E_h under the stated conditions, we can employ the equation,

$$E_h = E'_{\ominus}+\frac{2.303RT}{nF}\log\frac{[\text{ox}]}{[\text{red}]}, \quad \text{when, at 298 K } \frac{2.303RT}{nF} \text{ is } \frac{0.059}{n},$$

and since, for the $NAD^+|NADH$ couple, n equals 2,

$$\therefore \quad E_h = -0.32+\frac{0.059}{2}\log\frac{[NAD^+]}{[NADH]}$$

(a) *Thus when* $[NAD^+]|[NADH]$ *equals 10/90*

$$\log\frac{[NAD^+]}{[NADH]} = \log 0.111 = \bar{1}.0453 = -0.9547$$

Substituting this value for the log term in the equation derived above,

$$E_h = -0.32+0.0295(-0.9547) = -0.32-0.0282$$
$$= \underline{-0.348 \text{ V}}$$

(b) *When* [NAD⁺]|[NADH] *equals 90/10*

$$\log \frac{[\text{NAD}^+]}{[\text{NADH}]} = \log 9 = 0.9542$$

$$\therefore \quad E_h = -0.32+0.0295(0.9542) = \underline{-0.292 \text{ V}}$$

The Nernst equation $\left(E_h = E'_\ominus + \text{constant} \times \log \frac{[\text{oxidant}]}{[\text{reductant}]}\right)$, is strikingly similar in form to the Henderson-Hasselbalch equation, which relates the pH of a mixture of a weak acid and its conjugate base to the proportional composition of the mixture and the pK_a of the weak acid $\left(\text{i.e. pH} = \text{p}K_a + \log \frac{[\text{base}]}{[\text{acid}]}\right)$. In Chapter 5 we saw that when the pH of a solution containing a weak acid with its conjugate base is plotted against the fraction of the mixture that consists of the conjugate base, a characteristically sigmoid curve is obtained. Similarly, when the potential (E_h) of a redox couple is plotted against the percentage content of reductant, a sigmoid curve is obtained at whose inflection point the concentrations of oxidant and reductant are equal. At this point, the value of E_h equals the standard, midpoint potential of the couple, $E'_\ominus$ (see Fig. 12.3). Similarly to the way in which a mixture of a weak acid with its conjugate base can act as a pH buffer, a redox couple can 'poise' a medium by buffering its redox potential at a value close to the $E'_\ominus$ value of the couple.

Since the logarithmic term in the Nernst equation is multiplied by $2.303RT/nF$ which is a constant/n, the midsections of the curves shown in Fig. 12.3 have identical slopes for redox couples of the same valency. But the curve for the bivalent couple ($n=2$) is half as steep as the similar midportion of the curve relating to a univalent couple ($n=1$).

HOW THE pH MAY AFFECT THE REDOX POTENTIAL OF A REDOX COUPLE

The effect of pH when H⁺ is the oxidant

This situation is unique to the hydrogen redox couple, i.e.

$$\underset{\text{(oxidant)}}{H^+ + e} \rightleftharpoons \underset{\text{(conjugate reductant)}}{\tfrac{1}{2}H_2 \text{ (gas)}}$$

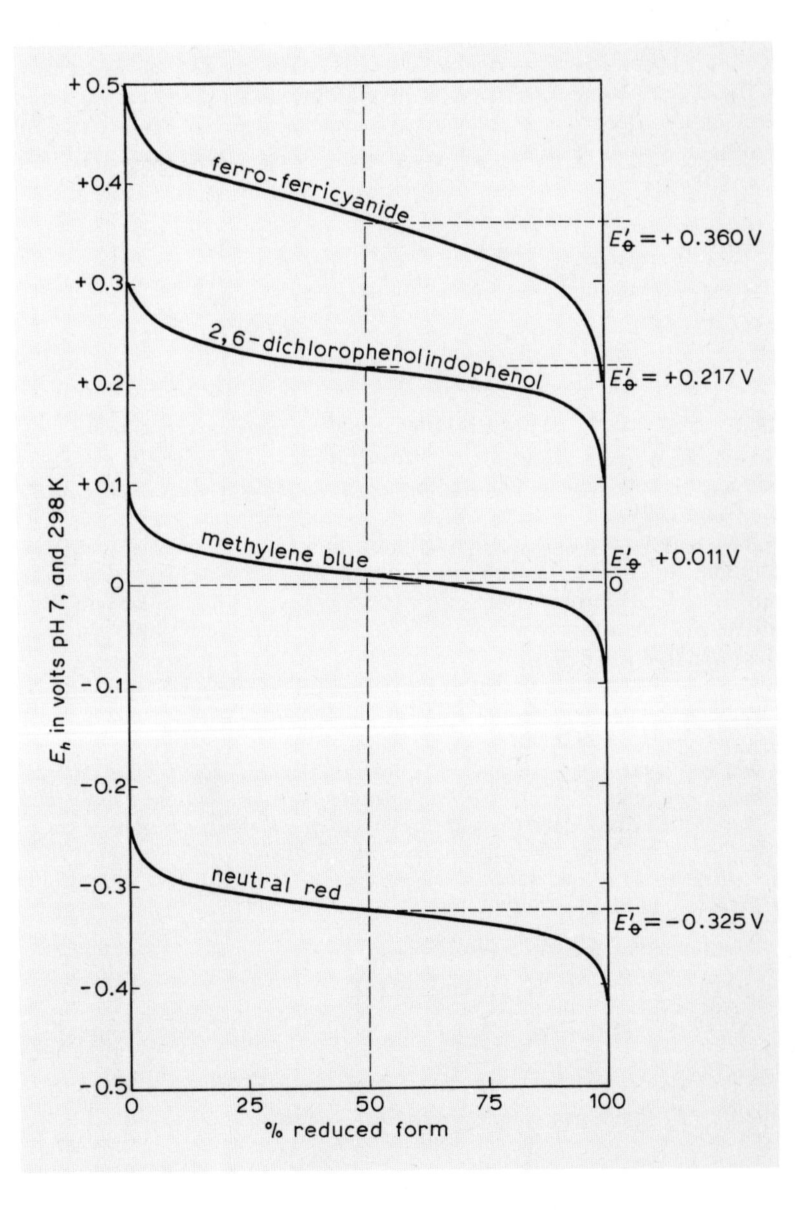

Fig. 12.3 Relationship between the redox potential of a couple and its percentage reduction.

In the standard hydrogen half cell, where the partial pressure of hydrogen gas (P_{H_2}) is maintained at 1 atmosphere (i.e. 101 325 Pa), and the pH is 0, the redox potential ($E^\ominus$) is by convention 0.00 V. If the partial pressure of hydrogen is maintained at 1 atmosphere, but the activity of hydrogen ions is changed (pH$\neq$0), then the new redox potential has the value E_h,

where, $$E_h = E^\ominus + \frac{RT}{nF} \ln \frac{(H^+)}{(P_{H_2})^{\frac{1}{2}}}$$

At 298 K, $E^\ominus = 0$ volt, $n = 1$, $\frac{2.303RT}{F} = 0.059$, $(P_{H_2})^{\frac{1}{2}} = 1$

$\therefore$ at 298 K, $$E_h = 0.059 \log (H^+) = -0.059 \text{ pH}$$

This means that the pH of a solution could be measured by equilibrating it with H_2 gas at 1 atmosphere, introducing an inert, metallic electrode and joining this half cell to a standard hydrogen electrode (or other reference half cell).

The measured e.m.f. of this cell will be a function of the pH of the solution under test. Though a pH meter could be constructed on this principle, hydrogen electrodes are not easily prepared, and the potentiometric pH meter that is commonly used in the laboratory employs a glass electrode (see p. 363).

EXAMPLE

Hydrogen gas at 1 atmosphere (i.e. 101 325 Pa) was equilibrated with an aqueous solution at the surface of a platinum electrode. This half cell joined to a saturated calamel electrode at 303 K gave a null-point e.m.f. of 0.572 V (the calomel electrode acting as the oxidizing half cell). Calculate the pH of the solution. (E_h of the saturated calomel electrode at 303 K is +0.242 V; 2.303RT/F at 303 K equals 0.06.)

The following cell was set up:

$$^{-}\text{Pt, } H_2 \text{ (1 atm), H (pH ?)} \parallel \text{S.C.E.}^{+}$$

The e.m.f. of this cell is $\Delta E_h = E_h(\text{S.C.E.}) - E_h$ (hydrogen couple)

$$\therefore \quad 0.572 = 0.242 - E_h \text{ (hydrogen couple)}$$

whence, E_h (hydrogen couple) $= (0.242 - 0.572) = -0.330$ V

Since for the hydrogen electrode $E_h = \frac{-2.303RT}{F}$ pH, at 303 K

$$E_h = -0.06 \text{ pH} \quad \text{and} \quad \text{pH} = \frac{-E_h}{0.06}$$

In the present example E_h is -0.33 V

$\therefore$ pH of the solution in the hydrogen half cell $= \dfrac{-(-0.33)}{0.06} = \underline{5.5}$

The effect of pH on the E_h of a redox couple whose half reaction involves H^+ ions

Many redox processes only involve the transfer of electrons, e.g. $Fe^{3+} + e \rightleftharpoons Fe^{2+}$, but a large number of those that occur in biological systems also involve the simultaneous transfer of H^+ ions, e.g.

$$\text{oxaloacetate} + 2H^+ + 2e \rightleftharpoons \text{malate}$$

or, more generally,

$$A + 2H^+ + 2e \rightleftharpoons AH_2$$

Because H^+ is a participant in this reaction, the E_h of this couple must be affected by the activity of hydrogen ions in the medium. In fact, in this example E_h will be related to $E_\ominus$ by the following equation,

$$E_h = E_\ominus + \frac{RT}{2F} \ln \frac{[A](H^+)^2}{[AH_2]}$$

whence, $$E_h = E_\ominus + \frac{RT}{2F} \ln \frac{[A]}{[AH_2]} + \frac{2.303RT}{2F} \times 2 \log (H^+)$$

and at 298 K $$E_h = E_\ominus + \frac{0.059}{2} \log \frac{[A]}{[AH_2]} - 0.059 \text{ pH}$$

At the 'midpoint' when $[A] = [AH_2]$, E_h equals $E'_\ominus$ and we obtain the following relationship,

$$E'_\ominus = E_\ominus - 0.059 \text{ pH (since } \log \frac{[A]}{[AH_2]} = \log 1 = 0)$$

This explains why, when calculating the value for E_h of a redox couple which involves H^+ ions, we use the 'modified' standard value $E'_\ominus$ in place of $E_\ominus$ in the Nernst equation. Thus to calculate the value of E_h for the redox couple $A|AH_2$ at, say, pH 5, we would use the following equation,

$$E_h = E'_\ominus + \frac{RT}{2F} \ln \frac{[A]}{[AH_2]}$$

where $E'_\ominus$ is the standard (midpoint) potential of this couple *at pH* 5, and $E'_\ominus = E_\ominus - 0.059 \times 5$ (see also the examples on pp. 341 and 352).

In those instances in which the reductant of the redox couple is capable of dissociating as a weak acid (as is the case for many biologically important substances) the situation will be further complicated by the fact that the manner and extent of its dissociation will be pH-determined. Suppose, for example, that the reductant AH_2 dissociates as a biprotic weak acid whose pK_a values are 8 and 11 at 298 K, i.e.

$$AH_2 \rightleftharpoons H^+ + AH^- \qquad pK_{a_1} = 8$$

$$AH^- \rightleftharpoons H^+ + A^{2-} \qquad pK_{a_2} = 11$$

This means that only in media whose pH is less than about pH 6 to 7 will AH_2 be wholly undissociated, and only at these 'acid' pH's is the net redox half reaction,

$$A + 2H^+ + 2e \rightleftharpoons AH_2$$

At pH's in the range pH 9 to 10, AH_2 is completely dissociated into H^+ and AH^- so that the net half reaction is,

$$A + H^+ + 2e \rightleftharpoons AH^-$$

while at pH's greater than pH 12 to 13, AH^- is itself completely dissociated and the redox half reaction is,

$$A + 2e \rightleftharpoons A^{2-}$$

In these very alkaline media, a small change in pH will have no effect on the redox potential of the couple since H^+ ions no longer appear in the net stoichiometric equation. Figure 12.4 shows how the dissociation of AH_2 is reflected in the way in which the value of $E'_\ominus$ of this couple changes with pH (assuming quite arbitrarily that $E'_\ominus$ at pH 7 is -0.3 V).

This curve is linear and most steep (slope $= -0.06$ at 303 K) when 2 protons are involved in the reduction of a single molecule of oxidant (at pH's <7). When only 1 proton is taken up per molecule of oxidant reduced (pH 9 to 10), then this plot is linear, but is only half as steep (slope $= -0.03$), while, as explained above, when no protons are involved, the redox potential is independent of pH and the curve ($>$pH 12) runs parallel to the pH axis (slope$=0$). In general then, the relationship between the value of E_h and the H^+ ion activity for a redox couple which involves the transfer of **a** protons and **n** electrons,

$$\text{ox} + aH^+ + ne \rightleftharpoons \text{red}$$

can be expressed as an equation of the following form,

$$E_h = E_\ominus + \frac{RT}{nF} \ln \frac{[\text{ox}]}{[\text{red}]} + \frac{RT}{nF} \ln (H^+)^a$$

For simplicity, we will not discuss the implications of Fig. 12.4 in any greater detail than to note that a single equation can be derived to relate the value of $E'_\ominus$ to the value of $E_\ominus$ over the complete range of pH values;

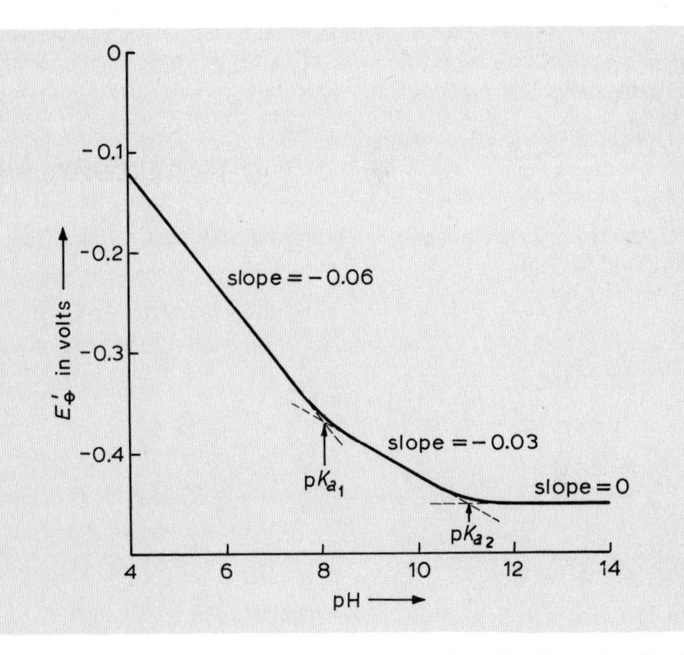

Fig. 12.4 Effect of pH change on the value of $E'_{\ominus}$ at 303 K for the bivalent redox couple $A|AH_2$ when AH_2 is a biprotic, very weak acid with pK_a values of 8 and 11.

this equation involves the H^+ activity and the acid dissociation constants of the reductant (see references [12, 21, 33]).

We need also only mention that the redox potential of a couple whose *oxidant* is a weak acid or a weak base will change characteristically as its pH is altered, for it to be obvious that it is generally desirable to conduct any oxidation-reduction experiment in a pH-buffered medium.

The following example illustrates the way in which it may be possible to determine the pH of a medium by measuring the redox potential of a suitable redox couple in whose half reaction H^+ ions participate.

EXAMPLE

Quinhydrone when dissolved in an aqueous solution of pH less than 8 yields equal concentrations of benzoquinone and benzoquinol. These compounds are respectively the reduced and oxidized members of a redox couple, which at pH <8 behaves according to the following half reaction:

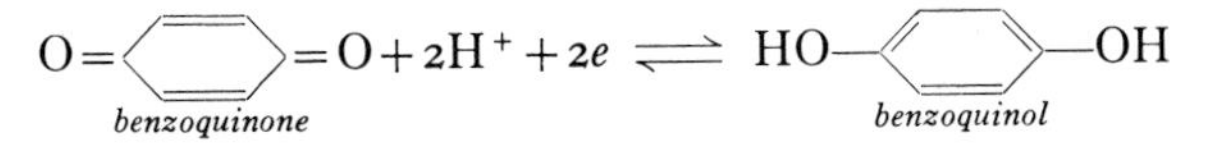

A quantity of quinhydrone was dissolved in a solution of acidic pH at 25°C, and a platinum electrode was immersed in the mixture. When joined to a saturated calomel electrode at this temperature, a cell of e.m.f. = 0.16 V was obtained, in which the calomel electrode formed the reducing half cell. Calculate the pH of the quinhydrone-containing solution, given that $E_\ominus$ *of the quinhydrone redox couple at 298 K = +0.700 V, and* E_h *of the saturated calomel electrode at 298 K = +0.244 V.*

Representing benzoquinone as Q, and benzoquinol as QH_2, then in a solution of acid pH,

$$Q + 2H^+ + 2e \rightleftharpoons QH_2$$

The redox potential E_h of this couple will be related to the prevailing pH by the equation

$$E_h = E_\ominus + \frac{RT}{nF} \ln \frac{[Q^-][H^+]^2}{[QH_2]}$$

$$= E_\ominus + \frac{2.303RT}{nF} \log \frac{[Q]}{[QH_2]} + \frac{2.303RT}{nF} \log [H^+]^2$$

Since quinhydrone yields equal concentrations of Q and QH_2,

$$\frac{2.303RT}{nF} \log \frac{[Q]}{[QH_2]} = \text{constant}.\log \frac{1}{1} = 0$$

$$\therefore \quad E_h = E_\ominus + \frac{2.303RT}{nF} \log [H^+]^2 \qquad \text{where} \begin{cases} E_\ominus = +0.700 \text{ V} \\ n = 2 \\ \dfrac{2.303RT}{F} = 0.059 \text{ at } 298 \text{ K} \end{cases}$$

$$\therefore \quad E_h = 0.7 + \frac{0.059}{2} \times 2.\log [H^+] = 0.7 - 0.059 \text{ pH}$$

The diagram for the cell formed between this quinhydrone electrode and a saturated calomel electrode is as follows:

$$^{-}\text{saturated calomel electrode} \parallel H^+ \text{ (pH = ?)}, \text{Q[C]}, \text{QH}_2\text{[C]}, \text{Pt}^+$$

$$E_h = 0.244 \text{ V} \qquad\qquad E_h = ? \text{ V}$$

$$\text{e.m.f.} = 0.16 = \Delta E_h = (E_h - 0.244) \text{ V}$$

$$\therefore \quad E_h = 0.16 + 0.244 \text{ V} = 0.404 \text{ V}$$

But $E_h = 0.70 - 0.059$ pH

$$\therefore \quad 0.404 = 0.70 - 0.059 \text{ pH, whence pH} = \frac{0.70 - 0.404}{0.059}$$

$$= \frac{0.296}{0.059} = \underline{5.0}$$

Thus the pH of the solution as estimated by use of this quinhydrone electrode is 5.0.

POTENTIOMETRIC TITRATION

Provided that the constituent redox couples are reversible and electromotively active (p. 353), the oxidation of a reducing agent by an oxidizing agent can be followed 'potentiometrically' by introducing an inert platinum electrode into the solution of the reducing agent, coupling this half cell with a reference electrode (e.g. calomel electrode) and continuously measuring the e.m.f. of this completed cell during the slow addition of a solution of the oxidizing agent.

Suppose one wished to follow the interaction between the two redox couples:

$$(1) \qquad \mathrm{ox}_1 + n_1 e \rightleftharpoons \mathrm{red}_1$$

$$\text{and } (2) \qquad \mathrm{ox}_2 + n_2 e \rightleftharpoons \mathrm{red}_2$$

If, at 298 K and pH 7, E_h of redox couple (2) is very much more positive than the value of E_h of redox couple (1), then $\mathbf{red}_1$ will be spontaneously oxidized by $\mathbf{ox}_2$ at this temperature and pH, in a reaction that proceeds virtually to completion (since ΔG will have a large negative value). Whether or not this oxidation-reduction reaction will be rapid cannot be predicted, but let us assume that either alone, or in the presence of a suitable catalyst, the reaction proceeds quickly. Starting with a solution of $\mathbf{red}_1$ into which is placed a platinum electrode, we could connect this to a reference electrode and measure the initial e.m.f. of the cell by the null-point potentiometric method. If the solution of $\mathbf{red}_1$ is now titrated with a solution containing an equivalent concentration of $\mathbf{ox}_2$, it is found that the e.m.f. of the cell slowly changes (Fig. 12.5). This change in e.m.f. reflects the increase in the redox potential of the experimental half cell, which rises as the value of $\log \frac{[\mathrm{ox}_1]}{[\mathrm{red}_1]}$ gradually increases due to the occurrence of the reaction,

$$\mathrm{red}_1 + \mathrm{ox}_2 \longrightarrow \mathrm{ox}_1 + \mathrm{red}_2$$

When just sufficient $\mathbf{ox}_2$ has been added to effect the oxidation of 50% of the $\mathbf{red}_1$ initially present, the measured redox potential of the experimental half cell equals the value of E'_0 of redox couple (1) since at this half-equivalence point,

$$[\mathrm{ox}_1] = [\mathrm{red}_1]$$

$$\text{and} \quad E_h = E'_{\ominus}$$

$$\left(\text{since } \frac{RT}{nF} \ln \frac{[\mathrm{ox}_1]}{[\mathrm{red}_1]} = 0\right)$$

As the titration is continued, the measured redox potential of the couple continues to increase slowly until the equivalence point is nearly reached. Then there is an abrupt change in redox potential that serves to indicate the end point of the titration (Fig. 12.5). Should the titration be continued beyond this end (equivalence) point, the increase in redox potential again becomes gradual once the phase of rapid change about the equivalence point has been passed. This continued slow rise in the electrode potential of the experimental half cell now reflects the increasing value of $\log \frac{[ox_2]}{[red_2]}$.

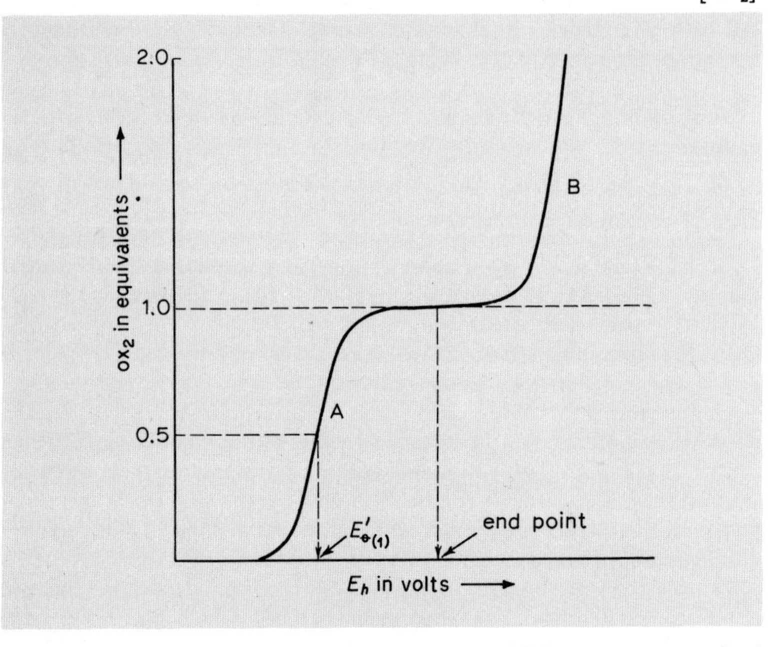

Fig. 12.5 Potentiometric titration of red_1 with the oxidizing agent ox_2 at a fixed temperature and pH. $E'_{\ominus(1)}$ is the standard, midpoint redox potential of the $red_1|ox_1$ couple at the given temperature and pH. (The e.m.f. measured potentiometrically during the course of the titration has been used to calculate the E_h of the mixture, and it is this value that is plotted on the abscissa.)

In summary, in segment A of the potentiometric titration curve shown in Fig. 12.5, the electrode potential of the experimental half cell is the redox potential of redox couple (1) whose value is determined by the proportion of **red**$_1$ that remains unoxidized,

$$\text{at 298 K, measured redox potential} = E'_{\ominus(1)} + \frac{0.059}{n_1} \log \frac{[ox_1]}{[red_1]}.$$

In segment B of the titration curve (Fig. 12.5), the electrode potential of the experimental half cell is the redox potential of redox couple (2) which increases as fresh $\mathbf{ox}_2$ is added, according to the equation,

$$\text{measured redox potential} = E'_{\ominus(2)} + \frac{0{\cdot}059}{n_2} \log \frac{[\text{ox}_2]}{[\text{red}_2]}.$$

Just as the apparent pK_a value of a weak acid is obtainable as the pH at the half-equivalence point during its titration with alkali (p. 117), so is the value of $E'_{\ominus}$ of a 'reducing' redox couple measurable as its redox potential at the half-equivalence point during its titration with a strong oxidizing agent (at a given temperature and pH).

Even when the oxidation of a reducing agent is accomplished by the loss of two electrons per molecule, it does not necessarily mean that these electrons are simultaneously donated (as a pair) to the oxidizing agent. They might be separately donated in two successive, univalent steps with the intermediary formation of a one electron donor or acceptor species (which could be a free radical of the semiquinone type). If the values of $E'_{\ominus}$ for each oxidation step are sufficiently different, the sequential oxidation of a reducing agent in a two-step reaction will be mirrored in the shape of its potentiometric titration curve with a strong oxidizing agent. The curve will then consist of two overlapping sigmoidal sections, the medial slope of each being that of a univalent oxidation reaction.

Biochemists have employed potentiometric titration methods for the identification and assay of many redox metabolites, e.g. the assay of thiol groups by titration with mercuric ions in the presence of a mercury electrode. Together with the amperometric methods of polarography, p. 364, that can be used to assay oxidants that are 'electroreducible' at a dropping mercury cathode, these potentiometric methods are likely to prove increasingly useful in the study of redox couples of biological importance.[33]

REDOX INDICATORS

Any intensely coloured dyestuff that behaves as a reversible redox couple may be used as a redox indicator if its oxidized and reduced forms are differently coloured. The colours must be sufficiently intense to allow the dye to be used in such small concentration that its addition to the test mixture insignificantly changes the redox potential. If the redox potential of the test mixture is E_h, then at 298 K the oxidized and reduced forms of the indicator will adjust themselves so that they will coexist at equilibrium in the mixture in concentrations given by the equation,

$$E_h = E'_{\ominus(\text{In})} + \frac{0{\cdot}059}{n_{(\text{In})}} \log \frac{[\text{In}_{\text{ox}}]}{[\text{In}_{\text{red}}]},$$

where $E'_{\ominus(\mathrm{In})}$ is the modified, midpoint redox potential of the indicator, and where $n_{(\mathrm{In})}$ for most organic, redox indicators is likely to equal 2.

This means that a redox indicator does not change its colour abruptly as the redox potential of its environment is gradually altered. Instead, the colour change is progressive, and extends over a range of values for the redox potential that is known as the 'transition range' or 'colour change interval' of the indicator. In a medium whose redox potential is equal to $E'_{\ominus(\mathrm{In})}$, the indicator shows its midpoint colour (or half-colour). Should $n_{(\mathrm{In})}$ equal 2, then at 298 K the indicator will show its fully reduced colour in any medium where the redox potential is more than about 0.06 V more negative than the value of $E'_{\ominus(\mathrm{In})}$. Similarly, the indicator will display its fully oxidized colour in any medium whose redox potential is more than 0.06 V more positive than its $E'_{\ominus(\mathrm{In})}$. It follows that the effective working range at 298 K of a redox indicator of this type is restricted to an E_h interval of just more than 0.1 V.

Most of the redox indicators in common use are colourless when fully reduced. Since in their case only the oxidized form is coloured, it is the *intensity*, and not the tint, of the colour that is indicative of the fraction present in the oxidized state.

EXAMPLE

The redox indicator methylene blue is an intense blue when oxidized but is colourless when reduced,

$$\underset{(blue)}{MB} + 2H^+ + 2e \rightleftharpoons \underset{(colourless)}{MBH_2}$$

If $E'_{\ominus}$ for methylene blue at 298 K and pH 7 equals 0.011 V, calculate the redox potentials in the media in which, at this temperature and pH, the indicator is (a) 2%, and (b) 90% in its coloured form. (Note: Reduced methylene blue is rapidly oxidized by O_2, and assays involving this indicator must be performed anaerobically.)

For any redox indicator,

$$E_h = E'_{\ominus(\mathrm{In})} + 2.303\,\frac{RT}{nF}\log\frac{[\mathrm{In_{ox}}]}{[\mathrm{In_{red}}]}$$

Thus for methylene blue at 298 K and pH 7,

$E'_{\ominus(\mathrm{In})} = 0.011$ V $\mathrm{In_{ox}} = \mathrm{MB}$

$n = 2$ $\mathrm{In_{red}} = \mathrm{MBH_2}$

$2.303\dfrac{RT}{F} = 0.059$

$$\therefore \qquad E_h = 0.011 + \frac{0.059}{2} \log \frac{[\text{MB}]}{[\text{MBH}_2]}$$

(a) *When 2% of methylene blue is present in its oxidized form*

$$\log \frac{[\text{MB}]}{[\text{MBH}_2]} = \log \frac{2}{98} = \bar{2}.3098 = -1.6902$$

$$\therefore \qquad E_h = 0.011 + \frac{0.059 \times (-1.6902)}{2} = \underline{-0.039 \text{ V}}$$

(b) *When 90% of methylene blue is present in its oxidized form*

$$\log \frac{[\text{MB}]}{[\text{MBH}_2]} = \log \frac{90}{10} = 0.9542$$

$$\therefore \qquad E_h = 0.011 + \frac{0.059 \times 0.9542}{2} = \underline{+0.039 \text{ V}}$$

Several redox indicators are also pH indicators (p. 133), and it is possible for these to undergo different colour transitions in acid and in alkaline media. For example, several substituted indophenol redox indicators, whose reduced forms are colourless at all pH values, possess oxidized forms that are bright red in solutions whose pH is lower than 6, but are blue if the pH is higher than 6. So long as these 'dual' indicators are used in pH-buffered media, their usefulness as redox indicators is not diminished.

SLUGGISH REDOX COUPLES

So far, we have assumed that all redox couples are *reversible* in the sense that they may be reduced or oxidized with equal ease, and *electromotively active* in the sense that they will communicate a stable redox potential via an inert platinum electrode when this is immersed in their solution. Yet some redox couples do not behave in this ideal fashion. For example, the tetrazolium salts that are often used in biological studies are readily reduced, passing easily from a colourless oxidized form to a distinctively coloured reduced state (the formazan). Yet, the reverse oxidation of the formazan is accomplished only with great difficulty under carefully controlled conditions, and the tetrazolium|formazan couple can be considered to be almost non-reversible. Indeed, although 'midpoint redox potentials' have been reported for many of these tetrazolium salts, the meaning of these values is uncertain, and is of questionable relevance in assessing the behaviour of these compounds in oxidation-reduction reactions. In general, redox potentials assigned to irreversible couples should be regarded with suspicion.

Sometimes a redox couple that behaves in a completely reversible manner in its interaction with another redox couple does not demonstrate a stable redox potential at a platinum electrode. This means in effect that the intrinsically reversible couple is sluggish in attaining electronic equilibrium with the metallic electrode. It is still possible to measure the redox potential of this sluggish couple by allowing it to react reversibly with a second couple of suitable E_h which *does* equilibrate rapidly with a Pt electrode. The resulting oxidation-reduction reaction will attain equilibrium when the interacting couples exhibit the same redox potential. If the components of the second couple are used in very much smaller concentrations than the members of the sluggish couple, the equilibrium redox potential of the mixture will be insignificantly different from the initial redox potential of the sluggish couple. Since the minor second couple is electromotively active, a platinum electrode immersed in the equilibrium mixture will register *its* redox potential and hence the equilibrium redox potential, which is virtually equal to the E_h of the sluggish couple. The minor couple in effect acts as a 'mediator' between the sluggish couple and the platinum electrode. A redox indicator makes a particularly convenient redox mediator since its colour in the solution of the sluggish couple immediately shows whether it is suitable for the role (for it can only act as a mediator of redox potentials which fall within its transition range). From the redox potential of the sluggish couple at the prevailing temperature and pH (E_h) and the concentrations at equilibrium of its oxidant and reductant, the value of its correspondingly modified, midpoint potential ($E'_\ominus$) can be calculated using the equation,

$$E_h = E'_\ominus + 2{\cdot}303\frac{RT}{nF}\log\frac{[\mathrm{ox}]}{[\mathrm{red}]}$$

Alternatively the $E'_\ominus$ of a sluggish couple can be measured by allowing the couple to interact (reversibly) with a comparable quantity of another redox couple whose $E'_\ominus$ value is known, and is suspected to lie close to the unknown $E'_\ominus$ of the sluggish couple. If necessary, this oxidation-reduction reaction may be speeded up by addition of a catalyst so that equilibrium is quickly achieved at a given temperature and pH. From the (measured) concentrations of the reaction components at equilibrium and the known value of $E'_{\ominus(2)}$ for the second couple, the value of $E'_{\ominus(1)}$ of the sluggish couple is easily calculated, as shown in the following example.

EXAMPLE

Succinate in buffered solution at pH 7 was incubated anaerobically at

310 K (i.e. 37°C) with methylene blue in the presence of the enzyme, succinic dehydrogenase. The dye was partially decolorized and some of the succinate was oxidized to yield fumarate:

$$succinate^{2-} + MB \rightleftharpoons fumarate^{2-} + MBH_2$$

The concentrations of these compounds at equilibrium were as follows:
succinate (36 mmol dm^{-3}); fumarate (6 mmol dm^{-3}); oxidized methylene blue (4.3 mmol dm^{-3}); and reduced methylene blue (6 mmol dm^{-3}).

Calculate the value of $E'_{\ominus}$ at 303 K and pH 7 for the fumarate|succinate couple, if $E'_{\ominus}$ at this temperature and pH for the methylene blue couple is +0.011 V.

The problem can be solved in two ways:

either (a) from the known value of $E'_{\ominus}$ for the methylene blue couple and the measured equilibrium concentrations of MB and MBH_2, the redox potential of the equilibrium mixture can be calculated ($E'_{\ominus}$). The value of this equilibrium redox potential and the measured equilibrium concentrations of succinate and fumarate can then be substituted in the following equation to yield the value of $E'_{\ominus}$ for the fumarate|succinate couple:

$$\text{Equilibrium redox potential} = E'_{\ominus} + 2.303\frac{RT}{2F}\log\frac{[\text{fumarate}^{2-}]}{[\text{succinate}^{2-}]};$$

or (b) from the equilibrium concentrations of the reaction components, the equilibrium constant of the reaction can be calculated. From the value of the equilibrium constant we can calculate the e.m.f. of the cell composed of the two couples in their modified standard states; taken in conjunction with the value of $E'_{\ominus}$ for the methylene blue 'half cell', the magnitude of this e.m.f. gives the value of $E'_{\ominus}$ of the fumarate | succinate 'half cell'.

Proceeding according to method (b), the calculation is as follows:
Equilibrium constant K'_{eq} for succinate oxidation by methylene blue

$$= \frac{[\text{fumarate}^{2-}][\text{MBH}_2]}{[\text{succinate}^{2-}][\text{MB}]} = \frac{6\times 6}{36\times 4.3} = 0.2326$$

Now, $$-2\ 303RT\log K'_{eq} = \Delta G^{\ominus\prime}$$

$$\therefore \quad \Delta G^{\ominus\prime} = -2.303\times 8.314\times 303\log 0.2326\ \text{J mol}^{-1}$$

$\therefore$ $\Delta G^{\ominus\prime}$ for oxidation of succinate by MB

$$= -2.303\times 8.314\times 303\times(-0.6335)\ \text{J mol}^{-1}$$
$$= \underline{+3675\ \text{J mol}^{-1}}$$

This means that it is the reverse reaction (reduction of fumarate by MBH_2) that is spontaneous under the modified standard conditions. In turn it follows that under these conditions the fumarate | succinate couple is more highly oxidizing than the methylene blue couple and possesses the more positive value of $E'_{\ominus}$;

i.e. e.m.f. $= \Delta E'_{\ominus} = E'_{\ominus}$(fumarate | succinate) $- E'_{\ominus}$(methylene blue couple)

For a spontaneous oxidation-reduction reaction,

$$\Delta G^{\ominus\prime} = -nF\Delta E'_{\ominus} \qquad \text{(p. 312)}$$

$$\therefore \qquad -3\,675 = -2 \times 96\,487 \times \Delta E'_{\ominus}$$

whence $$\Delta E'_{\ominus} = \frac{3\,675}{2 \times 96\,487}\ \text{V} = 0.019\ \text{V}$$

Substituting this value of $\Delta E'_{\ominus}$ in the equation given above,

$0.019 = E'_{\ominus}$ (fumarate | succinate couple) $- E'_{\ominus}$ (methylene blue couple)

But $E'_{\ominus}$ (methylene blue couple) equals $+0.011$ V

$\therefore$ $E'_{\ominus}$ (fumarate | succinate couple) $= 0.019 + 0.011$ V

$= \underline{+0.03\ \text{V}}$

Thus $E'_{\ominus}$ of the fumarate | succinate couple at pH 7 and 303 K is $\underline{+0.03\ \text{V}}$

USE OF TABLES OF STANDARD REDOX POTENTIALS

In Fig. 12.6 are listed values of $E'_{\ominus}$ at 298 K and pH 7 for several redox couples of biochemical interest. From these values we can predict the direction and extent of the oxidation-reduction reaction between members of any two of these couples under standard conditions (but at pH 7). For example, we see from this Figure that, under the prescribed reaction conditions, NADH is spontaneously oxidized by the flavin coenzyme FAD, succinate is spontaneously oxidized by nitrate, and hydrogen spontaneously reduces methylene blue (though whether these reactions will take place at a significant rate in the absence of suitable catalysts cannot be predicted). The value of $\Delta G^{\ominus\prime}$ for each of these reactions can also be calculated since it approximately equals $-nF\Delta E'_{\ominus}$ J mol^{-1} (see p. 198). When n equals 2, each 0.1 V difference between the values of $E'_{\ominus}$ of the couples represents an increment in $\Delta G^{\ominus\prime}$ of -19.25 kJ mol^{-1}.

EXAMPLE

Calculate the value of $\Delta G^{\ominus\prime}$ at 298 K and pH 7 for the reduction of fumarate by butyryl.CoA if, at this temperature and pH, $E'_{\ominus}$ of fumarate | succinate is $+0.03$ V and $E'_{\ominus}$ of crotonyl.CoA/butyryl.CoA is $+0.19$ V.

The required reduction of fumarate by butyryl.CoA is the resultant of the following 'half cell' reactions:

Electron donating reaction
(by reducing couple): butyryl.CoA $\rightleftharpoons$ crotonyl.CoA + $2H^+$ + $2e$

Electron accepting reaction
(by oxidizing couple): $fumarate^{2-}$ + $2H^+$ + $2e$ $\rightleftharpoons$ $succinate^{2-}$

Net: butyryl.CoA + $fumarate^{2-}$ $\rightleftharpoons$ crotonyl.CoA + $succinate^{2-}$

At 298 K and pH 7 with the components present in their standard states, this is the reaction of the following cell:

$^{-}$butyryl.CoA (1 mol dm^{-3}), crotonyl.CoA (1 mol dm^{-3}) ‖ fumarate (1 mol dm^{-3}), succinate (1 mol dm^{-3})$^{+}$

$$\begin{aligned} \text{whose e.m.f.} = \Delta E'_{\ominus} &= E'_{\ominus}\text{ (right)} - E'_{\ominus}\text{ (left)} \\ &= E'_{\ominus}\text{ (oxidizing couple)} - E'_{\ominus}\text{ (reducing couple)} \\ &= +0.03\text{ V} - (+0.19)\text{ V} = \underline{-0.16\text{ V}} \end{aligned}$$

(We can immediately conclude that since $\Delta E'_{\ominus}$ is −ve, the reaction will not occur spontaneously, i.e. $\Delta G^{\ominus\prime}$ will prove to be +ve.)
For this reaction at 298 K and pH 7,

$$\Delta G^{\ominus\prime} = -nF\Delta E'_{\ominus}$$

$$\text{where} \quad \begin{cases} n = 2 \\ F = 96\,487\text{ C mol}^{-1} = 96\,487\text{ J V}^{-1}\text{ mol}^{-1} \\ \Delta E'_{\ominus} = -0.16\text{ V} \end{cases}$$

$$\therefore \quad \Delta G^{\ominus\prime} = -2 \times 96\,487 \times (-0.16)\text{ J mol}^{-1} = \underline{+30.9\text{ kJ mol}^{-1}}$$

Again it must be emphasized that even if a reaction is thermodynamically spontaneous it need not take place at a measurable rate (p. 191). For example, although $\Delta G^{\ominus\prime}$ at 298 K and pH 7 for the oxidation of lactate by oxygen has a large negative value (approx. −195 kJ mol^{-1}), lactate is seemingly perfectly stable in air at this temperature and pH, and the rapid aerobic oxidation of lactate in a living cell requires the presence of several specific enzymes and intermediary redox couples (electron carriers). Furthermore, values of $E'_{\ominus}$ are only immediately relevant to the behaviour of redox couples when their oxidized and reduced forms are present together in equal concentrations. To predict the outcome of the interaction of two redox couples in the living cell we would wish to know the actual values of E_h that they exhibit in the cell, which would depend upon the local concentrations of their oxidized and reduced

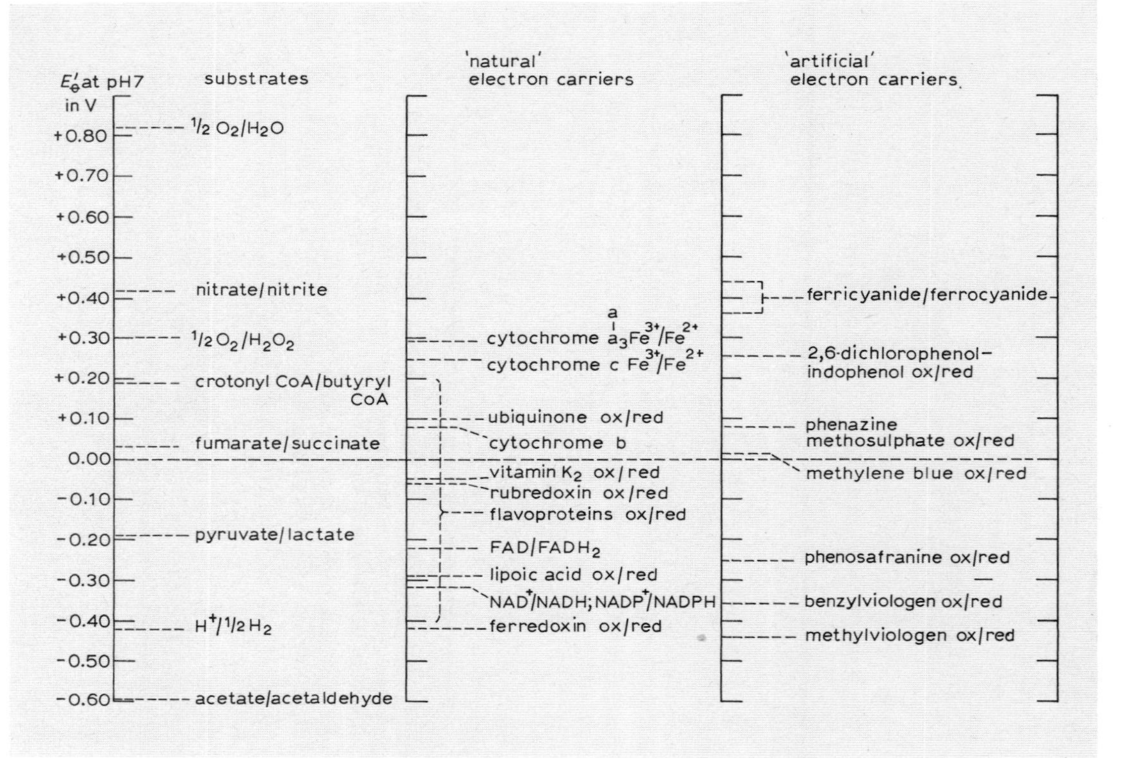

E'_0 at pH7 in V
substrates
'natural' electron carriers
'artificial' electron carriers
+0.80
+0.70
+0.60
+0.50
+0.40
+0.30
+0.20
+0.10
0.00
-0.10
-0.20
-0.30
-0.40
-0.50
-0.60
$\frac{1}{2}O_2/H_2O$
nitrate/nitrite
$\frac{1}{2}O_2/H_2O_2$
crotonyl CoA/butyryl CoA
fumarate/succinate
pyruvate/lactate
$H^+/\frac{1}{2}H_2$
acetate/acetaldehyde
cytochrome $a_3 Fe^{3+}/Fe^{2+}$
a
cytochrome c Fe^{3+}/Fe^{2+}
ubiquinone ox/red
cytochrome b
vitamin K_2 ox/red
rubredoxin ox/red
flavoproteins ox/red
FAD/$FADH_2$
lipoic acid ox/red
$NAD^+/NADH$; $NADP^+/NADPH$
ferredoxin ox/red
ferricyanide/ferrocyanide
2,6-dichlorophenol-indophenol ox/red
phenazine methosulphate ox/red
methylene blue ox/red
phenosafranine ox/red
benzylviologen ox/red
methylviologen ox/red

components at the actual site of interaction. One must therefore be particularly cautious when attempting to predict the outcome of the interaction in a living cell of two couples whose values of $E'_{\ominus}$ are very similar, since which of these couples will be the more oxidizing will very likely depend on the local concentrations of their components (which it might well be impossible to measure *in vivo*).

ELECTRON TRANSPORT AND THE RESPIRATORY CHAIN

Certain highly exergonic oxidation-reduction reactions are employed by living cells to accomplish spontaneous synthesis of ATP and so indirectly promote an immense number of biosynthetic and other metabolic processes (see Chapter 9).

Although the coupling of specific oxidation-reduction reactions to the synthesis of ATP from ADP and P_i is thermodynamically plausible, and although its occurrence has been demonstrated experimentally in extracts of living cells, the mechanism of this 'oxidative phosphorylation' is still not fully resolved. It is, however, well established that when a reduced metabolite is oxidized by oxygen during cellular respiration, a sequence of several distinct reactions is generally involved. and a series of electron carriers intervenes between the primary reductant (AH_2) and the terminal oxidant (O_2). These electron transport factors constitute redox couples of increasing redox potential many of which are usually located in sequence on a membranous matrix within the living cell.

In the respiring cells of higher organisms the components of the electron transport chain are localized in subcellular organelles called mitochondria. The most important of these electron carriers are shown in Fig. 12.7, which also indicates how the highly exergonic oxidation of NADH by O_2 ($\Delta E'_{\ominus} = 1.14$ V; $\Delta G^{\ominus\prime} = -220$ kJ mol^{-1}) is accomplished in a stepwise fashion. It has been shown experimentally that the oxidation of 1 molecule of NADH by O_2 in the mitochondrion enables 3 molecules of ATP to be synthesized from ADP plus P_i. On the other hand, the oxidation of 1 molecule of succinate by O_2 is associated with the synthesis of only 2 molecules of ATP. Since the primary electron (and 'hydrogen') acceptor in the oxidation of succinate is a flavoprotein, this suggests that the production of one of the three molecules of ATP formed by oxidation of NADH by O_2 must be associated with the initial oxidation of NADH by the flavoprotein. These and other experimental findings suggest that three segments of the electron transport chain are each concerned with the phosphorylation of 1 molecule of ADP. These segments (as shown in Fig. 12.7) are, (1) the oxidation of NADH by flavoprotein, (2) the oxidation of flavoprotein by cytochrome c_1 (or c), (3) the oxidation of cytochrome c by O_2 (via cytochrome $\begin{matrix} a \\ | \\ a_3 \end{matrix}$).

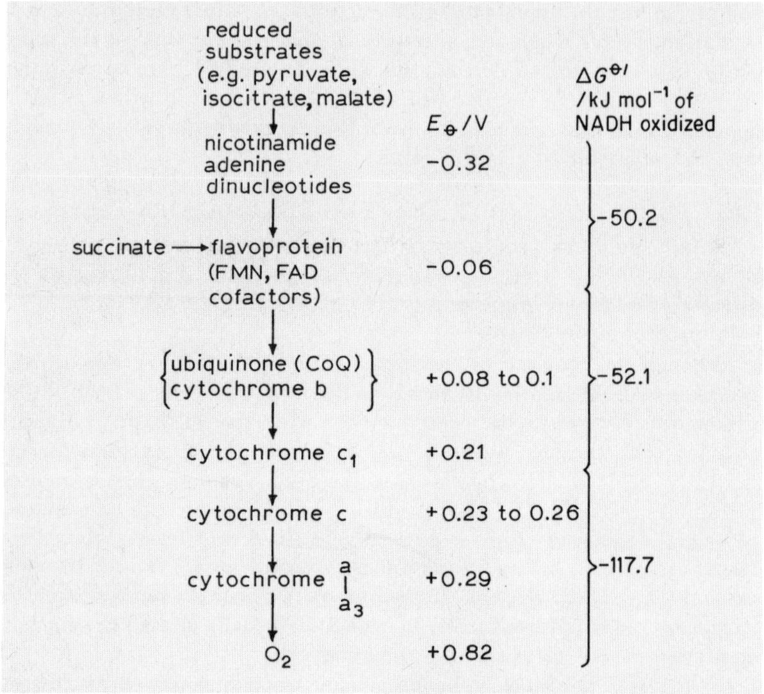

Component	$E_{\ominus}'$/V	$\Delta G^{\ominus\prime}$ /kJ mol^{-1} of NADH oxidized
reduced substrates (e.g. pyruvate, isocitrate, malate) ↓		
nicotinamide adenine dinucleotides ↓	−0.32	
		−50.2
succinate ⟶ flavoprotein (FMN, FAD cofactors) ↓	−0.06	
{ubiquinone (CoQ), cytochrome b} ↓	+0.08 to 0.1	−52.1
cytochrome c_1 ↓	+0.21	
cytochrome c ↓	+0.23 to 0.26	
cytochrome a–a_3 ↓	+0.29	−117.7
O_2	+0.82	

Fig. 12.7 Major components of the mitochondrial electron transport chain. (The values of $E_{\ominus}'$ and $\Delta G^{\ominus\prime}$ are necessarily approximate and refer to 298 K and pH 7.)

Current interest centres on discovering the location and operational sequence of these redox couples in the mitochondrion and on investigating the chemistry of oxidative phosphorylation.

PROBLEMS

(Use the following values: $R = 8.314$ J K^{-1} mol^{-1}; $F = 96\,487$ J V^{-1} mol^{-1}.)

1. At 310 K and pH 7.6, a solution containing compound A and its conjugate reductant AH_2 exhibited a redox potential of −0.06 V. A solution of compound B with its conjugate reductant BH_2 at the same temperature and pH had a redox potential of −0.12 V. If an electrochemical cell were constructed in which these solutions acted as half cells:

(i) which couple would act as the reducing half cell?

(ii) in the course of the spontaneous cell reaction would AH_2 be oxidized by B, or would BH_2 be oxidized by A?

(iii) what would be the e.m.f. of this cell, measured by the null-point, potentiometric method?

2. Platinum electrodes were introduced into solutions of three electromotively active redox couples at 303 K and pH 7, and these half cells were separately connected to a saturated calomel electrode (S.C.E.) at 303 K. The following cells were created:

(1) $^-$Redox couple 1 ‖ S.C.E.$^+$ e.m.f. = +0.54 V
(2) $^-$Redox couple 2 ‖ S.C.E.$^+$ e.m.f. = +0.11 V
(3) $^-$S.C.E. ‖ Redox couple 3$^+$ e.m.f. = +0.32 V

Calculate the values of E_h for the three redox couples at 303 K and pH 7, if E_h at 303 K of the saturated calomel electrode is +0.242 V.

3. Thioglycollate is often added to the culture media of anaerobic bacteria 'to remove dissolved oxygen and to poise the medium at the low redox potential required for growth of these micro-organisms'.

The thioglycollate redox couple behaves as follows:

$$\begin{array}{cc} CH_2.S.S.CH_2 \\ | \quad\quad\quad | \\ COO^- \quad COO^- \end{array} + 2H^+ + 2e \rightleftharpoons 2 \begin{array}{c} CH_2SH \\ | \\ COO^- \end{array}$$

If $E'_\ominus$ at 303 K and pH 7 of this couple is −0.34 V, calculate the values of E_h in the culture media in which thioglycollate is (i) 20%, and (ii) 75% reduced.

4. Calculate the value of $\Delta G^{\ominus\prime}$ at 303 K and pH 7 for the 'autoxidation' of a reduced flavoprotein by O_2, if, at this temperature and pH, $E'_\ominus$ of the flavoprotein ox|red couple is −0.06 V and $E'_\ominus$ of the $O_2|H_2O_2$ couple is +0.30 V. (The reaction may be assumed to be $FPH_2 + O_2 \rightleftharpoons FP + H_2O_2$.)

5. In many plant tissues and micro-organisms, glycollate is oxidized to glyoxylate in a reaction catalysed by the flavoprotein enzyme glycollate oxidase. In the laboratory, the activity of this enzyme is assayed by coupling the oxidation of glycollate to the reduction of added cytochrome c (using phenazine methosulphate as an electron carrier). The rate of reduction of cytochrome c is then followed spectrophotometrically.

If at 298 K and pH 7,

cytochrome $c_{ox} + e \rightleftharpoons$ cytochrome c_{red} $E'_\ominus = +0.25$ V

glyoxylate$^-$ $+ 2e + 2H^+ \rightleftharpoons$ glycollate$^-$ $E'_\ominus = -0.085$ V

calculate the values of (i) $\Delta G^{\ominus\prime}$, and (ii) the equilibrium constant, for the oxidation of glycollate by oxidized cytochrome c at 298 K and pH 7.

6. Vitamin C (ascorbic acid) is readily oxidized to yield dehydroascorbic acid. If $E'_\ominus$ at 303 K and pH 7 for this redox couple is +0.058 V (i.e. dehydroascorbate $+ 2H^+ + 2e \rightleftharpoons$ ascorbate);

(i) calculate the proportion of the vitamin that will be present in the reduced form in a solution whose E_h is +0.019 V;
(ii) which of the following dyes might be employed as an oxidizing agent

suitable for the assay of ascorbic acid by titrimetric analysis at 303 K and pH 7?
(a) 2,6-dichlorophenolindophenol, $E'_{\circ} = +0.217$ V
(b) thionine, $E'_{\circ} = +0.063$ V
(c) benzylviologen, $E'_{\circ} = -0.359$ V

(iii) since the ascorbate redox couple does not itself register a stable potential at a platinum electrode, which of the dyes listed under (ii) might be employed as a 'mediator' of its E_h at 303 K and pH 7?

7. What is the redox potential at 310 K of the hydrogen half cell that contains 0.001 mol dm^{-3} hydrochloric acid equilibrated with H_2 at standard atmospheric pressure?

8. The production of acid in the culture medium of a lactobacillus was followed by an intermittent sampling technique. A pinch of solid quinhydrone was added to each sample, a platinum electrode inserted, and the E_h of the half cell measured at 303 K against a standard calomel electrode (acting as the reducing half cell). When the e.m.f. of the cell so created was $+0.21$ V, calculate the pH of the culture medium. ($E'_{\circ}$ at 303 K of the saturated calomel electrode is $+0.242$ V; $E'_{\circ}$ of the 'quinhydrone' redox couple at 303 K is $+0.696$ V.)

9. In the oxidation of malate by NAD^+ that can be catalysed by malate dehydrogenase, the equilibrium constant at 298 K and pH 7 is 6×10^{-12} mol dm^{-3}:

$$\text{malate}^{2-} + NAD^+ \rightleftharpoons \text{oxaloacetate}^{2-} + NADH + H^+$$

Calculate $E'_{\circ}$ at 298 K and pH 7 for the oxaloacetate | malate couple, if $E'_{\circ}$ is -0.32 V for the NAD | NADH couple at this temperature and pH.

10. In the presence of the enzyme butyryl.CoA dehydrogenase, butyryl. CoA is rapidly oxidized by the redox dye pyocyanine:

$$\text{butyryl.CoA} + \text{pyc.}_{\text{ox}} \rightleftharpoons \text{crotonyl.CoA} + \text{pyc.}_{\text{red}}$$

In one experiment performed at 303 K and pH 7, the concentrations of the reaction components at equilibrium were as follows: butyryl.CoA (90 mmol dm^{-3}); crotonyl.CoA (42 mmol dm^{-3}); oxidized pyocyanine (131 mmol dm^{-3}); and reduced pyocyanine (62 mmol dm^{-3}). Calculate the value of $E'_{\circ}$ at 303 K and pH 7 of the crotonyl.CoA | butyryl.CoA couple, if $E'_{\circ}$ of the pyocyanine couple is -0.034 V at this temperature and pH.

Appendix

GLASS ELECTRODE pH METER

In essence, this consists of a glass membrane which encloses a solution of constant H^+ ion activity (0.1 mol dm^{-3}), into which dips an 'internal' silver/silver chloride electrode of constant electrode potential $E_{Ag/AgCl}$. This assembly, in conjunction with the solution of unknown pH into which it is dipped, forms a half cell with the following composition,

$$\text{Ag, AgCl(s), HCl (0.1 mol dm}^{-3}\text{)} \;\vdots\; \text{H}^+ \text{ (pH = ?)}$$

Glass membrane

Without going into details, the electrode potential of this half cell is the sum of three components:

1. $E_{Ag/AgCl}$ = a constant;
2. the 'asymmetric potential', E_{asym}, of the glass membrane. The significance of this need not concern us, for its value is not great and is constant for any one glass electrode.

$$\therefore \qquad E_{asym} = \text{a constant};$$

3. the difference between the potentials set up on each side of the glass membrane. This potential difference arises because the glass membrane is permeable to H^+ ions, but not to other ions. Whenever the membrane separates solutions whose H^+ activities are different, the tendency for H^+ ions to flow through the membrane from one solution to the other creates a proportional potential difference 'across the membrane'. If pH_x is the unknown pH of the solution *outside* the membrane, and pH_c is the constant pH of the contents of the glass electrode (i.e. *within* the membrane), this 'membrane potential' will equal 0.059 $(pH_x - pH_c)$ V at 298 K.

Thus the full electrode potential, E'_G, of a glass electrode which is immersed at 298 K in a solution of pH_x equals,

$$\underset{\text{(constant)}}{E_{Ag/AgCl}} + \underset{\text{(constant)}}{E_{asym}} - \underset{\text{(constant)}}{0.059\,pH_c} + \underset{(\textit{variable})}{0.059\,pH_x}$$

Therefore, $E_G = E_{constant} + 0.059\ pH_x$, and the pH of any solution into which the glass electrode is dipped, is given by the equation,

$$pH = \frac{E'_G - E_{const}}{0.059}$$

where E'_G is the measured electrode potential and E_{const} is the characteristic potential of the electrode, the value of which is determined for each electrode using standardized buffers of known pH.

The pH meters that employ glass electrodes are 'direct reading instruments', the pH of the solution in which the glass electrode is dipped being read off in pH units from a scale pre-set at reference pH values supplied by standard buffer solutions. To determine the pH of a solution with such an instrument therefore involves no calculation, but you should be aware that:

(*a*) the instrument measures H^+ ion *activities*;
(*b*) because of restrictions set by the chemical composition of the usual glass membranes, glass electrode instruments may usually be used only in the range of pH values of pH 1 to pH 10;
(*c*) the glass membrane must be kept clean; when used to determine the pH of cell extracts, etc., one must be careful not to allow a protein film to dry on its surface.

POLAROGRAPHY

(*I am indebted to my ex-colleague Dr. P. D. J. Weitzman for much information concerning this technique.*)

This is a technique in which the current produced at a polarizable micro-electrode, immersed in the solution under examination, is investigated as a function of the applied potential. The completed circuit contains a non-polarizable reference electrode, also in contact with the solution, and a sensitive instrument for measuring the tiny currents. The micro-electrode most commonly employed is the dropping mercury electrode, which consists of a very fine glass capillary from which mercury issues in the form of regular and identical drops; the reference electrode is usually a saturated calomel electrode. The current produced as a result of the applied potential derives from the reduction or oxidation of material at the micro-electrode.

Consider the polarography of a solution containing one electro-reducible species, S^+, as the potential of the micro-electrode is made progressively more negative. At a certain potential value, reduction of S^+ will begin to occur at the electrode, and a small current will flow. As the potential is made still more negative, the current increases sharply as more S^+ is reduced, until a point is reached at which S^+ is reduced as fast as it arrives at the electrode surface by diffusion from the body of the solution. At this point, a 'limiting current' is produced, and the electrode is said to be 'concentration polarized'. Figure A.1 shows these changes of current with applied potential; such a current-voltage curve is known as a polarographic wave and has two important properties. First, the potential at which reduction occurs (more precisely the midpoint of the wave, or half-wave potential, $E_{\frac{1}{2}}$) is characteristic of the material being reduced. Secondly, since the magnitude of the limiting current is controlled by the rate of diffusion of the electro-reducible species, it will be directly proportional to the concentration of the latter. These two features of such current–voltage relationships form the basis of the various applications of polarography. Thus the technique is of immense

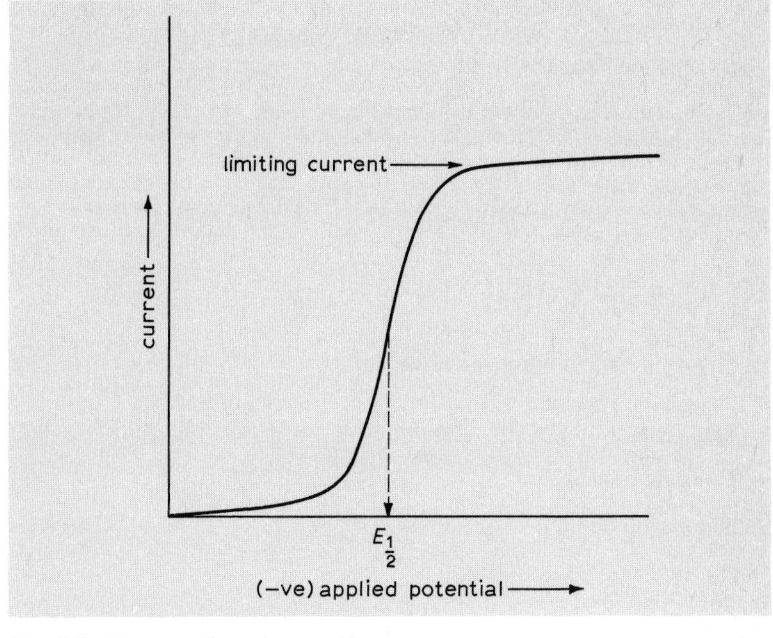

Fig. A.1 Diagram of a polarographic wave.

value in both the identification, and the estimation of trace amounts of metals and of numerous other materials.

More specifically biochemical and biological applications include the following:

1. The determination of the standard reduction potentials of substances of biological interest; in ideal cases, these are identical with the half-wave potentials.
2. The determination of SH (and, indirectly, of S-S) groups in proteins, by polarographic (amperometric) titration with heavy metal thiol reagents.
3. The assay of enzyme-catalysed reactions involving electro-active substrates, e.g. reactions in which acyl-thioesters are cleaved to yield free thiols.
4. A particular type of polarographic electrode may be used to measure oxygen concentration. Such 'oxygen electrodes' are most useful in biochemical and physiological studies.

Certain properties of the dropping mercury electrode are advantageous in the study of biological systems. Thus the fact that the electrode surface is continually and reproducibly renewed is important when studying proteins; the metal electrodes used in potentiometric studies, by contrast, may be rendered unresponsive by the adsorption of protein. It is also noteworthy that the high overvoltage of hydrogen on mercury extends the applicability of the electrode to substances whose reduction potentials are more negative than that of hydrogen.

ANSWERS TO PROBLEMS

As far as possible, I have attempted to use currently acceptable values for the various terms employed in examples and problems (e.g. dissociation constants, rate constants and thermodynamic functions). Yet it is likely that even while this book is in the press, some of these values will be altered in the light of new and more accurate experimental findings. Thus you should not employ these values for other purposes without first checking their authenticity in up-to-date reference tables. The only truly reliable values are those derived from sources which give details of the manner and conditions in which they were determined. Failing this, current standard reference books should be consulted.

Chapter 3. The Behaviour of Gases (p. 55)

1. Diameter = 2.284 m
2. (i) 411 cm^3 (ii) 264 cm^3 (iii) 18.9 cm^3
3. (i) 1101 cm^3 (ii) 52.7 cm^3 (iii) 472 cm^3
4. 31.4 mmol dm^{-3}
5. 366 cm^3 at s.t.p.
6. 66.9 cm^3 of air at 303 K
7. 598 kPa
8. 457 mg of anhyd. Na_2CO_3
9. 783 cm^3 of N_2
10. 275 kPa
11. Approx. 95% pure
12. Molecular weight = 16 g mol^{-1}; probably methane (CH_4)
13. (a) K for N_2 at 303 K = 2.13 (b) K for CO_2 at 298 K = 2.39
14. 20 mmol dm^{-3} pyruvate

Chapter 4. Some Properties of Aqueous Solutions (p. 96)

1. A = 92.8 g; B = 407.2 g
2. Molecular weight = 666 g mol^{-1}; stachyose is a tetrasaccharide
3. 1.17 g NaCl/100 cm^3
4. Activity coefficient = 0.85
5. Freezing point = −0.979°C
6. Molecular weight = 80 200 g mol^{-1}
7. (i) 0.03 (ii) 0.4 (iii) 1.5
8. In order of increasing ionic strength: 0.25 mol dm^{-3} ethanol < 0.1 mol dm^{-3} NaCl < 0.03 mol dm^{-3} $MgSO_4$ < 0.05 mol dm^{-3} $CaCl_2$ < 0.03 mol dm^{-3} $FeCl_3$
9. 0.96
10. 380.8 kPa
11. Activity of both K^+ and Cl^- ions = 9.91×10^{-6}
12. In order of increasing osmotic pressure: protein < NaCl < $CaCl_2$ < sucrose (i.e. (i) < (iii) < (iv) < (ii))
13. (i) Apparent $K_{a_1} = 1.24 \times 10^{-2}$
    (ii) van't Hoff factor = 1.93; osmotic pressure = 4.76 kPa
14. 1.63%

15. (i) (a) 1.04×10^{-5} (b) 1.04×10^{-5} mol dm^{-3}
(ii) (a) 1.04×10^{-5} (b) 2.81×10^{-5} mol dm^{-3}
(iii) (a) 6.75×10^{-9} (b) 4.22×10^{-8} mol dm^{-3}

Chapter 5. Acids, Bases and Buffers in Aqueous Solutions (p. 140)

1. (a) pH 3.0; pH 3.82; pH 5.51; pH 9.96
(b) 6.31×10^{-4} mol dm^{-3}; 2×10^{-8} mol dm^{-3}; 2.51×10^{-11} mol dm^{-3}; 3.16×10^{-14} mol dm^{-3}
2. pH 2.91
3. pH 8.65
4. 75 cm^3 of 0.05 mol dm^{-3} NaOH
5. (a) pH 3.78 (b) pH 4.26 (c) pH 4.74 (d) pH 5.22 (e) pH 5.69
6. (a) pH 4.74 (b) pH 5.04; hydroxyl ion concentrations are
(a) 5.5×10^{-10} mol dm^{-3} (b) 1.1×10^{-9} mol dm^{-3}
7. Apparent $K_b = 2 \times 10^{-5}$
8. pH 6.5
9. (a) pH 10.06 (b) pH 9.28 (c) pH 6.8
10. From 5.5×10^{-9} to 5.5×10^{-11} mol dm^{-3}
11. Mix 200 cm^3 of 0.1 mol dm^{-3} pyridine with 7 cm^3 of 2 mol dm^{-3} HCl and make up to 1 dm^3 with water; check pH on pH meter and adjust to pH 5.0 with small additions of pyridine aq. and/or HCl aq.
12. Dissolve 5.93 g $Na_2HPO_4.2H_2O$ and 2.3 g $NaH_2PO_4.H_2O$ in water and make up to 500 cm^3
13. Molecular weight = 206 g mol^{-1}
14. (a) pH 10.72 (b) pH 4.89
15. Methyl orange

Chapter 6. Biochemical Relevance of pH (p. 178)

1. (a) Cation (b) Zwitterion (c) Zwitterion (d) Anion
(i) pH 9.15 (ii) pH 9.45. Yes, they will be buffers
2. (a) 9.1×10^{-2} mol dm^{-3} (b) 0.1 mol dm^{-3} (c) 9.1 mmol dm^{-3}
3. (a) pH 6.01 (b) pH 5.67 (c) pH 9.73 (d) pH 5.06
4. $pK_{a_1} = 4.2$ and $pK_{a_2} = 10.4$
5. Histidine concentration = 0.72 mol dm^{-3}
6. Molecular weight = 75.5 g mol^{-1}; glycine ($CH_2NH_2.COOH$)
7. The aspartate will have migrated towards the anode, the histidine to the cathode, and the valine will have remained close to the starting line.
8. Salmine will be 'net' cationic at all three pH values.
9. Isoelectric point of egg albumin probably about pH 5 (i.e. pH value somewhat larger than the pK_{a_2} values of aspartic and glutamic acids but smaller than the pK_{a_2} of histidine). Not bound to β-lactoglobulin but probably bound to ribonuclease.
10. (a) β-lactoglobulin; load on DEAE-cellulose column in buffer at about pH 8, then elute with buffer using a gradient of decreasing pH and/or increasing ionic strength.
(b) cytochrome c: load on CM-cellulose at pH 7 to 8, elute with buffer, using a gradient of increasing pH and/or increasing ionic strength.
11. (a) 10^{-5} mmol dm^{-3} (b) 5 mmol dm^{-3}
12. 2×10^{-8} mol dm^{-3}

13. *iso*-Amyl alcohol is non-ionizable and the pH does not determine what proportion of its molecules are present in the anaesthetically most effective form; cocaine is a more effective anaesthetic in its basic form than in its cationic form; nembutal is more effective in its acid form than in its anionic form.
14. (a) If you calculate the concentration of undissociated benzoic acid present in the minimum fungistatic concentration at each pH, you will find that this is the same at all pH's (10^{-3} mol dm^{-3}). This suggests that it is only the undissociated acid that is fungistatic, i.e. that the anion is without fungistatic activity.
(b) No. Minimum fungistatic concentration of benzoic acid at pH 8 would be 6.3 mol dm^{-3}!

Chapter 7. Background Thermodynamics (p. 209)

1. (a) -103.5 kJ at constant pressure (b) -103.7 kJ at constant volume.
2. $+0.8$ kJ
3. $\Delta G^{\ominus} = -76$ kJ mol^{-1}
4. $\Delta G^{\ominus\prime}$ at pH $7 = -82.4$ kJ mol^{-1}
5. (a) (i) K_a acetic acid $= 1.74 \times 10^{-5}$ (ii) $\Delta G^{\ominus\prime}$ at pH $7 = -12.77$ kJ mol^{-1}
(b) $\Delta G^{\ominus}$ of dissociation of monochloroacetic acid $= +16.3$ kJ mol^{-1}
(c) $\Delta S^{\ominus}$ of dissociation of trichloroacetic acid $= -8.4$ J K^{-1} mol^{-1}
6. $\Delta G^{\ominus} = -7.66$ kJ mol^{-1}
7. $\Delta G^{\ominus\prime}$ at pH $7 = -25.1$ kJ mol^{-1}
8. ΔG at 298 K and pH $7 = -66.83$ kJ mol^{-1}
9. ΔG at 298 K and pH 7 for decarboxylation of malate $= -8.93$ kJ mol^{-1}
10. (a) In man at 310 K, $\Delta G^{\ominus\prime}$ at physiological pH $= -4.43$ kJ mol^{-1}
(b) In frog at 280 K, $\Delta G^{\ominus\prime}$ at same pH $= -2.56$ kJ mol^{-1}
11. $\Delta G^{\ominus}$ of ionization $= +11.97$ kJ mol^{-1}

Chapter 8. Chemical Equilibrium and the Coupling of Reactions (p. 232)

1. (a) 0 (b) 1 (c) >1 (d) (i) >1 (ii) Negative (iii) Yes
2. (a) No (b) No (c) Yes
3. (a) $K_{eq} = 19$; $\Delta G^{\ominus} = -7.32$ kJ mol^{-1}
(b) $\Delta G = -18.75$ kJ mol^{-1}
4. Glucose 1-phosphate 3.7 mmol dm^{-3}; glucose 6-phosphate 67.3 mmol dm^{-3}; fructose 6-phosphate 29 mmol dm^{-3}
5. (a) $\Delta G^{\ominus\prime}$ at pH $7 = +27.96$ kJ mol^{-1}
(b) $E_{\ominus}'$ at 298 K, pH 7 of oxaloacetate^{2-}|malate^{2-} couple $= -0.175$ V
6. Values of $\Delta G^{\ominus\prime}$ at pH 7.4 are: (a) $+8.58$ kJ mol^{-1} (b) -1.88 kJ mol^{-1} (c) $+6.69$ kJ mol^{-1}
7. At 310 K and pH 7, $\Delta G^{\ominus\prime} = -4.35$ kJ mol^{-1}, $\Delta H^{\ominus\prime} = -3.98$ kJ mol^{-1}, $\Delta S^{\ominus\prime} = +1.21$ J K^{-1} mol^{-1}
8. Malate:fumarate $= 5.14:1$ (i.e. 83.7%:16.3%)
9. $\Delta G^{\ominus\prime}$ at 310 K and pH $7.5 = -30.5$ kJ mol^{-1}
10. K_{eq}' at 310 K and pH $7.4 = 134$
11. Equilibrium concentration of APS $= 1.89 \times 10^{-3}$ μmol cm^{-3} (i.e. 1.89 μmol dm^{-3})

12. At 310 K and pH 8: (a) $\Delta G^{\ominus\prime} = +19.2$ kJ mol^{-1} (b) $\Delta G^{\ominus\prime} = -26.8$ kJ mol^{-1}

Chapter 10. The Kinetics of Chemical Reactions (p. 276)

1. (a) 0.02 mol dm^{-3} s^{-1} (b) 0.0223 s^{-1} (c) 0.025 dm^3 mol^{-1} s^{-1}
2. 0.1755 mol dm^{-3} of A; 0.0755 mol dm^{-3} of B
3. Equilibrium concentration of $B = 9.95 \times 10^{-5}$ mol dm^{-3}
4. (a) Energy of activation = 135.5 kJ mol^{-1}
   (b) Enthalpy of activation = 133.0 kJ mol^{-1}
5. Initial rate = 1.06 μmol cm^{-3} s^{-1}
6. Yes. The observed rate constant is a pseudo 1st order rate constant for the reaction which proceeds simultaneously by two mechanisms, viz.
   (a) uncatalysed (rate constant = 2×10^{-4} s^{-1})
   (b) catalysed by H^+ (catalytic constant of H^+ ions = 0.65 s^{-1}) i.e.
   $$k_{obs} = k_c[H^+] + k_u$$
   where k_c is the catalytic constant (0.65 s^{-1}) and k_u is the rate constant of the uncatalysed reaction (2×10^{-4} s^{-1}).

Chapter 11. The Kinetics of Enzyme-Catalysed Reactions (p. 319)

1. $K_m = 2.35$ mmol dm^{-3}
2. (iii) the rate at which EA decomposes to yield P (Probably only the value of k_2 is affected by the ionic strength.)
3. $V_{max} = 1.5$ μmol glucose min^{-1}. (Value of V_{max} is not affected by a competitive inhibitor.)
4. $K_m = 20$ mmol dm^{-3}. Urea acts as a non-competitive inhibitor (both slopes and intercepts on $1/v_0$ ordinate of double reciprocal plots are changed).
5. Ca^{2+} ions act as a competitive inhibitor of the binding of Mg^{2+}. K_i for $Ca^{2+} = 10^{-5}$ mol dm^{-3}.
6. Non-sequential (ping-pong) mechanism, since plots of $1/v_0$ versus 1/[aspartate] are parallel at different fixed concentrations of glyoxylate.
7. $E = 34.6$ kJ mol^{-1}, which is much lower than the energy of activation of the uncatalysed reaction (121 kJ mol^{-1})

Chapter 12. Oxidation and Reduction (p. 360)

1. (i) B | BH_2 (ii) BH_2 by A (iii) e.m.f. = +0.06 V
2. Values of E_h at 303 K and pH 7 are: (1) −0.298 V (2) +0.132 V (3) +0.562 V
3. (i) $E_h = -0.322$ V (ii) $E_h = -0.354$ V
4. $\Delta G^{\ominus\prime}$ at 303 K and pH 7 = −69.45 kJ mol^{-1}
5. (i) $\Delta G^{\ominus\prime}$ at pH 7 = −64.65 kJ mol^{-1} (ii) K'_{eq} at pH 7 = 2.14×10^{11}
6. (i) 95.2% (ii) 2,6-dichlorophenolindophenol (iii) thionine
7. E_h at 310 K = −0.185 V
8. pH = 4.1
9. $E'_{\ominus}$ at 298 K and pH 7 of the oxaloacetate | malate couple = +0.012 V
10. $E'_{\ominus}$ at 303 K and pH 7 of the crotonyl.CoA | butyryl.CoA couple = −0.014 V

References and Suggested Reading

TEXTBOOKS

Elementary

1. CONN, E. E. and STUMPF, P. K. (1972). *Outlines of Biochemistry*, 3rd edn. Wiley, New York and London.
2. CROCKFORD, H. D. and KNIGHT, S. B. (1964). *Fundamentals of Physical Chemistry*, 2nd edn. Wiley, New York and London.
3. CROWE, A. and CROWE, A. (1969). *Mathematics for Biologists*. Academic Press, London and New York.
4. GIESE, A. C. (1968). *Cell Physiology*, 3rd edn. Saunders, Philadelphia and London.
5. OPEN UNIVERSITY (1970). *The Handling of Experimental Data*. Science Foundation Course Unit E., Open University Press, Bletchley, U.K.
6. SPEAKMAN, J. C. (1966). *Molecules*. McGraw-Hill Book Co., New York and London.
7. SUTTIE, J. W. (1972). *Introduction to Biochemistry*. Holt Rinehart and Winston Inc., New York and London.

More Advanced

8. BARROW, G. M. (1966). *Physical Chemistry*, 2nd edn. McGraw-Hill Book Co., New York and London.
9. BULL, H. B. (1971). *An Introduction to Physical Biochemistry*, 2nd edn. F. A. Davis Co., Philadelphia.
10. DAWES, E. A. (1972). *Quantitative Problems in Biochemistry*, 5th edn. Churchill-Livingstone, Edinburgh and London.
11. LEHNINGER, A. L. (1970). *Biochemistry*, Worth Publishers Inc., New York.
12. MAHLER, H. R. and CORDES, E. H. (1971). *Biological Chemistry*, 2nd edn. Harper & Row, New York and London.
13. SEGEL, I. H. (1968). *Biochemical Calculations*, Wiley New York and London.
14. WILLIAMS, V. R. and WILLIAMS, H. B. (1967). *Basic Physical Chemistry for the Life Sciences*. W. H. Freeman, San Francisco and London.

Supplementary Reading

15. ANDERSON, T. F. (1951). *Transactions of the New York Academy of Sciences, B*, **13**, 130.

16. BAKER, J. J. W. and ALLEN G. E. (1970). *Matter, Energy and Life*, 2nd edn. Addison-Wesley, Reading Mass. and London.
17. BERNHARD, S. (1968). *The Structure and Function of Enzymes*. W. A. Benjamin Inc., New York.
18. BROWN, H. D. (1969). *Biochemical Microcalorimetry*. Academic Press, New York and London.
19. DAVIS, B. D. (1958). *Arch. Biochem. Biophys.*, **78**, 497.
20. GUTFREUND, H. (1965). *An Introduction to the Study of Enzymes*. Blackwell Sci., Oxford.
21. HEWITT, L. F. (1950). *Oxidation-Reduction Potentials in Bacteriology & Biochemistry*, 6th edn. Churchill-Livingstone, Edinburgh and London.
22. JENCKS, W. P. (1969). *Catalysis in Chemistry and Enzymology*. McGraw-Hill Book Co., New York and London.
23 KING, E. L. (1963). *How Chemical Reactions Occur*. W. A. Benjamin Inc., New York.
24. KING, E. L. and ALTMAN, C. (1956). *J. phys. Chem. Wash.*, **60**, 1375.
25. KITZINGER, C. and BENZINGER, T. H. (1960). *Meth. biochem. Analysis*, **8**, 309.
26. KLOTZ, I. (1967). *Energy Changes in Biochemical Reactions*. Academic Press Inc., New York and London.
27. KREBS, H. A. and KORNBERG, H. L. (1957) *Energy Transformations in Living Matter*. Springer-Verlag, Berlin.
28. LEHNINGER, A. (1971). *Bioenergetics*, 2nd edn. W. A. Benjamin, Inc., New York.
29. LINFORD, J. H. (1966). *An Introduction to Energetics with Applications to Biology*. Butterworths, London.
30. MAHAN, B. H. (1963). *Elementary Chemical Thermodynamics*. W. A. Benjamin Inc., New York.
31. MORRISON, J. F. (1965). *Aust. J. Sci.*, **27**, 317.
32. NASH, L. K. (1970). *Elements of Chemical Thermodynamics*, 2nd edn. Addison-Wesley, Reading Mass. and London.
33. PURDY, W. C. (1965). *Electroanalytical Methods in Biochemistry*. McGraw-Hill Book Co., New York and London.
34. STURTEVANT, J. M. (1959). In *Techniques of Organic Chemistry*, 3rd edn., **1**, 633. Interscience, New York.
35. UMBREIT, W. W., BURRIS, R. H. and STAUFFER, J. F. (1971). *Manometric and Biochemical Techniques*, 5th edn., Burgess, Minneapolis, Minn.
36. WALL, F. T. (1965). *Chemical Thermodynamics*. W. H. Freeman & Co., San Francisco and London.

REFERENCE BOOKS

37. CLARK W. M. (1960). *Oxidation-Reduction Potentials of Organic Systems*. Balliere, London; Williams & Wilkins, Baltimore, Md.
38. DAWSON, R. M. C., ELLIOTT, D. C., ELLIOTT, W. H. and JONES, K. M., Eds. (1969) *Data for Biochemical Research*, 2nd edn. Oxford University Press, London.
39. KAYE, G. W. C. and LABY, T. H. (1966). *Tables of Physical and Chemical Constants*, 13th edn. Longmans, London.
40. SOBER, H. A. Ed. (1970). *CRC Handbook of Biochemistry. Selected Data for Molecular Biology*, 2nd edn. C.R.C. Press, Cleveland, Ohio; Blackwell Sci., Oxford.
41. WEAST, R. C. Ed. (1972). *CRC Handbook of Chemistry and Physics*, 53rd edn. C.R.C. Press, Cleveland, Ohio; Blackwell Sci., Oxford.

LOGARITHMS

	0	1	2	3	4	5	6	7	8	9	1	2	3	4	5	6	7	8	9
10	0000	0043	0086	0128	0170	0212	0253	0294	0334	0374	4	8	12	17	21	25	29	33	37
11	0414	0453	0492	0531	0569	0607	0645	0682	0719	0755	4	8	11	15	19	23	26	30	34
12	0792	0828	0864	0899	0934	0969	1004	1038	1072	1106	3	7	10	14	17	21	24	28	31
13	1139	1173	1206	1239	1271	1303	1335	1367	1399	1430	3	6	10	13	16	19	23	26	29
14	1461	1492	1523	1553	1584	1614	1644	1673	1703	1732	3	6	9	12	15	18	21	24	27
15	1761	1790	1818	1847	1875	1903	1931	1959	1987	2014	3	6	8	11	14	17	20	22	25
16	2041	2068	2095	2122	2148	2175	2201	2227	2253	2279	3	5	8	11	13	16	18	21	24
17	2304	2330	2355	2380	2405	2430	2455	2480	2504	2529	2	5	7	10	12	15	17	20	22
18	2553	2577	2601	2625	2648	2672	2695	2718	2742	2765	2	5	7	9	12	14	16	19	21
19	2788	2810	2833	2856	2878	2900	2923	2945	2967	2989	2	4	7	9	11	13	16	18	20
20	3010	3032	3054	3075	3096	3118	3139	3160	3181	3201	2	4	6	8	11	13	15	17	19
21	3222	3243	3263	3284	3304	3324	3345	3365	3385	3404	2	4	6	8	10	12	14	16	18
22	3424	3444	3464	3483	3502	3522	3541	3560	3579	3598	2	4	6	8	10	12	14	15	17
23	3617	3636	3655	3674	3692	3711	3729	3747	3766	3784	2	4	6	7	9	11	13	15	17
24	3802	3820	3838	3856	3874	3892	3909	3927	3945	3962	2	4	5	7	9	11	12	14	16
25	3979	3997	4014	4031	4048	4065	4082	4099	4116	4133	2	3	5	7	9	10	12	14	15
26	4150	4166	4183	4200	4216	4232	4249	4265	4281	4298	2	3	5	7	8	10	11	13	15
27	4314	4330	4346	4362	4378	4393	4409	4425	4440	4456	2	3	5	6	8	9	11	13	14
28	4472	4487	4502	4518	4533	4548	4564	4579	4594	4609	2	3	5	6	8	9	11	12	14
29	4624	4639	4654	4669	4683	4698	4713	4728	4742	4757	1	3	4	6	7	9	10	12	13
30	4771	4786	4800	4814	4829	4843	4857	4871	4886	4900	1	3	4	6	7	9	10	11	13
31	4914	4928	4942	4955	4969	4983	4997	5011	5024	5038	1	3	4	6	7	8	10	11	12
32	5051	5065	5079	5092	5105	5119	5132	5145	5159	5172	1	3	4	5	7	8	9	11	12
33	5185	5198	5211	5224	5237	5250	5263	5276	5289	5302	1	3	4	5	6	8	9	10	12
34	5315	5328	5340	5353	5366	5378	5391	5403	5416	5428	1	3	4	5	6	8	9	10	11
35	5441	5453	5465	5478	5490	5502	5514	5527	5539	5551	1	2	4	5	6	7	9	10	11
36	5563	5575	5587	5599	5611	5623	5635	5647	5658	5670	1	2	4	5	6	7	8	10	11
37	5682	5694	5705	5717	5729	5740	5752	5763	5775	5786	1	2	3	5	6	7	8	9	10
38	5798	5809	5821	5832	5843	5855	5866	5877	5888	5899	1	2	3	5	6	7	8	9	10
39	5911	5922	5933	5944	5955	5966	5977	5988	5999	6010	1	2	3	4	5	7	8	9	10
40	6021	6031	6042	6053	6064	6075	6085	6096	6107	6117	1	2	3	4	5	6	8	9	10
41	6128	6138	6149	6160	6170	6180	6191	6201	6212	6222	1	2	3	4	5	6	7	8	9
42	6232	6243	6253	6263	6274	6284	6294	6304	6314	6325	1	2	3	4	5	6	7	8	9
43	6335	6345	6355	6365	6375	6385	6395	6405	6415	6425	1	2	3	4	5	6	7	8	9
44	6435	6444	6454	6464	6474	6484	6493	6503	6513	6522	1	2	3	4	5	6	7	8	9
45	6532	6542	6551	6561	6571	6580	6590	6599	6609	6618	1	2	3	4	5	6	7	8	9
46	6628	6637	6646	6656	6665	6675	6684	6693	6702	6712	1	2	3	4	5	6	7	7	8
47	6721	6730	6739	6749	6758	6767	6776	6785	6794	6803	1	2	3	4	5	5	6	7	8
48	6812	6821	6830	6839	6848	6857	6866	6875	6884	6893	1	2	3	4	4	5	6	7	8
49	6902	6911	6920	6928	6937	6946	6955	6964	6972	6981	1	2	3	4	4	5	6	7	8
50	6990	6998	7007	7016	7024	7033	7042	7050	7059	7067	1	2	3	3	4	5	6	7	8
51	7076	7084	7093	7101	7110	7118	7126	7135	7143	7152	1	2	3	3	4	5	6	7	8
52	7160	7168	7177	7185	7193	7202	7210	7218	7226	7235	1	2	2	3	4	5	6	7	7
53	7243	7251	7259	7267	7275	7284	7292	7300	7308	7316	1	2	2	3	4	5	6	6	7
54	7324	7332	7340	7348	7356	7364	7372	7380	7388	7396	1	2	2	3	4	5	6	6	7

LOGARITHMS

	0	1	2	3	4	5	6	7	8	9	1	2	3	4	5	6	7	8	9
55	7404	7412	7419	7427	7435	7443	7451	7459	7466	7474	1	2	2	3	4	5	5	6	7
56	7482	7490	7497	7505	7513	7520	7528	7536	7543	7551	1	2	2	3	4	5	5	6	7
57	7559	7566	7574	7582	7589	7597	7604	7612	7619	7627	1	2	2	3	4	5	5	6	7
58	7634	7642	7649	7657	7664	7672	7679	7686	7694	7701	1	1	2	3	4	4	5	6	7
59	7709	7716	7723	7731	7738	7745	7752	7760	7767	7774	1	1	2	3	4	4	5	6	7
60	7782	7789	7796	7803	7810	7818	7825	7832	7839	7846	1	1	2	3	4	4	5	6	6
61	7853	7860	7868	7875	7882	7889	7896	7903	7910	7917	1	1	2	3	4	4	5	6	6
62	7924	7931	7938	7945	7952	7959	7966	7973	7980	7987	1	1	2	3	3	4	5	6	6
63	7993	8000	8007	8014	8021	8028	8035	8041	8048	8055	1	1	2	3	3	4	5	5	6
64	8062	8069	8075	8082	8089	8096	8102	8109	8116	8122	1	1	2	3	3	4	5	5	6
65	8129	8136	8142	8149	8156	8162	8169	8176	8182	8189	1	1	2	3	3	4	5	5	6
66	8195	8202	8209	8215	8222	8228	8235	8241	8248	8254	1	1	2	3	3	4	5	5	6
67	8261	8267	8274	8280	8287	8293	8299	8306	8312	8319	1	1	2	3	3	4	5	5	6
68	8325	8331	8338	8344	8351	8357	8363	8370	8376	8382	1	1	2	3	3	4	4	5	6
69	8388	8395	8401	8407	8414	8420	8426	8432	8439	8445	1	1	2	2	3	4	4	5	6
70	8451	8457	8463	8470	8476	8482	8488	8494	8500	8506	1	1	2	2	3	4	4	5	6
71	8513	8519	8525	8531	8537	8543	8549	8555	8561	8567	1	1	2	2	3	4	4	5	5
72	8573	8579	8585	8591	8597	8603	8609	8615	8621	8627	1	1	2	2	3	4	4	5	5
73	8633	8639	8645	8651	8657	8663	8669	8675	8681	8686	1	1	2	2	3	4	4	5	5
74	8692	8698	8704	8710	8716	8722	8727	8733	8739	8745	1	1	2	2	3	4	4	5	5
75	8751	8756	8762	8768	8774	8779	8785	8791	8797	8802	1	1	2	2	3	3	4	5	5
76	8808	8814	8820	8825	8831	8837	8842	8848	8854	8859	1	1	2	2	3	3	4	5	5
77	8865	8871	8876	8882	8887	8893	8899	8904	8910	8915	1	1	2	2	3	3	4	4	5
78	8921	8927	8932	8938	8943	8949	8954	8960	8965	8971	1	1	2	2	3	3	4	4	5
79	8976	8982	8987	8993	8998	9004	9009	9015	9020	9025	1	1	2	2	3	3	4	4	5
80	9031	9036	9042	9047	9053	9058	9063	9069	9074	9079	1	1	2	2	3	3	4	4	5
81	9085	9090	9096	9101	9106	9112	9117	9122	9128	9133	1	1	2	2	3	3	4	4	5
82	9138	9143	9149	9154	9159	9165	9170	9175	9180	9186	1	1	2	2	3	3	4	4	5
83	9191	9196	9201	9206	9212	9217	9222	9227	9232	9238	1	1	2	2	3	3	4	4	5
84	9243	9248	9253	9258	9263	9269	9274	9279	9284	9289	1	1	2	2	3	3	4	4	5
85	9294	9299	9304	9309	9315	9320	9325	9330	9335	9340	1	1	2	2	3	3	4	4	5
86	9345	9350	9355	9360	9365	9370	9375	9380	9385	9390	1	1	2	2	3	3	4	4	5
87	9395	9400	9405	9410	9415	9420	9425	9430	9435	9440	0	1	1	2	2	3	3	4	4
88	9445	9450	9455	9460	9465	9469	9474	9479	9484	9489	0	1	1	2	2	3	3	4	4
89	9494	9499	9504	9509	9513	9518	9523	9528	9533	9538	0	1	1	2	2	3	3	4	4
90	9542	9547	9552	9557	9562	9566	9571	9576	9581	9586	0	1	1	2	2	3	3	4	4
91	9590	9595	9600	9605	9609	9614	9619	9624	9628	9633	0	1	1	2	2	3	3	4	4
92	9638	9643	9647	9652	9657	9661	9666	9671	9675	9680	0	1	1	2	2	3	3	4	4
93	9685	9689	9694	9699	9703	9708	9713	9717	9722	9727	0	1	1	2	2	3	3	4	4
94	9731	9736	9741	9745	9750	9754	9759	9763	9768	9773	0	1	1	2	2	3	3	4	4
95	9777	9782	9786	9791	9795	9800	9805	9809	9814	9818	0	1	1	2	2	3	3	4	4
96	9823	9827	9832	9836	9841	9845	9850	9854	9859	9863	0	1	1	2	2	3	3	4	4
97	9868	9872	9877	9881	9886	9890	9894	9899	9903	9908	0	1	1	2	2	3	3	4	4
98	9912	9917	9921	9926	9930	9934	9939	9943	9948	9952	0	1	1	2	2	3	3	4	4
99	9956	9961	9965	9969	9974	9978	9983	9987	9991	9996	0	1	1	2	2	3	3	3	4

ANTILOGARITHMS

	0	1	2	3	4	5	6	7	8	9	1	2	3	4	5	6	7	8	9
.00	1000	1002	1005	1007	1009	1012	1014	1016	1019	1021	0	0	1	1	1	1	2	2	2
.01	1023	1026	1028	1030	1033	1035	1038	1040	1042	1045	0	0	1	1	1	1	2	2	2
.02	1047	1050	1052	1054	1057	1059	1062	1064	1067	1069	0	0	1	1	1	1	2	2	2
.03	1072	1074	1076	1079	1081	1084	1086	1089	1091	1094	0	0	1	1	1	1	2	2	2
.04	1096	1099	1102	1104	1107	1109	1112	1114	1117	1119	0	1	1	1	1	2	2	2	2
.05	1122	1125	1127	1130	1132	1135	1138	1140	1143	1146	0	1	1	1	1	2	2	2	2
.06	1148	1151	1153	1156	1159	1161	1164	1167	1169	1172	0	1	1	1	1	2	2	2	2
.07	1175	1178	1180	1183	1186	1189	1191	1194	1197	1199	0	1	1	1	1	2	2	2	2
.08	1202	1205	1208	1211	1213	1216	1219	1222	1225	1227	0	1	1	1	1	2	2	2	3
.09	1230	1233	1236	1239	1242	1245	1247	1250	1253	1256	0	1	1	1	1	2	2	2	3
.10	1259	1262	1265	1268	1271	1274	1276	1279	1282	1285	0	1	1	1	1	2	2	2	3
.11	1288	1291	1294	1297	1300	1303	1306	1309	1312	1315	0	1	1	1	2	2	2	2	3
.12	1318	1321	1324	1327	1330	1334	1337	1340	1343	1346	0	1	1	1	2	2	2	2	3
.13	1349	1352	1355	1358	1361	1365	1368	1371	1374	1377	0	1	1	1	2	2	2	3	3
.14	1380	1384	1387	1390	1393	1396	1400	1403	1406	1409	0	1	1	1	2	2	2	3	3
.15	1413	1416	1419	1422	1426	1429	1432	1435	1439	1442	0	1	1	1	2	2	2	3	3
.16	1445	1449	1452	1455	1459	1462	1466	1469	1472	1476	0	1	1	1	2	2	2	3	3
.17	1479	1483	1486	1489	1493	1496	1500	1503	1507	1510	0	1	1	1	2	2	2	3	3
.18	1514	1517	1521	1524	1528	1531	1535	1538	1542	1545	0	1	1	1	2	2	2	3	3
.19	1549	1552	1556	1560	1563	1567	1570	1574	1578	1581	0	1	1	1	2	2	3	3	3
.20	1585	1589	1592	1596	1600	1603	1607	1611	1614	1618	0	1	1	1	2	2	3	3	3
.21	1622	1626	1629	1633	1637	1641	1644	1648	1652	1656	0	1	1	2	2	2	3	3	3
.22	1660	1663	1667	1671	1675	1679	1683	1687	1690	1694	0	1	1	2	2	2	3	3	3
.23	1698	1702	1706	1710	1714	1718	1722	1726	1730	1734	0	1	1	2	2	2	3	3	4
.24	1738	1742	1746	1750	1754	1758	1762	1766	1770	1774	0	1	1	2	2	2	3	3	4
.25	1778	1782	1786	1791	1795	1799	1803	1807	1811	1816	0	1	1	2	2	2	3	3	4
.26	1820	1824	1828	1832	1837	1841	1845	1849	1854	1858	0	1	1	2	2	3	3	3	4
.27	1862	1866	1871	1875	1879	1884	1888	1892	1897	1901	0	1	1	2	2	3	3	3	4
.28	1905	1910	1914	1919	1923	1928	1932	1936	1941	1945	0	1	1	2	2	3	3	4	4
.29	1950	1954	1959	1963	1968	1972	1977	1982	1986	1991	0	1	1	2	2	3	3	4	4
.30	1995	2000	2004	2009	2014	2018	2023	2028	2032	2037	0	1	1	2	2	3	3	4	4
.31	2042	2046	2051	2056	2061	2065	2070	2075	2080	2084	0	1	1	2	2	3	3	4	4
.32	2089	2094	2099	2104	2109	2113	2118	2123	2128	2133	0	1	1	2	2	3	3	4	4
.33	2138	2143	2148	2153	2158	2163	2168	2173	2178	2183	0	1	1	2	2	3	3	4	4
.34	2188	2193	2198	2203	2208	2213	2218	2223	2228	2234	1	1	2	2	3	3	4	4	5
.35	2239	2244	2249	2254	2259	2265	2270	2275	2280	2286	1	1	2	2	3	3	4	4	5
.36	2291	2296	2301	2307	2312	2317	2323	2328	2333	2339	1	1	2	2	3	3	4	4	5
.37	2344	2350	2355	2360	2366	2371	2377	2382	2388	2393	1	1	2	2	3	3	4	4	5
.38	2399	2404	2410	2415	2421	2427	2432	2438	2443	2449	1	1	2	2	3	3	4	4	5
.39	2455	2460	2466	2472	2477	2483	2489	2495	2500	2506	1	1	2	2	3	3	4	5	5
.40	2512	2518	2523	2529	2535	2541	2547	2553	2559	2564	1	1	2	2	3	4	4	5	5
.41	2570	2576	2582	2588	2594	2600	2606	2612	2618	2624	1	1	2	2	3	4	4	5	5
.42	2630	2636	2642	2649	2655	2661	2667	2673	2679	2685	1	1	2	2	3	4	4	5	6
.43	2692	2698	2704	2710	2716	2723	2729	2735	2742	2748	1	1	2	3	3	4	4	5	6
.44	2754	2761	2767	2773	2780	2786	2793	2799	2805	2812	1	1	2	3	3	4	4	5	6
.45	2818	2825	2831	2838	2844	2851	2858	2864	2871	2877	1	1	2	3	3	4	5	5	6
.46	2884	2891	2897	2904	2911	2917	2924	2931	2938	2944	1	1	2	3	3	4	5	5	6
.47	2951	2958	2965	2972	2979	2985	2992	2999	3006	3013	1	1	2	3	3	4	5	5	6
.48	3020	3027	3034	3041	3048	3055	3062	3069	3076	3083	1	1	2	3	4	4	5	6	6
.49	3090	3097	3105	3112	3119	3126	3133	3141	3148	3155	1	1	2	3	4	4	5	6	6

ANTILOGARITHMS

	0	1	2	3	4	5	6	7	8	9	1	2	3	4	5	6	7	8	9
.50	3162	3170	3177	3184	3192	3199	3206	3214	3221	3228	1	1	2	3	4	4	5	6	7
.51	3236	3243	3251	3258	3266	3273	3281	3289	3296	3304	1	2	2	3	4	5	5	6	7
.52	3311	3319	3327	3334	3342	3350	3357	3365	3373	3381	1	2	2	3	4	5	5	6	7
.53	3388	3396	3404	3412	3420	3428	3436	3443	3451	3459	1	2	2	3	4	5	6	6	7
.54	3467	3475	3483	3491	3499	3508	3516	3524	3532	3540	1	2	2	3	4	5	6	6	7
.55	3548	3556	3565	3573	3581	3589	3597	3606	3614	3622	1	2	2	3	4	5	6	7	7
.56	3631	3639	3648	3656	3664	3673	3681	3690	3698	3707	1	2	3	3	4	5	6	7	8
.57	3715	3724	3733	3741	3750	3758	3767	3776	3784	3793	1	2	3	3	4	5	6	7	8
.58	3802	3811	3819	3828	3837	3846	3855	3864	3873	3882	1	2	3	4	4	5	6	7	8
.59	3890	3899	3908	3917	3926	3936	3945	3954	3963	3972	1	2	3	4	5	5	6	7	8
.60	3981	3990	3999	4009	4018	4027	4036	4046	4055	4064	1	2	3	4	5	6	6	7	8
.61	4074	4083	4093	4102	4111	4121	4130	4140	4150	4159	1	2	3	4	5	6	7	8	9
.62	4169	4178	4188	4198	4207	4217	4227	4236	4246	4256	1	2	3	4	5	6	7	8	9
.63	4266	4276	4285	4295	4305	4315	4325	4335	4345	4355	1	2	3	4	5	6	7	8	9
.64	4365	4375	4385	4395	4406	4416	4426	4436	4446	4457	1	2	3	4	5	6	7	8	9
.65	4467	4477	4487	4498	4508	4519	4529	4539	4550	4560	1	2	3	4	5	6	7	8	9
.66	4571	4581	4592	4603	4613	4624	4634	4645	4656	4667	1	2	3	4	5	6	7	9	10
.67	4677	4688	4699	4710	4721	4732	4742	4753	4764	4775	1	2	3	4	5	7	8	9	10
.68	4786	4797	4808	4819	4831	4842	4853	4864	4875	4887	1	2	3	4	6	7	8	9	10
.69	4898	4909	4920	4932	4943	4955	4966	4977	4989	5000	1	2	3	5	6	7	8	9	10
.70	5012	5023	5035	5047	5058	5070	5082	5093	5105	5117	1	2	4	5	6	7	8	9	11
.71	5129	5140	5152	5164	5176	5188	5200	5212	5224	5236	1	2	4	5	6	7	8	10	11
.72	5248	5260	5272	5284	5297	5309	5321	5333	5346	5358	1	2	4	5	6	7	9	10	11
.73	5370	5383	5395	5408	5420	5433	5445	5458	5470	5483	1	3	4	5	6	8	9	10	11
.74	5495	5508	5521	5534	5546	5559	5572	5585	5598	5610	1	3	4	5	6	8	9	10	12
.75	5623	5636	5649	5662	5675	5689	5702	5715	5728	5741	1	3	4	5	7	8	9	10	12
.76	5754	5768	5781	5794	5808	5821	5834	5848	5861	5875	1	3	4	5	7	8	9	11	12
.77	5888	5902	5916	5929	5943	5957	5970	5984	5998	6012	1	3	4	5	7	8	10	11	12
.78	6026	6039	6053	6067	6081	6095	6109	6124	6138	6152	1	3	4	6	7	8	10	11	13
.79	6166	6180	6194	6209	6223	6237	6252	6266	6281	6295	1	3	4	6	7	9	10	11	13
.80	6310	6324	6339	6353	6368	6383	6397	6412	6427	6442	1	3	4	6	7	9	10	12	13
.81	6457	6471	6486	6501	6516	6531	6546	6561	6577	6592	2	3	5	6	8	9	11	12	14
.82	6607	6622	6637	6653	6668	6683	6699	6714	6730	6745	2	3	5	6	8	9	11	12	14
.83	6761	6776	6792	6808	6823	6839	6855	6871	6887	6902	2	3	5	6	8	9	11	13	14
.84	6918	6934	6950	6966	6982	6998	7015	7031	7047	7063	2	3	5	6	8	10	11	13	15
.85	7079	7096	7112	7129	7145	7161	7178	7194	7211	7228	2	3	5	7	8	10	12	13	15
.86	7244	7261	7278	7295	7311	7328	7345	7362	7379	7396	2	3	5	7	8	10	12	13	15
.87	7413	7430	7447	7464	7482	7499	7516	7534	7551	7568	2	3	5	7	9	10	12	14	16
.88	7586	7603	7621	7638	7656	7674	7691	7709	7727	7745	2	4	5	7	9	11	12	14	16
.89	7762	7780	7798	7816	7834	7852	7870	7889	7907	7925	2	4	5	7	9	11	13	14	16
.90	7943	7962	7980	7998	8017	8035	8054	8072	8091	8110	2	4	6	7	9	11	13	15	17
.91	8128	8147	8166	8185	8204	8222	8241	8260	8279	8299	2	4	6	8	9	11	13	15	17
.92	8318	8337	8356	8375	8395	8414	8433	8453	8472	8492	2	4	6	8	10	12	14	15	17
.93	8511	8531	8551	8570	8590	8610	8630	8650	8670	8690	2	4	6	8	10	12	14	16	18
.94	8710	8730	8750	8770	8790	8810	8831	8851	8872	8892	2	4	6	8	10	12	14	16	18
.95	8913	8933	8954	8974	8995	9016	9036	9057	9078	9099	2	4	6	8	10	12	15	17	19
.96	9120	9141	9162	9183	9204	9226	9247	9268	9290	9311	2	4	6	8	11	13	15	17	19
.97	9333	9354	9376	9397	9419	9441	9462	9484	9506	9528	2	4	7	9	11	13	15	17	20
.98	9550	9572	9594	9616	9638	9661	9683	9705	9727	9750	2	4	7	9	11	13	16	18	20
.99	9772	9795	9817	9840	9863	9886	9908	9931	9954	9977	2	5	7	9	11	14	16	18	20

Index

(Italicized page numbers refer to Figures)